T0188804

Local Structural Characterisation

Inorganic Materials Series

Series Editors:

Professor Duncan W. Bruce
Department of Chemistry, University of York, UK

Professor Dermot O'Hare
Chemistry Research Laboratory, University of Oxford, UK

Professor Richard I. Walton
Department of Chemistry, University of Warwick, UK

Series Titles

Functional Oxides
Molecular Materials
Porous Materials
Low-Dimensional Solids
Energy Materials
Local Structural Characterisation

Forthcoming Titles

Multi Length-Scale Characterisation
Structure From Diffraction Methods

Local Structural Characterisation

Edited by

Duncan W. Bruce
University of York, UK

Dermot O'Hare
University of Oxford, UK

Richard I. Walton
University of Warwick, UK

This edition first published 2014
© 2014 John Wiley & Sons, Ltd

Registered office
John Wiley & Sons Ltd, The Atrium, Southern Gate, Chichester, West Sussex, PO19 8SQ, United Kingdom

For details of our global editorial offices, for customer services and for information about how to apply for permission to reuse the copyright material in this book please see our website at www.wiley.com.

Library of Congress Cataloging-in-Publication Data

Local structural characterisation/edited by Duncan W. Bruce, Dermot O'Hare, Richard I. Walton.
    pages cm
  Includes bibliographical references and index.
  ISBN 978-1-119-95320-3 (cloth)
1. Chemical structure. 2. Materials. I. Bruce, Duncan W., editor of compilation.
II. O'Hare, Dermot, editor of compilation. III. Walton, Richard I., editor of compilation.
  QD471.L685 2014
  543′.5–dc23

                                                                      2013015036

A catalogue record for this book is available from the British Library.

ISBN: 9781119953203

Set in 10.5/13pt Sabon by Laserwords Private Limited, Chennai, India
Printed and bound in Malaysia by Vivar Printing Sdn Bhd

1 2014

Contents

Inorganic Materials Series Preface

Back in 1992, two of us (DWB and DO'H) edited the first edition of *Inorganic Materials* in response to the growing emphasis on and interest in materials chemistry. The second edition, which contained updated chapters, appeared in 1996 and was reprinted in paperback. The aim had always been to provide chapters that while not necessarily comprehensive, nonetheless gave a first-rate and well-referenced introduction to the subject for the first-time reader. As such, the target audience was from first-year postgraduate students upwards. In these two editions, we believe our authors achieved this admirably.

In the intervening years, materials chemistry has grown hugely and it now finds itself central to many of the major challenges that face global society. We felt, therefore, that there was a need for more extensive coverage of the area, and so Richard Walton joined the team and, with Wiley, we set about working on a new and larger project.

The *Inorganic Materials Series* is the result, and our aim is to provide chapters with a similar pedagogical flavour to the first and second editions, but now with much wider subject coverage. As such, the work will be contained in several volumes. Many of the early volumes concentrate on materials derived from continuous inorganic solids. Later volumes, however, will emphasise methods of characterisation as well as molecular and soft-matter systems, as we aim for a much more comprehensive coverage of the area than was possible with *Inorganic Materials*.

We are delighted with the calibre of authors who have agreed to write for us and we thank them all for their efforts and cooperation. We believe they have done a splendid job and that their work will make these volumes a valuable reference and teaching resource.

DWB, York
DO'H, Oxford
RIW, Warwick
June 2013

Preface

Inorganic materials show a diverse range of important properties that are desirable for many contemporary, real-world applications. Good examples include recyclable battery cathode materials for energy storage and transport, porous solids for capture and storage of gases and molecular complexes that can be used in electronic devices. Some of these families of materials, and many others, were reviewed in earlier volumes of the *Inorganic Materials Series*. When considering the property-driven research in this large field, it is immediately apparent that methods for structural characterisation must be applied routinely in order to understand the function of materials and thus optimise their behaviour for real applications. Thus, 'structure–property relationships' are an important part of research in this area. To determine structure effectively, advances in methodology are important: the aim is often rapidly to examine increasingly complex materials in order to gain knowledge of structure over length scales ranging from local atomic order, through crystalline, long-range order to the meso- and macroscopic.

No single technique can examine all levels of structural order simultaneously, and the chapters presented in this volume deal with recent advances in important techniques that allow investigation of the structures of inorganic materials on the local atomic scale. Such short-range order is concerned with local atomic structure – the arrangement of atoms in space about a central probe atom – and deals with bond distances, coordination geometry and the local connectivity of the simple building units of a complex structure. It is often by studying this shortest of structural length scales that information about the underlying behaviour of a material can be deduced. The techniques employed are usually spectroscopic in origin, involving observation of the effect of interaction of an appropriate energy source with the substance being studied, which supplies information about the probe atoms' environments. It should be noted that these methods have no requirement for any long-range order (translational symmetry) and so can be applied equally to poorly crystalline, glassy, amorphous or heterogeneous systems, as well as to crystalline substances. Another consideration of any characterisation study is the need to examine materials under real operating conditions in order to understand properly their function; here, spectroscopic, short-range probes of structure often provide the key.

Some of the techniques discussed in this volume may be familiar to the reader (such as NMR, EPR and XPS), but with recent advances broadening their applicability and making them available more routinely, it is timely to provide up-to-date overviews of their uses. Also included are techniques that require large-scale facilities, such as X-ray absorption spectroscopy (XAS) and inelastic neutron scattering (INS). With the investment by many countries in major facilities for X-ray and neutron science, such methods provide an important, and increasingly accessible, addition to the toolbox of techniques available to the scientist studying the structures of materials. We approached an international set of expert authors to write the chapters in this volume with the brief to provide an introduction to the principles of their technique, to describe recent developments in the field and then to select examples from the literature to illustrate the method under discussion. We believe they have done an excellent job in all respects and hope that the chapters provide a valuable set of references for those who wish to learn the principles of some important methods in the study of inorganic materials.

DWB, York
DO'H, Oxford
RIW, Warwick
June 2013

List of Contributors

Sharon E. Ashbrook School of Chemistry, University of St Andrews, St Andrews, UK

Daniel M. Dawson School of Chemistry, University of St Andrews, St Andrews, UK

Pieter Glatzel European Synchrotron Radiation Facility, Grenoble, France

John M. Griffin School of Chemistry, University of St Andrews, St Andrews, UK

Amélie Juhin Institut de Minéralogie et de Physique des Milieux Condensés (IMPMC), CNRS, Pierre-and-Marie-Curie University, Paris, France

Adam F. Lee Department of Chemistry, University of Warwick, Coventry, UK and School of Chemistry, Monash University, Melbourne, Australia

Tomasz Mazur Faculty of Chemistry, Jagiellonian University, Krakow, Poland

Philip C. H. Mitchell Department of Chemistry, University of Reading, Reading, UK

Piotr Pietrzyk Faculty of Chemistry, Jagiellonian University, Krakow, Poland

A. J. Ramirez-Cuesta Neutron Sciences Directorate, Oak Ridge National Laboratory, Oak Ridge, Tennessee, USA and ISIS Facility, Rutherford Appleton Laboratory, STFC, Oxford, UK

Zbigniew Sojka Faculty of Chemistry, Jagiellonian University, Krakow, Poland

Karen Wilson European Bioenergy Research Institute, School of Engineering and Applied Science, Aston University, Aston Triangle, Birmingham, UK

1

Solid-State Nuclear Magnetic Resonance Spectroscopy

Sharon E. Ashbrook, Daniel M. Dawson and John M. Griffin
School of Chemistry, University of St Andrews, St Andrews, UK

1.1 OVERVIEW

Although solution-state nuclear magnetic resonance (NMR) spectroscopy is one of the most widely applied analytical tools in chemistry, providing a sensitive probe of local structure for systems ranging from small molecules to large proteins, it is only relatively recently that solid-state NMR has been able to provide information of a similar quality. The anisotropic (*i.e.* orientation-dependent) interactions affecting NMR spectra, which ultimately provide valuable information about structure, symmetry and bonding, are averaged in solution by the rapid tumbling motion of the molecules, resulting in simplified spectra from which information can be more easily obtained. In contrast, NMR spectra of solids remain broadened by these interactions, hindering the extraction of structural information. This broadening poses significant challenges both in the acquisition of high-resolution NMR spectra for solids and in their interpretation and analysis. However, in recent years considerable developments in hardware (*e.g.* increasing magnetic field strengths) and in software (*e.g.* improvements in computational simulations and analysis packages) have enabled solid-state NMR to

Local Structural Characterisation, First Edition. Edited by Duncan W. Bruce, Dermot O'Hare and Richard I. Walton.
© 2014 John Wiley & Sons, Ltd. Published 2014 by John Wiley & Sons, Ltd.

develop to the point where it can play a central role in the atomic-level understanding of materials as diverse as zeolites, glasses, polymers, energy materials, pharmaceuticals and proteins.

The ability of NMR spectroscopy to probe the local atomic-scale environment, without any requirement for long- or short-range order, enables it to be used alongside more conventional diffraction-based approaches for the study of solids. The sensitivity of NMR to small changes in the local environment (and its element specificity) makes it an ideal approach for studying disorder in solids, be it positional or compositional, resulting in numerous applications to the study of glasses, gels and ceramics. NMR is also an excellent probe of dynamics, sensitive to motion over a wide range of timescales, depending upon the exact experiment used. However, despite this wealth of information, the interpretation of solid-state NMR spectra and the extraction of relevant structural detail remain a challenge. In recent years there has been growing interest in the use of computational methods alongside experimental measurement. While there has been a long tradition in quantum chemistry of the calculation of NMR parameters from first principles, much of the development has been focused on molecules (either in vacuum or in solution), rather than on the extended and periodic structures found in the solid state. Recent methods utilising periodic approaches to recreate the three-dimensional (3D) structure from a high-symmetry small-volume unit have found great favour with experimentalists, and are currently being applied to a wide range of different systems, helping to interpret complex NMR spectra, improve structural models and provide new insight into disorder and/or dynamics.

At first sight, the vast array of NMR experiments in the literature can seem daunting to the non-specialist; however they can be easily categorised by their overall aim. Many experiments are designed to improve resolution and/or sensitivity, typically through more efficient removal of anisotropic broadening – an enduring theme in solid-state NMR spectroscopy. Experiments have also been developed to measure the magnitudes of individual interactions, providing information on local geometry or symmetry, for example. Further experiments are concerned with the transfer of magnetisation between different nuclei, probing their through-bond or through-space connectivity. In many cases, the exact experimental detail is not of vital importance; it is more useful to understand the type of information available from a particular NMR spectrum and how it can be extracted. In this chapter, we will give an overview of solid-state NMR spectroscopy, focusing in particular upon its application to inorganic solids. We briefly introduce the theoretical basis of the technique and the interactions that affect NMR spectra (and ultimately provide information). We describe the basic and routinely used experimental techniques, and the information that is available from solid-state NMR spectra. We then review the nuclear species most commonly studied and provide a range of examples of the application of NMR spectroscopy for a wide variety of materials, demonstrating the versatility and promise of the technique.

1.2 THEORETICAL BACKGROUND

A brief description of the theoretical basis of NMR spectroscopy is provided here. For a detailed description, see references [1, 2].

1.2.1 Fundamentals of NMR

Atomic nuclei possess an intrinsic spin angular momentum, $\mathbf{I}$, described by the nuclear spin quantum number, I, which may take any positive integer or half-integer value. The projection of $\mathbf{I}$ onto a specified axis, arbitrarily the z-axis, is quantised in units of $m_I \hbar$, where m_I is the magnetic quantum number, and takes values between $+I$ and $-I$ in integer steps, leading to $2I + 1$ degenerate spin states. Nuclei with $I > 0$ possess a magnetic dipole moment, $\boldsymbol{\mu}$, related to $\mathbf{I}$ by the gyromagnetic ratio, γ, which is characteristic of a given nuclide. Therefore, $\boldsymbol{\mu}$ is quantised along the (arbitrary) z-axis in units of $\gamma m_I \hbar$. When an external magnetic field, B_0, is present, the axis of quantisation is defined and the degeneracy of the nuclear spin states is removed. The field-induced splitting of nuclear energy levels is known as the Zeeman interaction, with the Zeeman energy of a state, m_I, given by:

$$E_{m_I} = -\gamma m_I \hbar B_0 \tag{1.1}$$

as shown in Figure 1.1. Only transitions with $\Delta m_I = \pm 1$ are observable in NMR spectroscopy and, therefore, all observable transitions are degenerate, with a frequency:

$$\omega_0 = -\gamma B_0 \tag{1.2}$$

where ω_0 is the Larmor frequency, with units of rad s^{-1} (or $v_0 = \omega_0/2\pi$, in Hz). In a macroscopic sample at thermal equilibrium, nuclei occupy energy levels according to the Boltzmann distribution. The equilibrium

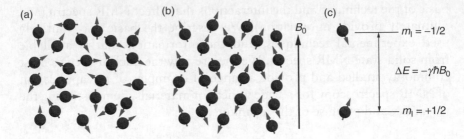

Figure 1.1 (a) In the absence of an external magnetic field, all orientations of the nuclear magnetic moment are degenerate. (b) An external magnetic field, B_0, aligns the nuclear spins and lifts the degeneracy of the nuclear spin energy levels through the Zeeman interaction. (c) For a nucleus with spin quantum number I (here $I = 1/2$), this gives rise to $2I + 1$ spin states of energy $m_I \hbar \gamma B_0$, and $2I$ degenerate transitions with frequency ω_0.

population difference gives rise to a bulk nuclear magnetisation, which may be represented by a vector, M, aligned with the field. The magnitude of M is exponentially dependent on ω_0, so that, at a given field strength, M is much larger for nuclei with higher γ and, for a given nucleus, the magnitude of M will increase with field strength. Typically, therefore, high magnetic field strengths are employed in NMR spectroscopy (usually between 4 and 24 T) to ensure sufficient sensitivity.

1.2.2   Acquisition of Basic NMR Spectra

In the simplest NMR experiment, a short 'pulse' of high-power radiofrequency (rf) electromagnetic radiation is applied to the sample, exciting transitions with energies corresponding to its frequency, ω_{rf}. While all NMR experiments are performed in the static or 'laboratory' frame (*i.e.* a static Cartesian coordinate system); it is convenient to consider the effects of a pulse in the rotating frame; a coordinate system in which the z-axis remains aligned with B_0 and the xy-plane rotates around the z-axis at a frequency of ω_{rf}. In the laboratory frame, the pulse appears as two counter-rotating magnetic fields, with angular frequencies $+\omega_{rf}$ and $-\omega_{rf}$. In the rotating frame, the first of these components appears static and the second rotates at $-2\omega_{rf}$. The static component interacts with nuclear spins, while the rotating component has no effect.

The static field, B_1, supplied by the pulse causes nutation of M about B_1 at a frequency $\omega_1 = -\gamma B_1$ for the duration of the pulse, τ_p. Pulses are generally described by a 'flip angle', $\beta = \omega_1 \tau_p$, the angle through which M nutates during the pulse. The phase, ϕ, of a pulse indicates the direction along which B_1 lies in the rotating frame, and a pulse of flip angle β and phase ϕ is described using the notation β_ϕ. The simplest (sometimes termed 'one-pulse') NMR experiment begins by applying a $90°$ pulse to the system and so creating magnetisation in the transverse (xy) plane (as shown in Figure 1.2). After the pulse, M precesses about the z-axis with a frequency $\Omega = \omega_0 - \omega_{rf}$. This precession is recorded, typically by two orthogonal detectors in the xy-plane, leading to a complex time-dependent signal, $S(t)$, known as a free induction decay or FID. Fourier transformation of $S(t)$ yields the frequency domain signal, $S(\omega)$, or spectrum. In most NMR experiments, 'signal averaging' is carried out, with an experiment repeated N times, and the resulting FIDs co-added. As shown in Figure 1.3, this enables an improvement in the signal-to-noise ratio (SNR) of the resulting spectrum, as the true signal increases linearly with N, whereas random noise increases with $\sqrt{N}$, giving a $\sqrt{N}$ increase in the SNR. Signal averaging is extensively used in NMR, especially in cases where sensitivity is low.

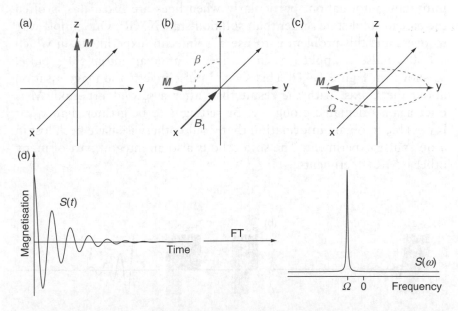

Figure 1.2 (a) Vector model representation of the bulk magnetisation vector, M, aligned along the z-axis of the rotating frame. (b) A pulse applied along the x-axis causes nutation of M in the yz-plane. (c) M then undergoes free precession (and relaxation) in the xy-plane at a frequency Ω. (d) Fourier transformation (FT) of the resulting time domain signal, $S(t)$, yields the frequency domain spectrum, $S(\omega)$.

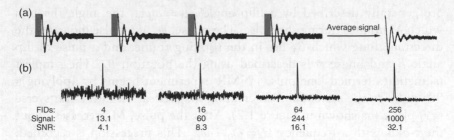

FIDs:	4	16	64	256
Signal:	13.1	60	244	1000
SNR:	4.1	8.3	16.1	32.1

Figure 1.3 (a) Schematic representation of signal averaging. An NMR experiment is repeated several times, with the FIDs co-added to improve the SNR. (b) ^{2}H NMR spectrum of H_2O (natural abundance), acquired with signal averaging of 4, 16, 64 and 256 FIDs. The relative integrated intensity (or 'signal') and SNR are indicated.

While it would be preferable to begin acquiring the FID immediately after the pulse, for a short time any detected signal will contain remnants of the pulse itself (called 'pulse ringing'), which cause distortion and artefacts in the spectrum; thus there must be a short delay or 'dead time' (τ_D) before the FID is acquired. However, this may lead to the loss of important information, particularly when lines are broad (as is often the case in solid-state rather than solution-state NMR). One simple way to overcome this problem is to use a 'spin echo' experiment, in which a $180°_x$ pulse is applied a short time, τ, after an initial $90°_x$ pulse, as shown in Figure 1.4.[3] This second pulse inverts the magnetisation about the x-axis, with the result that, after a second τ period, M is once again orientated along $-y$. By setting τ to be greater than τ_D, it is possible to obtain information that would otherwise have been lost in a one-pulse experiment. The spin echo is also an integral part of many other NMR experiments.

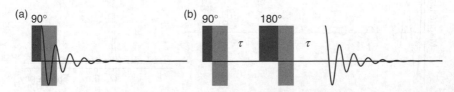

Figure 1.4 Pulse sequences for (a) one-pulse and (b) spin echo experiments. Pulses are shown as dark grey blocks and the dead time, τ_D, is marked in light grey. By refocusing the magnetisation at a time, τ (where $\tau > \tau_D$), after the $180°$ pulse, the spin echo experiment allows acquisition of the whole FID, including any information that would be lost during τ_D in a one-pulse experiment.

1.2.3 Relaxation

In order for signal averaging to be successful, the nuclear spin population must return to thermal equilibrium prior to acquisition of successive FIDs. The return of the magnetisation to equilibrium, termed 'relaxation', is described by a time constant, T_1 (the longitudinal relaxation constant). It is often assumed that equilibrium has been restored after $\sim 5T_1$, and usually it is best to wait for this time between the acquisition of successive FIDs. Although T_1 relaxation times in solution-state NMR can be rapid (typically a few milliseconds), in the solid state they can be much longer (typically a few seconds, but up to many minutes or even hours). Therefore, the acquisition of NMR spectra of solids can be a time-consuming process, requiring, in some cases, very long experiments to achieve an acceptable SNR. In addition to longitudinal relaxation, various processes also attenuate the transverse magnetisation. Transverse relaxation can have a number of different contributions, but is generally described by the time constant T_2; typically, in solids, $T_1 \gg T_2$. Transverse relaxation can alter the width and shape of the line observed in the spectrum, with the shape dependent on the nature of the distribution of frequencies (usually described by a mixture of Gaussian and Lorentzian behaviour) and the width related to $1/T_2$.

1.2.4 Interactions in NMR Spectroscopy

NMR spectroscopy provides a valuable analytical tool, as, in addition to the Zeeman interaction, nuclear spins are also affected by a variety of other interactions, either between two spins or between the spin and its local environment. These provide a sensitive probe of the local structure, symmetry and bonding in a molecule or a solid. Table 1.1 summarises these interactions, their origin and magnitude and the effect they have upon NMR spectra of liquid and solid samples.

1.2.4.1 Chemical Shielding

Although the Larmor frequency, ω_0, depends in principle only upon γ and B_0, in most NMR spectra multiple resonances are observed in the spectrum for any one nuclear species. This is a result of the circulation of

Table 1.1 Summary of the interactions affecting NMR spectra of liquid and solid samples.

Interaction	Description	Magnitude/ Hz	Solution	Solid
Zeeman	Interaction of magnetic dipole moments with external magnetic field	10^7-10^9	yes	yes
Shielding	Alteration of local magnetic field by surrounding electrons	10^2-10^5	isotropic	anisotropic
Dipolar coupling	Through-space magnetic spin–spin coupling	10^3-10^5	0	anisotropic
J coupling	Spin–spin interaction mediated by the bonding electrons	$1-10^3$	isotropic	anisotropic
Quadrupolar	Interaction of nuclear quadrupole moment with electric field gradient (EFG)	10^3-10^7	0	anisotropic
Paramagnetic	Interaction with isolated unpaired electrons in the sample	10^2-10^5	isotropic	anisotropic
Knight shift	Interaction with electrons at the Fermi level in metals	10^3-10^6	no	anisotropic

electrons around the nucleus when in an atom or molecule, generating a magnetic field, B', proportional to B_0. In isolated atoms, B' will always oppose B_0 (*i.e.* it will 'shield' the nucleus from the external magnetic field), but in molecules, B' may oppose or augment B_0 (*i.e.* provide a shielding or deshielding effect[4]). The effective magnetic field experienced by a nucleus, B_{eff}, is given by:

$$B_{eff} = B_0 - B' = B_0(1 - \sigma) \qquad (1.3)$$

where σ is a field-independent shielding constant. The effect of this local magnetic field is to alter the observed precession frequency, ω_{obs}, of a spin:

$$\omega_{obs} = -\gamma B_{eff} = -\gamma B_0(1 - \sigma) \qquad (1.4)$$

resulting in different resonances in the NMR spectrum for magnetically inequivalent nuclei. As an example, Figure 1.5 shows a ^{13}C NMR spectrum of solid L-alanine, where the three distinct chemical environments

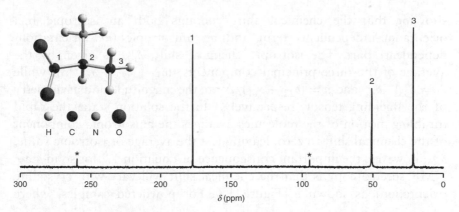

Figure 1.5 ^{13}C MAS NMR spectrum of solid L-alanine. Asterisks denote the 'spinning sidebands' of C1 (see Section 1.3.1.1). Three resonances are observed, arising from the three inequivalent carbons in the molecule. The integrated intensity ratio (1.02 : 1.02 : 1.00) matches that expected from the structure (1 : 1 : 1).

result in three resonances in the spectrum, with relative intensities of 1 : 1 : 1, as expected. In practice, the absolute value of σ is hard to measure, and instead a chemical shift, δ, is measured relative to the known frequency of a reference compound, ω_{ref}. As generally $\sigma \ll 1$, δ is typically reported in parts *per* million (ppm):

$$\delta = 10^6 (\omega_{\text{obs}} - \omega_{\text{ref}})/\omega_{\text{ref}} \tag{1.5}$$

It should be noted that δ is opposite in sign to σ, so that while σ is a measure of shielding and increases with increasing B', δ is a measure of deshielding and increases with increasing B_{eff}.

In general, the electron distribution around a nucleus is rarely perfectly spherical and, therefore, rather than using the scalar constant, σ, the shielding must be described by a shielding tensor, $\boldsymbol{\sigma}$ (and corresponding shift tensor, $\boldsymbol{\delta}$). The observed chemical shift, δ, of a resonance can be shown to be:

$$\delta = \delta_{11} \sin^2\theta \cos^2\phi + \delta_{22} \sin^2\theta \sin^2\phi + \delta_{33} \cos^2\theta \tag{1.6}$$

where δ_{11}, δ_{22} and δ_{33} are the three principal components of $\boldsymbol{\delta}$ when expressed in its principal axis system (see reference [4] for further details), and the angles θ and ϕ describe the orientation of the tensor relative to the external field B_0. Equation 1.6 can be rewritten as:

$$\delta = \delta_{\text{iso}} + (\Delta/2) \left[(3\cos^2\theta - 1) + \eta \left(\sin^2\theta \cos 2\phi\right) \right] \tag{1.7}$$

showing that the chemical shift contains both an isotropic (*i.e.* orientation-independent) term and an anisotropic (*i.e.* orientation-dependent) part. The isotropic chemical shift, δ_{iso}, is given by the average of the three principal components $((\delta_{11} + \delta_{22} + \delta_{33})/3)$, while $\Delta = \delta_{33} - \delta_{iso}$ and $\eta = (\delta_{22} - \delta_{11})/\Delta$ are the magnitude and asymmetry of the shielding tensor, respectively.[a] In the solution state, the rapid tumbling motion of the molecules averages the anisotropic component of the chemical shift to zero, leaving just the average or isotropic value, δ_{iso}; however, the important consequence of Equation 1.7 for solid-state NMR spectroscopy is that the chemical shift will vary with crystallite orientation, as shown in Figure 1.6. For powdered samples, where

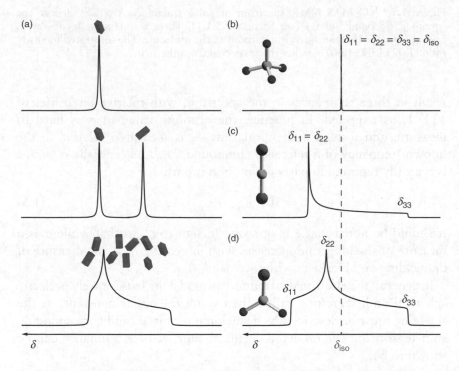

Figure 1.6   (a) The anisotropic nature of the shielding results in a single orientation-dependent resonance for a single crystallite, multiple resonances for chemically equivalent sites in different crystallites and a powder-pattern lineshape in a polycrystalline sample. (b–d) Powder-pattern lineshapes simulated for (b) spherical ($\Delta = 0, \eta = 0$), (c) axially symmetric ($\Delta \neq 0, \eta = 0$) and (d) axially asymmetric ($\Delta \neq 0, \eta \neq 0$) shielding tensors. In each case, the isotropic shielding $((\delta_{11} + \delta_{22} + \delta_{33})/3)$ is marked.

---

[a] Note there are several (often confusing) conventions for shielding in the literature; see reference [4].

crystallites have all possible orientations, the result is a broadened or 'powder-pattern' lineshape, with the centre of gravity at δ_{iso}. The width and shape of the line are determined primarily by Δ and η, respectively, providing information on local structure and symmetry. This can be seen in Figure 1.6, in which simulated lineshapes corresponding to sites with spherical, axially symmetric and axially asymmetric shielding are shown.

1.2.4.2 Internuclear Interactions

In addition to the shielding effects of nearby electrons, the position of a spectral resonance is often affected by interactions with other nuclei. Nuclear dipole moments may couple either directly through space, as in the dipolar interaction, or indirectly (mediated by electrons), as in the through-bond scalar or J coupling.[5] In the dipolar interaction, one spin is affected by the small, localised magnetic fields resulting from another. For an isolated spin pair, this results in an orientation-dependent splitting in the spectrum, proportional to:

$$\omega_D = -\frac{\mu_0}{4\pi}\frac{\gamma_I\gamma_S\hbar}{r_{IS}^3}\frac{1}{2}\ (3\cos^2\theta - 1) \tag{1.8}$$

where r_{IS} is the internuclear distance between spins I and S and θ is the angle between the internuclear vector and the external magnetic field, as shown in Figure 1.7. Therefore, for a powdered sample, where all crystallite orientations are present, the result is a 'Pake doublet' powder-pattern lineshape. However, in most solids there is a virtually infinite number of $I - S$ dipolar interactions present, and the orientation and distance dependence of ω_D leads to a Gaussian-like broadening of the spectrum, as shown in Figure 1.7d. The dipolar interaction is strongest for high-γ nuclei that are close in space, such as ^{1}H and ^{19}F, and can significantly broaden the spectral lines, often over many kHz in the solid state. In solution, however, the dipolar interaction is averaged to zero by the rapid molecular tumbling.

Unlike solution-state NMR spectroscopy, J coupling is rarely observed in solid-state NMR spectra, as it is typically much smaller than the other anisotropic interactions, as shown in Table 1.1. However, as J coupling acts exclusively through regions of shared electron density (e.g. covalent or hydrogen bonds), transfer of magnetisation using this interaction can be used to probe connectivity in solids, as discussed in Section 1.3.1.3.

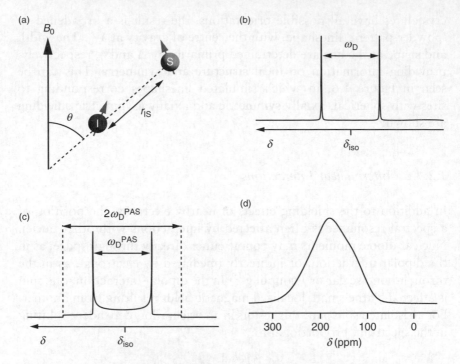

Figure 1.7 (a) Schematic representation of the dipolar interaction between two spins I and S. (b–c) Schematic NMR spectrum for a dipolar-coupled heteronuclear two-spin $I = S = 1/2$ system for (b) a single crystallite and (c) a powdered sample. (d) ^{13}C NMR spectrum of $2[^{13}C]$-glycine, showing the Gaussian-like broadened lineshape observed for many solids where a variety of different dipolar interactions are present.

1.2.4.3 The Quadrupolar Interaction

Around 75% of NMR-active nuclei are quadrupolar (*i.e.* have spin quantum number $I > 1/2$), and their spectra are additionally broadened by the anisotropic interaction of the nuclear quadrupole moment, Q, with the surrounding electric field gradient (EFG). This interaction is usually described by its magnitude, $C_Q = eQV_{zz}/h$, and its asymmetry (or shape), $\eta_Q = (V_{yy} - V_{xx}/V_{zz})$, where V_{ii} are the principal components of the tensor describing the EFG (see reference [6] for further details). The coordinating atoms provide a large contribution to the EFG (although more remote atoms do, of course, have an effect in real materials). As the surroundings vary from a highly symmetric environment, such as

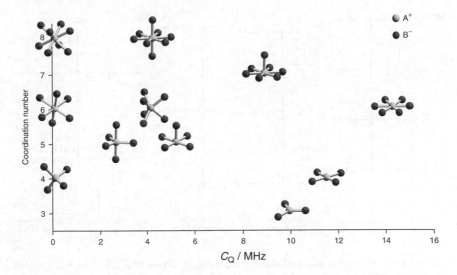

Figure 1.8 Calculated C_Q values for a number of different (idealised) coordination geometries using a point-charge model. After Koller *et al.* (1994) [7].

octahedral coordination where the EFG is spherically symmetrical and C_Q is zero, to a less symmetric one, such as square planar, the value of C_Q shows a corresponding increase, as shown in Figure 1.8 for a number of (idealised) coordination geometries.

In many cases, the quadrupolar interaction can be very large: sometimes many MHz in magnitude. However, in most practically relevant cases it remains smaller than the Zeeman interaction, and its effect can be treated as a perturbation to the Zeeman energy levels. A spin I nucleus has $2I + 1$ allowed orientations of the nuclear magnetic moment with respect to B_0, giving rise to $2I + 1$ Zeeman energy levels, as shown in Figure 1.9a for a spin $I = 3/2$ nucleus. This results in $2I$ degenerate transitions at the Larmor frequency, ω_0. The effect of the quadrupolar interaction (to a first-order approximation) is to perturb the energy levels and lift the degeneracy of the transitions, resulting (for nuclei with half-integer spin quantum number) in a central transition (CT) unaffected by the quadrupolar interaction and satellite transitions (STs) with transition frequencies that depend upon the quadrupolar splitting parameter:

$$\omega_Q = (\omega_Q{}^{PAS}/2)\,[(3\cos^2\theta - 1) + \eta_Q\,(\sin^2\theta\,\cos 2\phi)] \qquad (1.9)$$

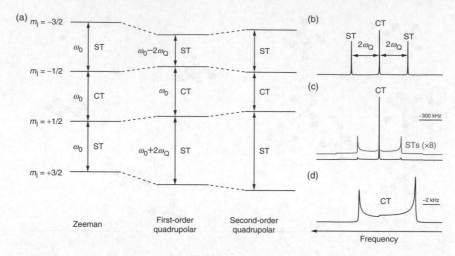

Figure 1.9 (a) Perturbation of the Zeeman energy levels of a spin $I = 3/2$ nucleus by the quadrupolar interaction. (b–c) Resulting spectra showing the effect of the first-order quadrupolar interaction for (b) a single crystallite and (c) a powdered sample. (d) Anisotropic broadening of the central transition (CT) by the second-order quadrupolar interaction.

where ω_Q^{PAS}, in rad s^{-1}, is given by:

$$\omega_Q^{PAS} = 3\pi \; C_Q/(2I(2I - 1)) \tag{1.10}$$

as shown in Figure 1.9. For a single crystal, this would result in $2I$ resonances, as shown in Figure 1.9b. However, in a powdered sample the orientation dependence of ω_Q results in a broadened powder-pattern lineshape for the STs, while the CT remains unaffected, as in Figure 1.9c. In many cases the STs are so broad that spectral acquisition is only concerned with the CT. For larger EFGs, this first-order approximation is insufficient to describe the spectrum and a second-order perturbation must also be considered. The second-order quadrupolar interaction affects all transitions within the spectrum, as shown in Figure 1.9a, and is also orientation dependent (although the dependence is more complex than that shown in Equation 1.10). This has the result that the CT lineshape is also anisotropically broadened, as shown in Figure 1.9d. In general, the second-order quadrupolar broadening is much smaller than the first-order quadrupolar interaction (as it is proportional to $(\omega_Q^{PAS})^2/\omega_0$, rather than ω_Q^{PAS}), and it often results in linebroadening over tens of kHz. It should be noted that for integer spins there is no CT, and all transitions are affected by the first-order quadrupolar interaction, resulting in broadened lineshapes that can be difficult to acquire experimentally, unless C_Q is small.

1.3   BASIC EXPERIMENTAL METHODS

1.3.1   Spin $I = 1/2$ Nuclei

While all of the anisotropic interactions discussed above are present in solution, rapid tumbling of the molecules averages these interactions to their isotropic values. Such motional averaging is absent in most solids, and solid-state NMR spectra of polycrystalline samples contain information on both the isotropic and the anisotropic components of all of the interactions present. This wealth of information leads to very broad, often overlapping lines, from which very little useful information can be obtained. Many of the basic experimental approaches in solid-state NMR spectroscopy are therefore concerned with improving spectral resolution and sensitivity.[1,2]

1.3.1.1   MAS and Decoupling

One widely used approach to obtaining high-resolution (isotropic) spectra is to mimic the orientational averaging that occurs in solution. As described above, the anisotropic parts of the dipolar, chemical shielding, J coupling (and first-order quadrupolar) interactions all have a similar orientational dependence, of the form $(3\cos^2\theta - 1)/2$. These interactions will therefore have a magnitude of zero when $\theta = 54.736°$. It is obviously not practically possible in a powdered sample to orient all crystallites at this angle simultaneously. However, a similar effect can be achieved using a physical rotation of the sample about an axis inclined at an angle, χ, of $54.736°$ to B_0, in a technique called magic angle spinning (MAS),[8–10] shown schematically in Figure 1.10a. While all possible crystallite orientations (β) are still present, if sample rotation is sufficiently rapid the *average* orientation for every crystallite is the same, *i.e.* aligned along the rotor axis at $\chi = 54.736°$. It is possible to describe this mathematically by:

$$\langle (1/2)(3\cos^2\theta - 1) \rangle = (1/2)(3\cos^2\chi - 1) \times (1/2)(3\cos^2\beta - 1)$$

$$(1.11)$$

where χ is the angle of the axis about which the sample is rotated and $\langle\rangle$ denotes the average orientation. The dramatic effect of MAS upon the ^{31}P NMR spectrum of the aluminophosphate, SIZ-4,[11] is shown in Figure 1.10c.

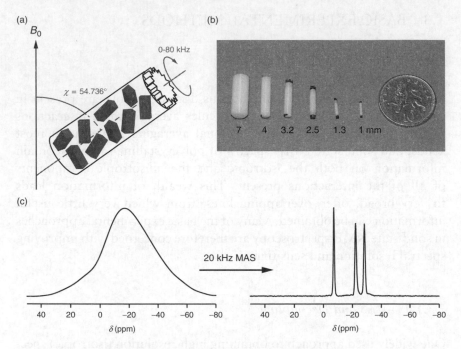

Figure 1.10 (a) Schematic depiction of the MAS experiment, in which a polycrystalline sample is rotated about an axis inclined at the magic angle, χ, of 54.736° to B_0. (b) Rotors of varying outer diameters, as described in Table 1.2. (c) The effect of MAS (20 kHz) upon the ^{31}P NMR spectrum of the aluminophosphate, SIZ-4,[11] which contains three crystallographically distinct phosphorus environments.

Practically, MAS is performed by packing the sample into a holder or 'rotor', typically machined from ZrO_2 (a strong material that can withstand the high forces associated with MAS), which is then rotated at rates of up to 80 kHz. Rotors of varying diameter are available, with the maximum possible MAS rate increasing as the rotor diameter decreases, as shown in Table 1.2 and Figure 1.10. The increase in rotation rate comes with the compromise of sample volume and, therefore, sensitivity. However, in order for anisotropic interactions to be efficiently removed, the rotation must be 'fast' (relative to the magnitude of the interaction that is to be removed).[9,10] Therefore, for ^{1}H and ^{19}F NMR, for example, where the homonuclear dipolar interaction is large, it may be desirable to spin at rapid rotation rates, *i.e.* 70–80 kHz, at the expense of sample volume. If the rotation rate is not sufficiently fast, the powder-pattern lineshape is broken into a series of 'spinning sidebands' (SSBs), separated by integer multiples of the spinning rate, ω_R, from the

Table 1.2 Practical considerations for experimental implementation of MAS NMR experiments.

Rotor diameter/ mm	Maximum rotation rate/kHz	Sample volume/ µl
14	~5	1000–3000
7	~7	300–500
4	~15	50–90
3.2	~23	20–40
2.5	~35	~11
1.3	~65	~2
1	~80	0.8

isotropic peak. At slow MAS rates, the intensity of the SSB manifold follows the static lineshape, but at higher MAS rates this resemblance is lost as the isotropic peak becomes more intense.[9,10] The effect of MAS on a lineshape broadened by the chemical shielding anisotropy (often referred to as the CSA) can be seen in Figure 1.11. For spin $I = 1/2$ nuclei, the CSA is usually the dominant interaction, and it is relatively straightforward to obtain information on the isotropic and anisotropic components from the SSB intensities in a slow MAS NMR spectrum. MAS has the added benefit of partially removing the heteronuclear dipolar coupling and anisotropic J interactions, increasing the resolution and sensitivity of the spectrum.[9,10]

Rather than averaging crystallite orientations in real space, it is possible to carry out averaging in 'spin space', i.e. by manipulating the nuclear spins using rf pulses. This approach, known as decoupling,[12,13] is able to remove the dipolar coupling, which may, as we have seen, be a large interaction (and MAS rates may not be sufficient to remove it completely). Decoupling will also remove J couplings, although these are typically much smaller. Heteronuclear decoupling,[12] i.e. the removal of the dipolar (or J) interaction between two different spins, I and S (e.g. ^{1}H and ^{13}C), is relatively straightforward and, at its most basic, takes the form of continuous rf irradiation at the Larmor frequency of S while the FID is acquired for I. In order to remove strong interactions, high-power pulses are often required, although care must be taken to avoid excessive heating of the sample. Over the years a vast range of more complicated multiple-pulse decoupling schemes has been developed in order to remove heteronuclear dipolar interactions with greater efficiency.[12] Figure 1.12 shows ^{13}C NMR spectra of the carbonyl resonance in glycine, acquired without and with ^{1}H

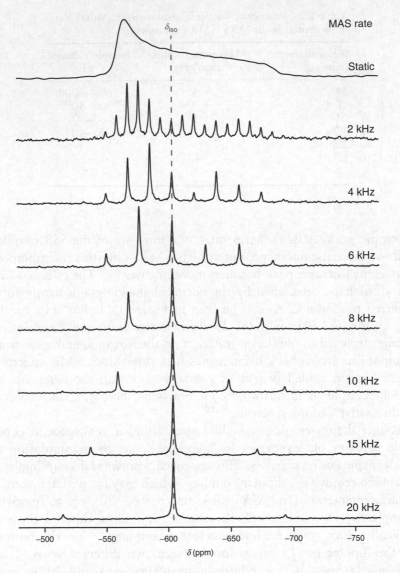

Figure 1.11 Effect of MAS upon the ^{119}Sn (14.1 T) NMR spectrum of SnO_2, containing a lineshape anisotropically broadened by the anisotropic chemical shift interaction. At low MAS rates, the SSB manifold mirrors the static lineshape. As the MAS rate increases, the sidebands decrease in intensity and the isotropic peak dominates the spectrum.

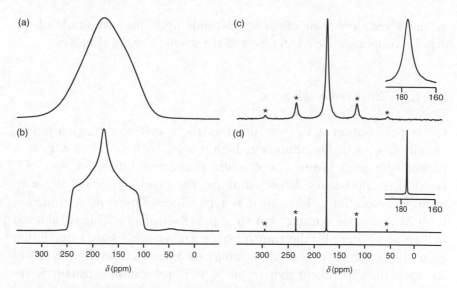

Figure 1.12 Effect of MAS and ^{1}H decoupling upon a ^{13}C (9.4 T) NMR spectrum of the carbonyl resonance of 2[^{13}C]-glycine (enriched to ~99% in ^{13}C). In (a), a broad, featureless lineshape is observed, while in (b) ^{1}H decoupling reveals a powder-pattern lineshape as a result of the ^{13}C CSA. The effect of MAS (6 kHz) upon the lineshapes in (a) and (b) can be seen in (c) and (d), respectively. SSBs are marked with asterisks, and insets in parts (c) and (d) show the narrowing of the isotropic peak (here by a factor of ~16) upon decoupling. The combination of MAS and decoupling narrows the carbonyl resonance by a factor of ~400.

decoupling. A featureless lineshape broadened by dipolar interactions (between ^{1}H and ^{13}C) is observed when no decoupling is applied, while ^{1}H decoupling results in a characteristic CSA powder-pattern lineshape. Decoupling cannot remove the CSA, and so to obtain truly high-resolution spectra a combination of MAS and decoupling is often used. Figure 1.12 shows that the use of MAS significantly improves the resolution of the ^{13}C spectrum of glycine, both without and with decoupling. In favourable cases, the combination of MAS and decoupling can lead to linewidths comparable to those obtained in the solution phase.

Homonuclear decoupling,[13] *i.e.* the removal of dipolar interactions between two I spins, is considerably more difficult, as it requires simultaneous manipulation and observation of the I spins. This is usually achieved with 'windowed' acquisition schemes, where decoupling pulses alternate with 'windows' in which decoupling is not applied, enabling acquisition of FID data points. As with heteronuclear decoupling, there are a variety of different approaches, or sequences of pulses, that can be employed, and their efficiency depends upon the magnitude of the dipolar couplings, the MAS rate and the strength of the rf pulses.

1.3.1.2 Cross Polarisation

Cross polarisation (CP)[14,15] involves the transfer of magnetisation, usually from a highly abundant, high-γ spin, such as ^{1}H or ^{19}F, to a second spin with lower γ and lower abundance (often ^{13}C or ^{29}Si). Unlike the approaches described above, the main aim of CP is not to improve resolution (although it is typically employed in conjunction with MAS and decoupling) but to improve sensitivity. This is achieved in two ways: first, by the transfer of magnetisation (with a maximum gain of a factor of γ_I/γ_S in favourable cases); and second, by the ability to repeat the experiment more rapidly, as T_1 relaxation is usually faster for high-γ, high-abundance spins (*e.g.* a factor of 15–20 faster in the case of ^{1}H/^{13}C). CP has revolutionised the acquisition of NMR spectra for ^{13}C and ^{15}N in particular, opening up the study of organic systems in the solid state, from small molecules to large proteins.[15]

The transfer of magnetisation in CP takes place *via* heteronuclear dipolar coupling, thereby also 'editing' the spectrum on the basis of spatial proximity to the heterospin. In this respect, not only does CP result in increased sensitivity, it also provides structural information. The pulse sequence used for CP is shown in Figure 1.13a,[15] where magnetisation initially created by a 90° pulse on spin *I* is transferred to spin *S* in a 'contact time', during which low-power pulses are applied to both spins to 'lock' the magnetisation along a particular direction while transfer takes place. The duration of this period is chosen to maximise the transferred signal intensity, which depends upon both the transfer rate (proportional to the dipolar coupling between the spins) and the relaxation of each spin during the spin–lock pulses (described by a time constant, $T_{1\rho}$). After the contact time, the *S* spin FID is acquired, using *I* spin decoupling if necessary. Variation of the CP contact time can help assign a spectrum and provide information on the material, *e.g.* magnetisation will build up more quickly for -CH$_2$- or {Si(OSi)$_3$(OH)} groups than for quaternary carbons or {Si(OSi)$_4$} species. This is shown in Figure 1.13c, where the intensities of the resonances in the ^{13}C CP MAS NMR spectrum of L-alanine vary with contact time. Although CP can provide structural information in this way, the dependence upon dipolar coupling does result in non-quantitative spectra, and care must

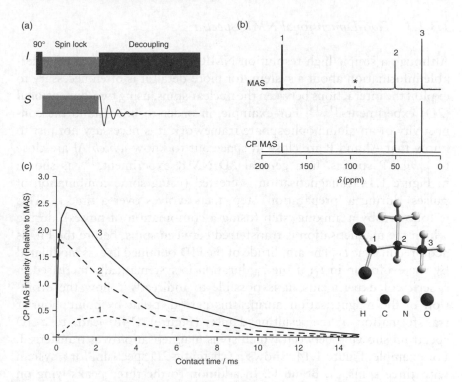

Figure 1.13 (a) Pulse sequence for a CP experiment transferring magnetisation from spin I to spin S. (b) ^{13}C MAS and CP MAS NMR spectra of L-alanine (contact time $=$ 1 ms), showing the non-quantitative nature of CP; the asterisk denotes a SSB. (c) Plot of CP MAS peak intensities (relative to the MAS NMR spectrum) as a function of contact time.

be taken when considering the relative intensities of resonances in a CP spectrum.

In the CP experiment, magnetisation transfer only occurs when the rf fields applied during the contact time fulfil the Hartmann–Hahn condition:[16]

$$\gamma_I B_{1I} = \gamma_S B_{1S} \tag{1.12}$$

This condition must be adapted if the experiment is performed under MAS conditions to:

$$\gamma_I B_{1I} = \gamma_S B_{1S} \pm n\omega_R \tag{1.13}$$

where ω_R is the MAS rate and n is an integer (typically 1 or 2). Any rf field strengths that satisfy these match conditions may, in principle, be used in the experiment. Practically, however, lower powers are preferred, in order to minimise detrimental effects on the hardware.

1.3.1.3   Two-Dimensional NMR Spectra

Although a simple high-resolution NMR spectrum can reveal considerable information about a system, for more detail it is often necessary to exploit the interactions between the nuclear spins, using two-dimensional (2D) experiments.[17,18] For example, in order to understand the connectivity of an aluminophosphate framework it is necessary not just to know that Al and P are close in space but to know *which* Al are close to *which* P species. In a general 2D NMR experiment,[18] as shown in Figure 1.14a, magnetisation is created (using some combination of pulses) during a 'preparation' step, then evolves over a time t_1. This is followed by a 'mixing' step (using a combination of pulses), during which the magnetisation is transferred between spins, before the FID is acquired in time t_2. The amplitude of the FID obtained in t_2 is modulated by the evolution in t_1; if the t_1 duration is systematically increased in a series of experiments, it is possible to indirectly follow the evolution of the magnetisation during this period, point by point. Fourier transformation of the resulting dataset (Figure 1.14b) leads to a 2D spectrum showing between which spins magnetisation was transferred. For example, Figure 1.14c shows a schematic 2D spectrum for a system with three spins, A, B and C. In addition to the three peaks lying on the diagonal (corresponding to magnetisation that was not transferred between spins), off-diagonal 'cross peaks' are observed. These reveal that magnetisation has been transferred between A and B and also between B and C, demonstrating the presence of an interaction or connection between these spins. In contrast, there is no transfer between A and C, showing that these two spins do not have such an interaction.[18]

Correlation experiments can be typically classified into two types: heteronuclear (*i.e.* between two different nuclear species) and homonuclear (*i.e.* between nuclear species of the same type). In the former case, it is necessary to apply pulses to both species in order to enable magnetisation to be transferred from one type of nucleus to the other. It is also possible to design 2D experiments (by careful choice of the pulses applied) such that the magnetisation transfer proceeds *via* either through-bond J coupling or through-space dipolar interaction; cross peaks will then demonstrate that two spins are either connected by covalent bonds or close in space, respectively. Note that although the J coupling is typically very small and is often unresolved in solid-state NMR spectra, it can still be exploited for the transfer of magnetisation. If this were the case in Figure 1.14c, it would indicate that A and B were connected by covalent bonds, as were B and C, but that the

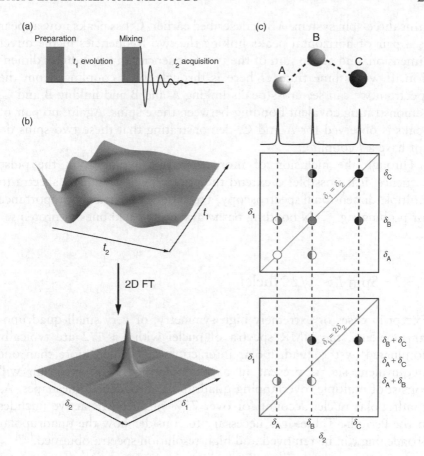

Figure 1.14 (a) Pulse sequence for a generic 2D experiment in which the t_1 duration is incremented. (b) The 2D FID and spectrum resulting from Fourier transformation. (c) Schematic 2D homonuclear correlation spectra for a system with three spins, A, B and C, showing cross peaks (and interactions) between A and B and between B and C. No cross peaks are observed between A and C. The spectra result from a conventional 2D experiment (upper) and a 'double-quantum/single-quantum' experiment (lower).

number of bonds between A and C was too large for a J coupling to be present. In some cases, the particular type of pulse(s) employed can result in a 2D spectrum that looks very different but essentially contains the same information. The theory behind these 'double-quantum/single-quantum' experiments in the solid state[19] is beyond the scope of the current discussion, but they are widely used, and it is worthwhile to consider the appearance of the resulting spectrum. Figure 1.14c also shows a schematic 'double-quantum/single-quantum' spectrum for the same three-spin system, ABC, described earlier. Cross peaks now appear as a pair of horizontal peaks linking the two frequencies in the direct dimension, and at the sum of the two frequencies in the indirect dimension. If we assume transfer here is through the *J* coupling, from the spectrum we can see cross peaks linking A and B and linking B and C, demonstrating covalent bonding between these spins. Again, no pair of peaks is observed for A and C, demonstrating that these two spins do not have a *J* coupling.

Through the inclusion of more than one evolution in the pulse sequence, it is possible to extend the principles of 2D NMR spectra to multiple-dimensional spectroscopy, a technique of particular importance for probing *e.g.* C–N bonding pathways in the backbones of proteins.

1.3.2   Spin *I* > 1/2 Nuclei

Except in cases of extremely high symmetry, or very small quadrupolar moments, the NMR spectra of nuclei with *I* > 1/2 are typically dominated by the quadrupolar interaction.[6,20] When more than one inequivalent site is present in a material, the NMR spectrum will consist of multiple, overlapping quadrupolar-broadened lineshapes. As quadrupolar nuclei account for over 75% of all NMR-active nuclides in the Periodic Table, it is necessary to consider how the quadrupolar broadening can be removed and high-resolution spectra obtained.[20]

1.3.2.1   MAS

As described previously, the orientation dependence of the first-order quadrupolar interaction is similar to that of the CSA or dipolar coupling, and so is, in principle, removed by MAS. In practice, however, this interaction is often many hundreds of kHz or MHz in magnitude, and it is usually impossible to spin sufficiently rapidly. Therefore, the broad ST lineshapes for nuclei with half-integer spin (and all lineshapes for nuclei with integer spin) exhibit extensive sideband manifolds.[6,20] However, for nuclei with half-integer spin quantum number, the CT lineshape is not affected by this large first-order quadrupolar interaction, but is broadened by the much smaller second-order quadrupolar broadening. Unfortunately, the second-order quadrupolar interaction has a more complex angular dependence than the other interactions and cannot be removed completely by MAS. The resonance frequency (assuming $\eta_Q = 0$ for simplicity) is given (under MAS) by:[6,21]

$$\omega = \frac{(\omega_Q^{PAS})^2}{\omega_0} \{A^0 + A^2 d_{00}^2(\beta)d_{00}^2(\chi) + A^4 d_{00}^2(\beta)d_{00}^4(\chi)\} \quad (1.14)$$

where A^n are spin-dependent coefficients, given in the literature, and β describes the orientation of the crystallite and χ the angle of the rotor axis. Equation 1.14 shows that the second-order quadrupolar interaction consists of an orientation-independent shift (proportional to A^0) and two anisotropic terms (proportional to A^2 and A^4, respectively). As:

$$d_{00}^2(\chi) = \frac{1}{2}(3\cos^2\chi - 1) \quad (1.15)$$

$$d_{00}^4(\chi) = \frac{1}{8}(35\cos^4\chi - 30\cos^2\chi + 3) \quad (1.16)$$

it can be seen that MAS will only remove one of these terms. The last term in Equation 1.14 is removed when $\chi = 30.56$ or $70.12°$, but there is no one angle about which the sample could be rotated that would remove all of the anisotropic broadening. Hence, for quadrupolar nuclei, MAS NMR spectra contain lines that are narrowed, but are not truly high resolution, and are shifted away from the isotropic chemical shift, as shown in Figure 1.15a. In principle, these lineshapes can be fitted to extract δ_{iso} and the quadrupolar parameters, C_Q and η_Q, providing information on the local environment or symmetry. However, if a number of inequivalent species are present, the spectrum may contain a number of overlapped quadrupolar lineshapes and it can once again be difficult to extract any useful information. This can be seen in Figure 1.15b – the ^{17}O MAS NMR spectrum of $MgSiO_3$[22] – where it is difficult to obtain any information due to the presence of six overlapping broadened lineshapes. It can be seen from Equation 1.14 that the quadrupolar broadening is reduced at high magnetic field strength (as it is proportional to $1/\omega_0$); however, in most cases significant broadening remains at the conventional B_0 fields that are available. It is desirable, therefore, to develop methods by which to remove the quadrupolar broadening completely and achieve truly isotropic spectra.

1.3.2.2 DOR and DAS

The earliest approaches to removing quadrupolar broadening used composite sample rotation, *i.e.* rotation around two different angles.

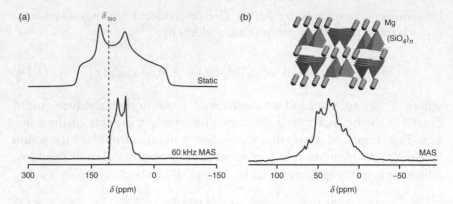

Figure 1.15 (a) Effect of MAS (60 kHz) on the ^{71}Ga (20.0 T) quadrupolar-broadened CT lineshape of GaPO$_4$ berlinite. A powder-pattern lineshape is still observed (although it is narrowed in comparison to the static case), shifted from δ_{iso} by the isotropic quadrupolar shift. (b) ^{17}O (9.4 T) MAS NMR spectrum of orthoen-statite (MgSiO$_3$),[22] containing six distinct O sites and exhibiting a complicated overlapping lineshape. Reprinted with permission from Ashbrook *et al.* (2007) [22]. Copyright (2007) American Chemical Society.

In double rotation (DOR),[20,23] the two rotations take place simultaneously, with an inner rotor spinning inside a much larger outer rotor, as shown in Figure 1.16a. The outer rotor is inclined at 54.74° to B_0 *i.e.* the magic angle, in order to remove the broadening proportional to $d^2_{00}(\chi)$ (and also dipolar and CSA interactions), while the inner rotor spins at 30.56° to the outer rotor, in order to remove the anisotropic broadening proportional to $d^4_{00}(\chi)$. In principle, this enables the complete removal of all quadrupolar broadening in a simple one-dimensional (1D) experiment, as shown in Figure 1.16c. A major limitation of DOR, however, is the restricted rotation rate (typically 1–2 kHz) of the bulky outer rotor, producing a series of SSBs in some cases that may complicate spectral analysis.

Like DOR, dynamic angle spinning (DAS)[20,25] removes quadrupolar broadening by spinning around two angles; however, this occurs sequentially rather than simultaneously. DAS is a 2D experiment, in which the magnetisation evolves in t_1 with the rotor inclined at one angle and then, before t_2, the angle of the rotor is changed so that, after a certain length of time, the first- and second-order quadrupolar broadening is refocused and only the isotropic peak remains. Between t_1 and t_2, the magnetisation is 'stored' without evolution, meaning that reasonably

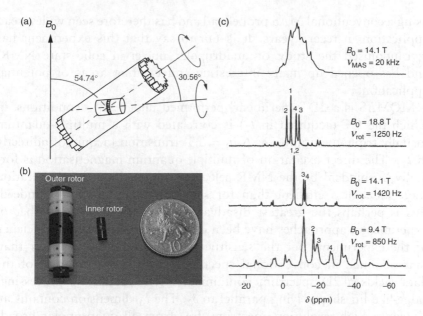

Figure 1.16 (a) Schematic depiction of a DOR experiment, with a large outer rotor (inclined at $54.74°$ to B_0) and a smaller inner rotor (inclined at $30.56°$ to the outer rotor). (b) Inner and outer DOR rotors. (c) ^{23}Na MAS (top) and DOR NMR spectra of $Na_4P_2O_7$ acquired at varying Larmor frequencies and rotation rates (of the outer rotor). Adapted with permission from Engelhardt *et al.* (1999) [24]. Copyright (1999) Elsevier Ltd.

slow T_1 relaxation is required. For ease of processing, the FID may be acquired with the rotor spinning at the magic angle (to remove dipolar and CSA interactions), although this requires an additional storage and angle-setting period.

Both DOR and DAS require specialist probeheads, capable of rotating two rotors simultaneously or of switching the rotor axis during the experiment, which increases the cost of both of these approaches and limits their widespread applicability.

1.3.2.3 MQMAS and STMAS

In contrast to DOR and DAS, multiple-quantum magic angle spinning (MQMAS)[20,26] uses spin manipulation rather than spatial manipulation to achieve an isotropic spectrum. The experiment can be performed using a conventional MAS probehead and has therefore seen widespread application in recent years. It is fair to say that this experiment has revolutionised the study of quadrupolar nuclei in solid-state NMR, and has opened up many interesting and exciting areas of potential application.

MQMAS is a 2D experiment, performed under MAS conditions, in which the CT (acquired in t_2) is correlated with a multiple-quantum (usually triple-quantum, *i.e.* $\Delta m_I = \pm 3$) transition (acquired indirectly in t_1). The direct excitation of multiple-quantum magnetisation is formally 'forbidden' by the NMR selection rule, making the preparation stage much less efficient than for single-quantum excitation; indeed, this is perhaps the greatest disadvantage of MQMAS. A number of experimental approaches have been developed to increase the efficiency of the experiment, but the sensitivity is still significantly lower than that obtained in conventional spectra. Fourier transformation of the data yields a 2D spectrum containing (after appropriate processing) ridge-like lineshapes lying parallel to δ_2. The δ_1 dimension contains an isotropic, high-resolution spectrum, free from all quadrupolar broadening. This can be seen in the ^{87}Rb MAS and MQMAS NMR spectra of RbNO$_3$ shown in Figure 1.17. The three quadrupolar-broadened lineshapes are overlapped in the MAS NMR spectrum, but can be separated out in the 2D spectrum, producing a high-resolution spectrum containing three sharp, narrow resonances. In addition, it is also possible to study each of the quadrupolar lineshapes individually, by extracting cross-sections parallel to δ_2. Figure 1.17c shows how the three lineshapes in the MQMAS NMR spectrum of RbNO$_3$ can be fitted to extract the quadrupolar parameters, C_Q and η_Q. Information on the quadrupolar interaction and the isotropic chemical shift can also be obtained from the position of the lineshape in the 2D spectrum.

The satellite transition magic angle spinning (STMAS) experiment[21,27] is conceptually very similar to MQMAS, in that the CT lineshape is correlated, in a 2D spectrum, with a spectrum corresponding to a different transition within the spin system. Rather than using a forbidden multiple-quantum transition, STMAS correlates the spectrum from the ST, increasing significantly the sensitivity of the experiment (typically by a factor of 4–8). However, as the STs are also affected by the very large first-order quadrupolar broadening, the experiment is technically more challenging to perform. It is important to ensure that the magic angle is accurately adjusted (to $\pm 0.002°$), the rotation rate is very stable and the pulses applied are timed extremely accurately.

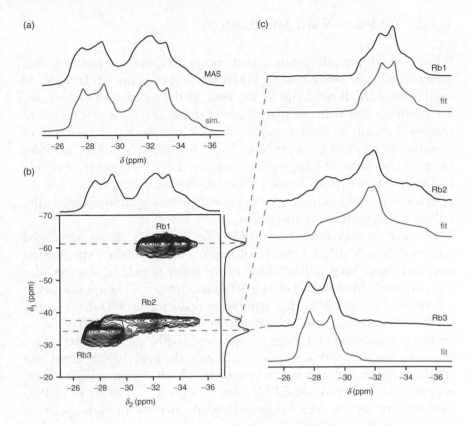

Figure 1.17 ^{87}Rb (14.1 T) (a) MAS and (b) MQMAS NMR spectra of RbNO$_3$, with the latter showing the presence of three distinct Rb species in the isotropic spectrum. (c) Fitting of the three quadrupolar lineshapes obtained from the spectrum in (b). In (a), the simulated MAS NMR spectrum uses the parameters obtained from the fits shown in (c).

The appearance of an STMAS NMR spectrum is very similar to that of a MQMAS NMR spectrum; however, the increased difficulty of implementation has limited its popularity in comparison to MQMAS, and it is typically only used for cases where sensitivity is a problem.

It should be noted that it is possible to utilise all of the experimental methods described above for spin $I = 1/2$ nuclei (*e.g.* CP, decoupling and 2D experiments) with quadrupolar nuclei, although in many cases there is more of a challenge in their experimental implementation. However, if high-resolution spectra are required, these approaches must be combined with experiments such as MQMAS, *e.g.* in order to remove the quadrupolar broadening prior to transfer of magnetisation.[20]

1.3.3   Wideline NMR Spectroscopy

Large anisotropic interactions, such as quadrupolar or paramagnetic interactions, can often lead to NMR lineshapes many MHz wide. In such cases, MAS is often not of any help, as the rotation rates available are much smaller than the magnitude of the interactions one wishes to remove. Instead, it is often simplest to acquire a 'wideline' spectrum of a static sample, using experiments based around spin echoes in order to ensure undistorted lineshapes are acquired.[20,28] However, it is only really feasible to extract detailed information from such spectra when the number of crystallographic species present is low, as multiple lineshapes will be overlapped in the spectrum.

In order to overcome the considerable sensitivity issues associated with wideline NMR, a number of improvements to the experimental approach have been utilised. One of the most popular is the use of a Carr–Purcell–Meiboom–Gill (CPMG) echo train.[29] The application of a series of 180° pulses during acquisition results in an FID that consists of multiple echoes, and subsequent Fourier transformation yields a spectrum consisting of a series of 'spikelets', with intensities that reflect the static lineshape; this therefore increases the peak-height signal and results in significant increases in sensitivity. This is shown in Figure 1.18, where ^{71}Ga spin echo and CPMG spectra of a static sample of GaPO$_4$ berlinite are shown. The integrated signal intensity in each spectrum is the same, but the peak-height sensitivity is vastly improved when the CPMG echo train is employed, as shown in Figure 1.18b. If, as is frequently true in the case of wideline NMR, the linewidth exceeds the excitation bandwidth (*i.e.* the frequency range that can be efficiently and uniformly excited by the rf pulses), the spectrum must be acquired in

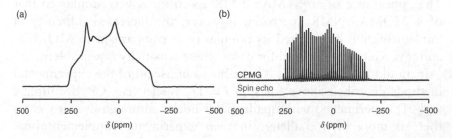

Figure 1.18   Static ^{71}Ga (14.1 T) NMR spectra of GaPO$_4$ berlinite acquired using (a) a spin echo pulse sequence and (b) a CPMG echo train with a spikelet spacing of 2 kHz. In part (b), the spin echo spectrum is plotted on the same vertical scale as the CPMG spectrum.

a stepwise fashion, varying the frequency offset of the transmitter and co-adding the individual subspectra to generate the final lineshape. It has also been recently shown that the use of so-called 'shaped' pulses, designed to excite over a much broader frequency range, and previously employed in solution-state NMR, can be used in either a simple echo experiment or a CPMG echo train in order to improve the excitation bandwidth. These can be implemented in a single experiment to acquire a whole spectrum in one step or can be combined with stepped acquisition to reduce the number of different steps required to acquire very broad lineshapes (and hence the overall experimental time).[20]

1.4   CALCULATION OF NMR PARAMETERS

Unlike solution-state NMR, the assignment of solid-state NMR spectra poses a considerable challenge. This is a result of the lack of the extensive databases of parameters that exist for the solution state, and also a consequence of the crystallographic inequivalence of chemically similar species resulting from crystal packing in the solid. Furthermore, for inorganic solids in particular, a variety of nuclides are typically studied and a vast array of local and longer-range coordination environments is possible, making database-based assignment methods extremely impractical. In recent years, there has been growing interest in the use of alternative tools to aid spectral assignment and interpretation, and first-principles calculations provide one such option. Owing to the latest developments in hardware and software, these approaches may be applied to ever more complicated materials. While a full understanding of density functional theory (DFT) is beyond the scope of this chapter (see references [30, 31] for more detail), we provide a brief introduction to the key steps involved in calculating the parameters relevant to solid-state NMR.

1.4.1   Introduction to Density Functional Theory

Electronic structure calculations aim to provide a description of the properties of a material using only the fundamental assumptions of quantum mechanics (*i.e.* from 'first principles'), starting from the Schrödinger equation ($H\psi = E\psi$) for a system of electrons and nuclei. From the total energy, E, it is possible to derive all fundamental NMR parameters.

However, the Schrödinger equation is too complicated to solve exactly for many-body systems. One approach to simplifying this problem is to treat part of the interaction between the electrons (termed their 'correlation') in an average way. This 'Hartree–Fock' approach provides an accurate answer, but any attempt to improve the treatment of electron correlation quickly becomes computationally too costly to apply to extended solids.[30,31] An alternative approach, DFT, exploits the fact that the total energy is known to be a function of the electron density.[30,31] As this is simply a function of position, it is much simpler to calculate than a many-body wavefunction and, therefore, computationally cheaper for larger systems. While most contributions to the energy can be computed exactly, the form of the 'exchange correlation' energy, which describes the interaction between electrons, is not known exactly. A number of simple approximations have been introduced to overcome this, including the local density approximation (LDA), which assumes that, for a small unit of space, the electron density is constant and equal to that of a uniform electron gas. Despite its simple nature, LDA has been shown to provide a good approximation for the energy and structure of solids. For other properties, it is possible to improve accuracy by adding terms based on the gradient of the density, *i.e.* the generalised gradient approximation (GGA). Although further adaptations of GGA can result in increased accuracy for small molecules, application to solids generally proves too computationally expensive.

1.4.2   Basis Sets and Periodicity

Although there has been considerable application of the calculation of NMR parameters in quantum chemistry, much of the effort has been focused on small isolated molecules, rather than the extended structures found for most solids. A solid, therefore, has to be approximated as a 'cluster', centred on the atom of interest with the termination of any 'dangling' bonds, usually with H. The accuracy of the calculations increases with the size of the cluster used; however, this also increases the computational cost of the calculation. For solids, a more efficient approach is to exploit the inherent translational symmetry, recreating the 3D structure from a small high-symmetry volume unit using periodic boundary conditions. This reduces the number of distinct atoms that must be considered in the calculation to manageable levels, again decreasing the cost.[31]

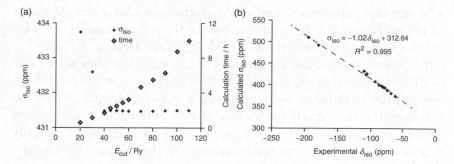

Figure 1.19   (a) Plot showing the variation of the calculated ^{29}Si isotropic shielding of SiO_2 coesite as a function of the plane wave cut-off energy, E_{cut}. The calculation time or 'computational cost' as a function of E_{cut} is also shown, and continues to increase above the 'converged' E_{cut} value of $\sim$45 Ry, with no improvement in the accuracy of the calculation. (b) Comparison of ^{29}Si calculated isotropic shielding and experimental isotropic shifts for a series of simple inorganic solids. Analytical fitting enables the value of the reference, σ_{ref}, to be determined from the y intercept.

It is usual to express the wavefunction as a linear combination of a group of simple functions, or 'basis set', for ease of calculation. In quantum chemical calculations of molecules, atom-centred basis sets, consisting *e.g.* of Gaussian functions, are often employed. However, for the periodic systems of interest in the solid state, a more natural choice of basis set is plane waves,[31] which are inherently periodic. In principle, the accuracy of the calculation will increase with the size of the basis set used, although in practice a finite number of functions are usually chosen, such that the calculated results do not change or improve significantly if more are added, *i.e.* the calculation is said to be 'converged'. This is straightforward for a plane wave basis set, where the NMR parameters can be calculated as a function of the maximum energy of the plane waves included (the cut-off energy, E_{cut}), as shown in Figure 1.19a for the ^{29}Si isotropic shielding of SiO_2 coesite. Here, any increase in E_{cut} past 45 Ry (612.3 eV) results only in an increase in cost, with no significant improvement in accuracy.

1.4.3   Reducing the Computational Cost of Calculations

The explicit description of core electrons is computationally costly and generally unnecessary, as the valence electrons are responsible for most chemical phenomena. It is possible to reduce computational cost by using a 'frozen-core' approximation, where electrons within a defined radius of the nucleus, r_{core}, are treated as part of the static potential provided by the nuclei, while electrons outside r_{core} are treated explicitly.[31] Furthermore, the oscillatory nature of the valence wavefunctions close to the nucleus may require a large number of plane waves to reproduce. It is possible to reduce this cost by using a 'pseudopotential', where the wavefunction is artificially smoothed close to the nucleus, thus saving on both cost and time. However, many NMR parameters are crucially dependent upon the electron density close to the nucleus; fortunately, this information can be easily recovered using schemes such as the projected augmented wavefunction (PAW)[30] method or the gauge-including projected augmented wavefunction (GIPAW)[32] method (for properties that involve the response of a system to magnetic field). A wide range of codes using a variety of different approaches are available, although the recent introduction of GIPAW into periodic codes has proven particularly popular.

1.4.4   Application of First-Principles Calculations

The resurgence of interest in the calculation of NMR parameters in the solid state has led to investigations in areas as diverse as biomaterials, minerals, microporous frameworks, energy materials and organic solids. In general, calculations provide support for the assignment of spectra and confirmation of the NMR parameters for a system (particularly important if sensitivity is poor). Calculations may also be used to predict spectra prior to experiment, in order to guide experimental acquisition for materials that are particularly challenging to study. For systems where the structure is unknown or under debate, calculations can enable the evaluation or validation of structural models and predictions against an experimental observable. It is usual to calculate both the EFG (and therefore the quadrupolar interaction) and the shielding; the former can be calculated in a fraction of the time required for the latter. More recent code development has also enabled the calculation of J couplings.[33] The isotropic and anisotropic components of all interactions are typically calculated, and the latter can be particularly useful as it can be averaged either by the NMR methods used or by dynamics in a system. It should be noted that absolute values of the shielding are calculated, and for comparison to experiment these generally have to be referenced in some way. Usually it is assumed that:

$$\delta_{iso}^{calc} = (-\sigma_{iso}^{calc} + \sigma_{ref})$$ (1.17)

where σ_{ref} is a reference shielding, determined either by matching the experimental and calculated data from a simple reference compound or (more usually) from a plot comparing experimental shifts and calculated shieldings for a range of simple compounds. An example is shown in Figure 1.19b, where ^{29}Si calculated shieldings and experimental shifts are compared for a series of Si-containing inorganic compounds. Analytical fitting enables σ_{ref} (312.64 ppm) to be determined from the y intercept of the plot.

One crucial consideration when calculating NMR parameters is the accuracy (or otherwise) of the initial structure or structural model. In many cases, structures are obtained from diffraction experiments, although many are generated from computational approaches. For structures determined from experiment, the quality of the data may be variable depending upon the type of approach (*e.g.* laboratory X-ray, synchrotron X-ray or neutron) and on whether a single crystal or powdered sample is available. In particular, the positions of lighter atoms, such as hydrogen, can be difficult to determine accurately from X-ray data, and even small errors in these positions, or those of other atoms, can significantly affect the calculated NMR parameters. Figure 1.20 shows an example from work on calcined AlPO-14, a microporous aluminophosphate.[34] NMR parameters calculated for the structure obtained directly from diffraction were in poor agreement with those obtained experimentally. However, after the optimisation of the atomic positions using DFT and subsequent calculation of the NMR parameters, much better agreement was observed. The structural changes are difficult to observe by visual inspection, as shown in Figure 1.20a, but result in a considerable change in the ^{27}Al and ^{31}P calculated NMR parameters, as shown in Figure 1.20b.

As a consequence of the improvements in software, the calculation of NMR parameters from first principles is no longer the realm of specialists, and many experimentalists now utilise them almost routinely alongside experiment to provide a detailed understanding of local structure. The real power of this approach, however, has been demonstrated in more recent work, where calculations have been used to provide insight into both disorder and dynamics in the solid state.

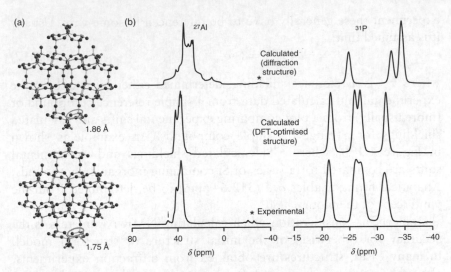

Figure 1.20   (a) Structure of calcined AlPO-14 from diffraction (top) and after geometry optimisation of the atomic positions (bottom). (b) Experimental (14.1 T) ^{27}Al and ^{31}P MAS NMR spectra for AlPO-14, and lineshapes simulated using the NMR parameters calculated with DFT prior to and post structural optimisation. Asterisks denote SSBs and i denotes a minor impurity phase observed in the experimental spectrum. Reproduced (in part) with permission from Ashbrook *et al.* (2008) [34]. Copyright (2008) PCCP Owner Societies.

1.5   APPLICATIONS OF SOLID-STATE NMR SPECTROSCOPY

1.5.1   Local and Long-Range Structure

The local magnetic environment at a nucleus is determined by a variety of structure-dependent interactions and NMR spectra therefore contain a wealth of information on the structure of crystalline materials. As in solution, the number of isotropic resonances corresponds to the number of distinct nuclei, and their relative intensities can be used to determine the proportion of each species (care may have to be taken for more complex experiments, where intensities may be non-quantitative, but this can be taken into account by appropriate spectral analysis). For molecular solids, NMR can provide information on the number of distinct molecules in the asymmetric unit, while for extended solids it may be possible to distinguish between two possible space groups of a material, or to rule out a proposed space group or model. NMR is also an excellent method for identifying or distinguishing between different

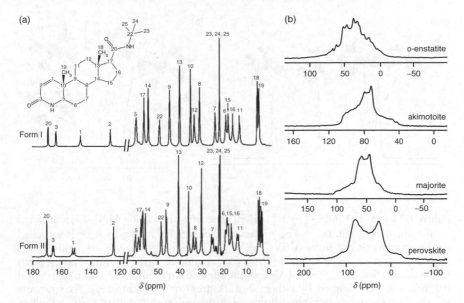

Figure 1.21 (a) ^{13}C CPMAS NMR spectra of two polymorphs of finasteride (Form I and Form II), with one and two distinct molecules in the asymmetric unit, respectively. Reproduced with permission from Othman *et al.* (1997) [36]. Copyright (2007) John Wiley & Sons Inc. (b) ^{17}O (9.4 T) MAS NMR spectra of four polymorphs of MgSiO$_3$: orthoenstatite, akimotoite, majorite and perovskite. Reproduced with permission from Ashbrook *et al.* (2007) [22]. Copyright (2007) American Chemical Society.

polymorphic forms of a material, often aided by DFT calculations. Figure 1.21a shows the ^{13}C CP MAS NMR spectra of two forms of finasteride, a molecular solid.[35,36] In Form I, the presence of one signal for each carbon in the molecule confirms there is a single molecule in the asymmetric unit. However, for Form II, the splitting of each resonance reveals there are two distinct molecules, confirming a difference in symmetry between the two polymorphs. In Figure 1.21b, ^{17}O MAS NMR spectra of four polymorphs of MgSiO$_3$ are shown – all high-pressure minerals of importance in the inner Earth.[22] Although the second-order quadrupolar broadening prevents the resolution of distinct species, the four spectra are clearly very different, reflecting the different structures. The lineshapes are composed of six, one, six and two overlapping signals for orthoenstatite, akimotoite, majorite and perovskite, respectively.

Although, in general, multiple complex factors determine the exact values of the NMR parameters, in many cases a particularly strong dependence on one factor may enable structural information to be

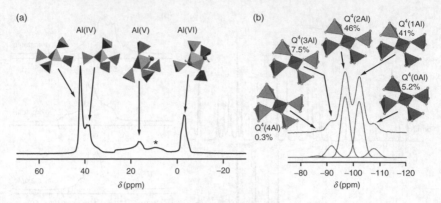

Figure 1.22   (a) ^{27}Al MAS (14.1 T) NMR spectrum of as-made AlPO-14, templated by isopropylammonium.[34] The presence of charge-balancing hydroxyls attached to the framework produces Al(IV), Al(V) and Al(VI) species. The asterisk denotes an impurity phase. (b) ^{29}Si MAS NMR spectrum (and corresponding analytical fit) for analcime ($Na_xAl_xSi_{3-x}O_6 \cdot H_2O$), containing one tetrahedrally-coordinated cation site that can be occupied by either Si (dark grey) or Al (light grey). Assignments for the five resonances are also shown. Reproduced with permission from Phillips (2009) [37]. Copyright (2009) John Wiley & Sons Ltd.

extracted from a single measurement, or from a series of measurements on related compounds. The isotropic chemical shift depends on the shielding effects of nearby electrons and, hence, on any factors affecting the electron density at the atom. For example, coordination number typically has a large effect on chemical shift, an effect exploited in ^{27}Al MAS NMR,[2] where four-, five- and six-coordinate Al are easily distinguished, as shown in Figure 1.22a, for as-made AlPO-14.[34] The charge on the structure-directing template (isopropylammonium) in the pores is balanced by hydroxyl groups that attach to the framework Al, increasing their coordination number. The spectrum shows the presence of Al(IV), Al(V) and Al(VI) species (where the number in brackets describes the coordination of the Al atom), and their relative proportions can be used to gain insight into the position and ordering of the hydroxyls. Upon calcination, a purely tetrahedral framework is produced, as previously shown by the ^{27}Al MAS NMR spectrum in Figure 1.20.[34] Silicon chemical shifts are also very dependent upon the coordination number, with very clear differences between tetrahedral SiO_4 (-60 to -120 ppm) and octahedral SiO_6 (-180 to -210 ppm) environments, a fact that has been widely exploited in the study of silicate minerals.[2]

In addition to the coordination number, the nature of the coordinated atoms can also have a significant impact on the chemical shift. For example, for SiO_4, SiN_4 and SiC_4 environments, chemical shifts are typically -110, -50 and -20 ppm, respectively.[2] Similarly, the chemical shift is also affected by the degree of condensation (or polymerisation) in the material, *e.g.* in phosphate and silicate chemistry. This is usually expressed using the Q^n nomenclature, where n is the number of other Q units attached to the unit in question. It has been shown that, for both ^{29}Si and ^{31}P, δ_{iso} becomes increasingly negative as the condensation increases.[2] This sensitivity provides an excellent tool for probing the network connectivity, particularly in glasses and minerals. The nature of the next-nearest neighbour (NNN) cations can also affect the isotropic chemical shift. This has been utilised in the study of aluminosilicates (both minerals and zeolites), enabling the distribution of Si and Al to be probed (see Section 1.7 for a more detailed description). As Figure 1.22b shows for the aluminosilicate mineral analcime ($Na_xAl_xSi_{3-x}O_6 \cdot H_2O$), there is a change in the ^{29}Si chemical shift of $\sim$6 ppm for each NNN Al substituted.[37] This substitution is usually denoted by $Q^n(m\,Al)$, where m represents the number of NNN Al species.

More subtle variations in the local environment can also influence the isotropic shift. For example, the isotropic shift of a number of nuclei has been shown to depend on the detailed local geometry (*i.e.* bond lengths or bond angles).[2] For a series of similar compounds (where other structural changes are inherently limited), the variation in chemical shift can be used to measure the geometrical changes directly. This has been shown to be particularly useful when probing phenomena such as hydrogen and halogen bonding,[20] where the shifts of the hydrogen- or halogen-bonded atoms are dependent on the geometry of the bond in question. DFT calculations can also provide an excellent tool for investiging the nature and magnitude of the dependence of the chemical shift (and other NMR parameters) upon the detailed local environment, owing to the ease of structural manipulation.

In solids, of course, one is not restricted to considering the isotropic chemical shift and, as demonstrated by reference,[38] even when two nuclei have the same isotropic shift, their shielding anisotropies are not necessarily identical. Measurement of the principal components of the shift tensor can often provide information that is not immediately apparent from the isotropic shift alone. It has already been shown in Figure 1.6 that the magnitude and asymmetry of the shielding tensor depends upon the local symmetry; however, smaller changes in local

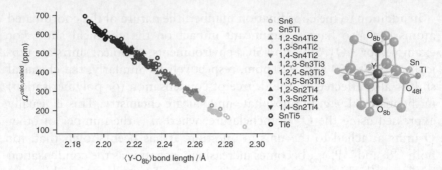

Figure 1.23 Plot showing the dependence of the ^{89}Y span, Ω (calculated using DFT), upon the average Y–O$_{8b}$ bond distance for a series of Y$_2$(Sn, Ti)$_2$O$_7$ pyrochlore materials. The inset shows the local environment, with Y located on the eight-coordinate A site, with six bonds to 48f and two to 8b oxygens. The symbols indicate the exact distribution of Sn/Ti NNN on the surrounding B sites. Reproduced with permission from Mitchell *et al.* (2012) [39]. Copyright (2012) American Chemical Society.

geometry can also affect the shielding anisotropy. Figure 1.23 shows a recent example, where the ^{89}Y shielding anisotropy (defined as the 'span', $\Omega = \delta_{11} - \delta_{33}$) exhibits a clear dependence upon the average Y–O$_{8b}$ bond distance (where 8b indicates the Wyckoff position) in a series of pyrochlore materials, proposed for the encapsulation of radioactive waste.[39]

For quadrupolar nuclei, small changes in δ_{iso} may be masked by the presence of significant quadrupolar broadening in MAS NMR spectra. However, the use of high-resolution approaches (*e.g.* MQMAS or DOR) results in isotropic spectra in which the number of distinct species can be identified. Figure 1.24 shows ^{23}Na NMR spectra of two samples of a perovskite, NaNbO$_3$, prepared using molten salt and solid-state synthetic approaches.[40] Although the two MAS NMR spectra look very similar, MQMAS reveals that the second sample contains two polymorphs, with four, rather than the expected two, ^{23}Na resonances. These were subsequently shown to be the *Pbcm* and *P2$_1$ma* polymorphs, which exhibit different tilting of the NbO$_6$ octahedra, as shown in Figure 1.24a. The presence of this second (polar) phase has a significant impact upon the physical properties of the material.

The quadrupolar interaction itself is, of course, an important source of structural information. It was shown in Figure 1.8 that, in simple cases, C_Q might be largely determined by the nature and arrangement of the coordinating atoms. This is widely exploited in ^{11}B NMR spectroscopy,

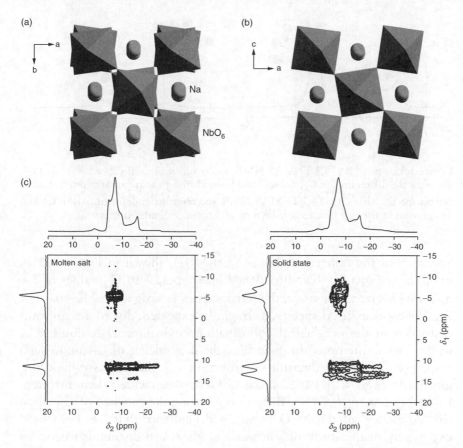

Figure 1.24 Crystal structures of (a) *Pbcm* and (b) *P2₁ma* polymorphs of NaNbO₃, showing different tilting modes of the NbO₆ octahedra. (c) ²³Na (9.4 T) MAS and MQMAS NMR spectra of NaNbO₃ prepared using two different synthetic approaches. The sample prepared using a molten-salt method contains a single polymorph, *Pbcm*, but that prepared using a solid-state synthesis contains a mixture of two polymorphs, *Pbcm* and *P2₁ma*, with four rather than the expected two resonances present in the MQMAS NMR spectrum. Reproduced with permission from Johnston *et al.* (2010) [40]. Copyright (2010) American Chemical Society.

where there is a large difference in C_Q for tetrahedral (0.0–0.5 MHz) *versus* trigonal (2.0–2.5 MHz) boron.[2] This is clearly seen in the ¹¹B MAS NMR spectrum of lithium diborate (Li₂B₄O₇) in Figure 1.25a, which shows the increased C_Q (and increased linewidth) of the trigonal boron species. Care must be taken, however, as the magnitude of the quadrupolar interaction is also dependent upon the exact local geometry and distortions from ideal symmetry. For example, the ²⁷Al MAS NMR

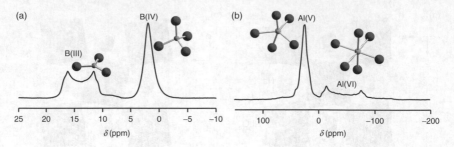

Figure 1.25 (a) ^{11}B (14.1 T) MAS NMR spectrum of lithium diborate ($Li_2B_4O_7$), showing the differences in C_Q values (and linewidth) for the trigonal and tetrahedral boron species. (b) ^{27}Al (14.1 T) MAS NMR spectrum of andalusite (Al_2SiO_5) and assignment of the resonances for the five- and six-coordinate Al species.

spectrum of the mineral andalusite (Al_2SiO_5), shown in Figure 1.25b, exhibits two quadrupolar-broadened lineshapes, with C_Q values of 5.8 and 15.3 MHz.[2] However, the latter species is assigned to the six-, not the five-coordinate Al species, as might be expected, due to a significant distortion in the octahedral coordination environment. A number of authors have attempted to determine the dependence of C_Q upon such distortions in the coordination environment (*e.g.* the shear strain or longitudinal strain), with varied success.[2] It has been widely demonstrated, however, that for ^{17}O NMR there is a strong dependence of C_Q upon the 'covalency' of the X−O bonds; a dependence that has been used extensively in the study of silicates and glasses to distinguish between 'bridging' (Si−O−Si) and 'nonbridging' (*e.g.* Si−O−Mg) oxygens, where typical C_Q values are 4−6 and 2−3 MHz, respectively.[41]

In simple cases, the asymmetry of the quadrupolar interaction is also able to provide support for structural models, owing to its dependence on both local and long-range symmetry. For example, if a nucleus lies on an *n*-fold rotation axis (with $n = 3 - 6$) in the structure, η_Q will be zero, due to the axial symmetry. For the anatase form of TiO_2, Ti sits on a C_4 rotation axis and $\eta_Q = 0$; however, in rutile the point symmetry at Ti is *mmm*, and this is reflected in the η_Q value of 0.19.[42] In many cases, it can be very difficult to predict the magnitude and asymmetry of the quadrupolar interaction accurately (owing to their dependence on long-range structure) and DFT calculations can prove invaluable in the interpretation and assignment of NMR spectra. For example, for the case of andalusite, described above, the C_Q values (calculated using a plane wave periodic code) are 5.1 and 16.0 MHz for Al(V) and Al(VI) species, respectively, in excellent agreement with the experimental values, which confirms their unexpected assignment.[43]

1.5.2 Measuring Internuclear Interactions

The internuclear interactions discussed in Section 1.2.4.2 can also provide structural information, although both dipolar and J coupling can be difficult to measure easily from solid-state NMR spectra. Owing to its strong distance dependence (proportional to r_{IS}^{-3}), the dipolar interaction encodes information on spatial proximity and internuclear separation, but is averaged to zero by the fast MAS used to resolve distinct resonances. A number of techniques have been proposed to reintroduce or 'recouple' the dipolar interaction, using combinations of pulses and delays to disrupt the effect of MAS.[44] It is possible to choose the nature and timings of these sequences such that only the (heteronuclear or homonuclear) dipolar interaction is recoupled and all other interactions are efficiently removed. Many different recoupling sequences exist and can be used for the transfer of magnetisation between nuclei in 2D experiments (as described in Section 1.3.1.3). In many cases, only qualitative information is required and experiments such as CP or 2D correlation spectra are sufficient to determine whether two nuclei are close in space, or which nuclei are closest. It is, however, possible to obtain quantitative information on interatomic distances using the rotational echo double resonance (REDOR) experiment.[45] In this approach, a spin echo experiment is performed (under MAS) for spin I, and the signal intensity is measured as a function of echo duration. The experiment is then repeated with the inclusion of a series of 180° pulses on a second spin, S, recoupling the $I - S$ heteronuclear dipolar interaction and leading to a modulation of the intensity of the FID. Owing to this recoupling, the intensity of any I spin resonance that has a significant dipolar coupling to S will be reduced. Qualitative information can be obtained, therefore, by comparing the two sets of spectra (acquired with and without the 180° pulses). However, if the signal intensity (or intensity difference) is plotted as a function of the echo duration, it is possible to measure the dipolar coupling and thus extract accurate information on the internuclear distance. An example is shown in Figure 1.26, where $^{27}Al/^{19}F$ REDOR was used to extract the Al–F distance in the microporous aluminophosphate AlPO-5. An Al–F distance of 1.92 Å

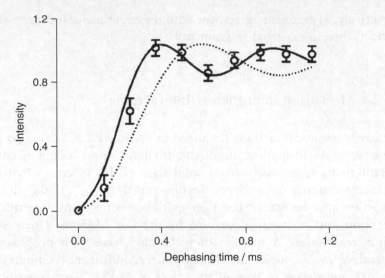

Figure 1.26 ^{27}Al/^{19}F REDOR signal as a function of echo duration (dephasing time) for AlPO-5. The calculated curves correspond to internuclear distances of 1.92 Å (solid line) and 2.19 Å (dotted line), with the latter being the distance in the crystal structure proposed initially. Reproduced with permission from Gougeon *et al.* (2001) [46]. Copyright (2001) American Chemical Society.

was obtained, confirming the presence of an Al–F bond, correcting the structure proposed from diffraction data.[46] REDOR generally works best for isolated spin pairs, *i.e.* where each *I* spin is only coupled to a single *S* spin. While this can be achieved in some cases by isotopic labelling, this is not practically possible for many materials. However, in most instances more complex analysis can lead to both qualitative and quantitative information in multiple-spin systems. In addition, although REDOR can be applied to quadrupolar nuclei, as in Figure 1.26, the analysis of the resulting data is more complicated, and there are a number of related techniques (*e.g.* REAPDOR and TRAPDOR) that are specifically designed to probe dipolar couplings involving quadrupolar nuclei.[2,20]

In contrast to the dipolar interaction, the *J* coupling has an isotropic component and so survives MAS. In principle, therefore, it should be measurable directly from the MAS NMR spectrum, and should provide information on the covalent bonding in the system. In practice, for most solids the large dipolar interactions result in linewidths that, even under MAS, are too broad to enable simple measurement of *J*

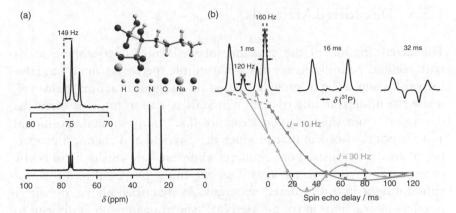

Figure 1.27 (a) ^{13}C (14.1 T) CP MAS NMR spectrum of monosodium alendronate, showing the presence of a 149 Hz one-bond J coupling to ^{31}P. Two-bond J couplings are not resolved due to the inherent linewidths observed. (b) Plot of the ^{31}P spin echo signal intensity for $(MoO_2)_2P_2O_7$ as a function of the echo duration, demonstrating the presence of two different ^{31}P/^{31}P homonuclear J couplings, of 10 and 30 Hz. This information was used to investigate structure and phase transitions in this material. Reproduced with permission from Apperley *et al.* (2012) [1]. Copyright (2012) Momentum Press LLC. From original data in Lister *et al.* (2010) [47].

couplings, unless they are particularly large, as is the case in Figure 1.27a, where the one-bond ^{13}C/^{31}P J coupling in monosodium alendronate (NaC$_4$H$_{18}$NO$_{10}$P$_2$) is 149 Hz: substantially larger than the linewidth (~40 Hz). However, it is not possible to resolve any two-bond couplings in this spectrum. It is possible to measure J couplings indirectly using a spin echo experiment, as the change in signal intensity as a function of the echo duration is modulated by any homonuclear or heteronuclear J couplings present. This is illustrated in Figure 1.27b for molybdenum pyrophosphate, $(MoO_2)_2P_2O_7$, in which ^{31}P/^{31}P J couplings of 10 and 30 Hz are observed, despite spectral linewidths that are of an order of magnitude larger.[1,47] It is possible to extend this approach and perform a Fourier transform of the oscillating signal, in order to obtain lineshapes in the indirect dimension of a 2D (J resolved) experiment from which the couplings (or their distribution, in more complex materials) can be measured directly. Despite not being resolved in many cases, it is possible to utilise J coupling for magnetisation transfer within 2D correlation experiments, as described previously, enabling through-bond, rather than through-space, connectivities to be investigated.

1.5.3   Disordered Materials

The determination of the structure of well-ordered crystalline solids with regular, periodic arrays of known atomic species is usually straightforward when using approaches based on Bragg diffraction. However, while the understanding of such materials is important, it is often the deviations from this periodic structure (*i.e.* compositional, positional and temporal disorder) that produce the physical and chemical properties of greatest industrial or commercial interest. The sensitivity of NMR to the local environment makes it an extremely powerful tool for providing insight into disordered materials. As described above, the effect of many interactions in solid-state NMR spectra can make it difficult to extract the detailed information required, and this is particularly true for disordered materials. In recent years, however, great progress has been made through the use of DFT calculations alongside experiment in helping spectral interpretation and assignment.

The presence of disorder in a material can have a range of possible effects upon the NMR spectrum. If the environment changes substantially (*e.g.* a change in coordination number) there are usually significant differences in the chemical shift, and resonances will be well separated in the spectrum. This is seen in the ^{89}Y CPMG MAS NMR spectrum of $Y_2Zr_2O_7$ shown in Figure 1.28a, where anion disorder of oxygen (each position 7/8 occupied) results in three resonances assigned to six-, seven- and eight-coordinate Y. The relative (integrated) intensities of each of these resonances can be used to provide insight into the nature of the anion disorder.

If the changes to the structural environment are less marked, the spectral resonances may not be completely resolved but instead overlap, producing a complex lineshape. Providing the components are sufficiently resolved, it is usually possible to deconvolute the lineshape and obtain information on each of the various contributions. An example of this can be seen in Figure 1.28b, an ^{89}Y MAS NMR spectrum of the pyrochlore, $Y_2Ti_{1.2}Sn_{0.8}O_7$, displaying resonances corresponding to different numbers of Sn/Ti on the six NNN B sites.[39] Once again, it is possible to extract information on cation disorder, *i.e.* the distribution of Sn/Ti throughout the material, from the relative spectral intensities (see Section 1.7.1). A similar composite lineshape is observed in the ^{29}Si MAS NMR spectrum of analcime in Figure 1.22b, with five overlapped resonances present, corresponding to 0–4 NNN Al. There has been considerable work over many years on the analysis of ^{29}Si NMR spectra

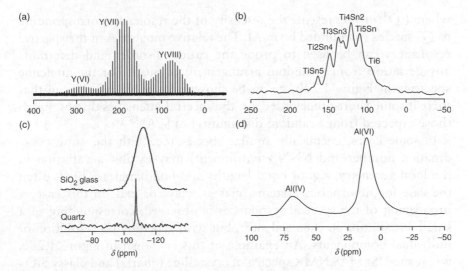

Figure 1.28 (a) ^{89}Y (14.1 T) CPMG MAS NMR spectrum of $Y_2Zr_2O_7$, with six-, seven- and eight-coordinate Y, resulting from anion disorder in the defect fluorite structure. (b) ^{89}Y (14.1 T) MAS NMR spectrum of $Y_2Ti_{1.2}Sn_{0.8}O_7$ pyrochlore, with distinct resonances resulting from different numbers of NNN Sn/Ti. Reproduced with permission from Mitchell *et al.* (2012) [39]. Copyright (2012) American Chemical Society. (c) ^{29}Si (8.45 T) MAS NMR spectra of crystalline (quartz) and glassy SiO_2. The increased linewidth can be related to a distribution in bond angles. Reproduced with permission from High-resolution ^{23}Na, ^{27}Al, and ^{29}Si NMR Spectroscopy of Framework Aluminosilicate Glasses by R. Oestrike *et al.*, *Geochimica et Cosmochimica Acta*, **51**, 2199–2209. Copyright (1987) Elsevier Ltd. (d) ^{27}Al (14.1 T) MAS NMR spectrum of γ-Al_2O_3, with the resonances attributed to Al(IV) and Al(VI) species exhibiting lineshapes characteristic of a distribution of quadrupolar parameters.

of aluminosilicate minerals and zeolites in order to extract information on the Si/Al ratio and the cation order/disorder. The latter is particularly difficult to achieve using X-ray diffraction, owing to the similar form factors of Si^{4+} and Al^{3+}. It has long been assumed that Al–O–Al linkages in aluminosilicates are disfavoured; if 'Lowenstein's rule'[48] holds (*i.e.* such linkages are completely absent), it is possible to determine the Si/Al ratio from ^{29}Si MAS NMR spectra by:

$$\left(\frac{Si}{Al}\right) = \frac{\sum\limits_{m} I(Q^4(m))}{0.25\sum\limits_{m} m\, I(Q^4(m))} \tag{1.18}$$

where $I(Q^4(m))$ represents the intensity of the resonance corresponding to Q^4 species surrounded by m Al. The relative intensities of the spectral resonances can be used to probe the cation disorder and determine any deviation from a random arrangement of Si/Al. For the analcime spectrum in Figure 1.22b, it can be shown that Si/Al = 2.55 and that there is some short-range order, *i.e.* the spectral intensities do not match those expected from a random distribution of Si/Al.[37]

In some cases, chemically similar species (*i.e.* with the same coordination number and NNN environment) may exhibit a variation in the local geometry, *e.g.* of bond lengths and bond angles. This is often the case for amorphous systems, glasses, clays or gels. In this case, a broadening of the spectral resonances is observed, corresponding to a small distribution in chemical shift, but with generally no resolution of individual components. An example of this is shown in Figure 1.28c, where the ^{29}Si MAS NMR spectra of crystalline (quartz) and glassy SiO_2 are compared. The considerable increase in linewidth for the glass can be related to a variation in the Si−O−Si bond angles, from 130 to 170° in this material.[49]

For quadrupolar nuclei, disorder results not only in a distribution of chemical shifts from the changes in the structural environment but also in a distribution in the quadrupolar parameters. In some cases these distributions are correlated, *i.e.* changes in one parameter are directly related to changes in the others, while in others the distributions appear to be independent. The resulting spectral lineshapes are broadened, but usually display an asymmetry, with a characteristic 'tail' to low frequency. The ^{27}Al MAS NMR spectrum of γ-Al_2O_3, shown in Figure 1.28d, has two distinct resonances, from Al(IV) and Al(VI) species, both exhibiting such asymmetric lineshapes, indicating a distribution of NMR parameters. In many cases it is possible to gain further insight into the magnitude and nature of the distributions either from analytical fitting of the lineshapes (preferably at more than one B_0 field strength) or by analysis of lineshapes in resulting 2D (*e.g.* MQMAS) spectra.[20]

In some materials there may be underlying periodicity or order, but one or more components of the material may not conform to this and may exhibit disorder. In Bragg-based diffraction, these components are either not refined at all or may be placed in the unit cell with a 'general' atom type or with fractional occupancy. In microporous materials, for example, the framework structure is ordered but the template, water or guest molecules incorporated within the pores are often disordered. Figure 1.29a shows ^{13}C CP MAS NMR spectra of the aluminophosphate framework SIZ-4, with 1-methyl-3-ethylimidazolium

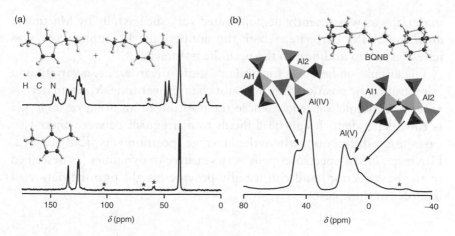

Figure 1.29 (a) ^{13}C CP MAS NMR spectra of as-made SIZ-4 templated with 1-methyl,3-ethylimidazolium and 1,3-dimethylimidazolium cations. In one case, the template appears disordered in the material, while in the other, only a single resonance is observed for each distinct carbon, indicating a more ordered material. Reproduced with permission from Griffin *et al.* (2012) [11]. Copyright (2012) Royal Society of Chemistry. (b) ^{27}Al (14.1 T) MAS NMR spectrum of as-made STA-2 (prepared with a BQNB template). The charge-balancing hydroxyls that attach to the framework are disordered; however, the relative intensities of the Al(IV) and Al(V) resonances reveal the hydroxyls are bridging rather than terminal. Reproduced with permission from Castro *et al.* (2010) [50]. Copyright (2010) American Chemical Society. In both parts, asterisks denote SSBs.

and 1,3-dimethylimidazolium templates.[11] In the former case, there are many resonances corresponding to each carbon in the template molecule, revealing its disordered nature. However, the latter template appears considerably more ordered in the material, with only one resonance *per* chemically distinct carbon, indicating that all template molecules are crystallographically equivalent. The charge-balancing anions (*e.g.* F$^-$ or [OH]$^-$) often attached to the framework atoms in microporous materials are not always ordered with the same symmetry or periodicity as the framework. Even if disorder is present, NMR spectra can still provide information on local structure, as in the ^{27}Al MAS NMR spectrum of the aluminophosphate STA-2 (Figure 1.29b).[50] STA-2 contains a disordered array of [OH]$^-$ anions, balancing the charge of the dicationic 1,4-bis-*N*-quinuclidiniumbutane (BQNB) template. Integrating the intensities of the Al(IV) and Al(V) resonances (1.5 : 1) reveals that the [OH]$^-$ species are bridging, rather than terminal (*i.e.* Al–OH–Al rather than Al–OH). In some cases, it is possible to determine more detailed information on the ordering of different components of a material – as was recently demonstrated very successfully by Martineau *et al.* for AlPO cloverite, where the nonperiodic F^- subnetwork was investigated, in addition to the periodic framework.[51]

The discussion here has focused on 'static' disorder, *i.e.* a variation in composition or position that does not change over time. Many materials exhibit 'dynamic' disorder, where the position of an atom or molecule is time dependent. It can be difficult to distinguish between these two cases using diffraction, where the 'average' position is typically refined. However, NMR spectroscopy is very sensitive to dynamics (as described in the next section) and can usually provide insight into the nature of disorder in these cases.

1.5.4   Studying Dynamics

NMR spectroscopy is a sensitive probe of dynamics over timescales spanning 15 orders of magnitude, as shown in Figure 1.30. Fast motion (*i.e.* correlation times, τ_c, of 10^{-12} to 10^{-9} seconds) is usually detected by changes in relaxation processes, which rely on fluctuations in the spin interactions induced by motion. To affect T_1 relaxation, any motion must have a correlation time of $\sim 1/\omega_0$, and information on the exact timescale can be obtained from the dependence of T_1 upon temperature. Motion on the correct timescale can have a dramatic effect, as demonstrated by the ^{13}C T_1 relaxation times of diamond (~ 24 hours) and adamantane (~ 5 seconds).[1] The latter is a so-called 'plastic crystal', in which the molecules are fixed on lattice points but exhibit isotropic motion around them, providing efficient relaxation. A detailed description of relaxation theory is beyond the scope of this chapter, however (see references [1, 2] for more detail).

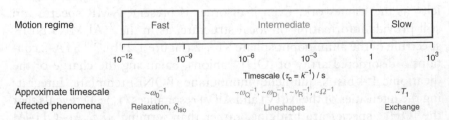

Figure 1.30 Schematic plot showing the sensitivity of NMR to dynamics over a range of timescales.

Slow motion is usually measured by 2D exchange experiments, in which 'mixing times' (typically 10^{-1} to 10^3 s) can be introduced to measure the physical motion of an atom. This results in a change in frequency for a particular spin, due to the change in local environment, which can be detected either as a cross peak in MAS-based experiments (if the motion exchanges different spins) or by analysis of a 2D lineshape (typically for static samples) in order to determine the geometry of a motional jump.[1,2,20]

Dynamic processes with a timescale intermediate to these two extremes can also be investigated using NMR, owing to the presence of anisotropic interactions. Motion typically results in a change in orientation and, therefore, in the anisotropic shielding or dipolar or quadrupolar coupling. The resulting changes in lineshapes or linewidths can then be followed as a function of temperature to provide information on the type and rate of the motional process. The timescales that can be probed are determined by the inverse of the magnitude of the interaction that is varied, e.g. 10^{-7} to 10^{-3} s for the quadrupolar interaction. The effect of motion upon solid-state NMR lineshapes can be understood by first considering a single spin that can hop between two positions, leading to two resonances in the NMR spectrum (labelled A and B in Figure 1.31a). If exchange between the two positions is slow, two distinct resonances will be observed. As the rate of exchange increases (with increased temperature), the two lines broaden and eventually coalesce. In the limit of fast exchange, a single, narrow line is observed at the average of the two resonance positions. In a powdered solid, such exchange-broadened 'pairs' must be averaged over all possible crystallite orientations to obtain the corresponding powder-pattern lineshapes. The lineshape contains information on the geometry, i.e. the change in orientation, of the dynamic process and on τ_c (and the rate constant $k = 1/\tau_c$). For example, Figure 1.31b shows a quadrupolar-broadened ^{2}H ($I = 1$) lineshape simulated as a function of k, for a nucleus undergoing a $180°$ jump (e.g. around the C_2 axis of a water molecule). When reorientation is slow, a typical Pake doublet lineshape is observed. As the rate increases, there are changes to the shape and width of the line, until an averaged lineshape is observed in the limit of fast motion. For different jumps (e.g. around a C_3 or C_6 axis), different lineshapes and widths will be observed. For isotropic motion, there will be full averaging of the interaction and a sharp, narrow line will be observed (in the fast motion limit).[1,2,20]

If experiments are performed under MAS, the timescale of rotation (typically $10–100\,\mu$s) is also relevant. If motion occurs during the

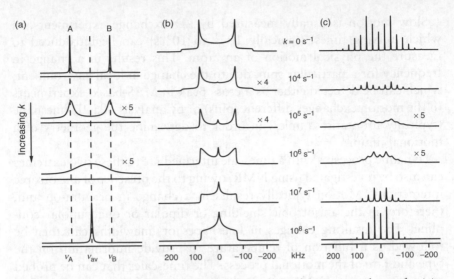

Figure 1.31 Simulated lineshapes showing the effect of dynamic exchange for (a) a single crystal and (b,c) a powder distribution of crystallites for a ^{2}H ($I = 1$) nucleus undergoing a 180° rotation around the C_2 axis of a water molecule, as a function of the rate constant, k. In (b), the sample is static, whereas in (c), a MAS rate of 25 kHz is used.

time it takes for the sample to rotate, with a corresponding change in the magnitude or orientation of the shielding, dipolar or quadrupolar interactions, this will affect the averaging that MAS endeavours to achieve. The incomplete averaging manifests itself as a broadening of the SSBs, as shown in Figure 1.31c. At slow rates, sharp sidebands are observed with a manifold that mirrors the static lineshape (in Figure 1.31b). As the rate increases, a broadening of the sidebands is observed, until at one point they are wider than their separation, with very little signal in the NMR spectrum. In the fast motion regime, sharp sidebands are again present, now with intensities that reflect the averaged lineshape. As an example, Figure 1.32a shows both ^{1}H and ^{2}H MAS NMR spectra of deuterated oxalic acid dihydrate.[52] Each spectrum contains two resonances (one from the hydroxyl and one corresponding to water); however, in the ^{2}H spectrum the resonance from D_2O is broadened as a result of molecular reorientation and the corresponding change in the quadrupolar interaction. Such a change is not observed for ^{1}H ($I = 1/2$). In many cases it is possible to measure the dependence of the NMR linewidth on temperature and extract information on the activation energy, E_A. This can be seen in Figure 1.32b, where a temperature-dependent linewidth is observed in the ^{2}H MAS NMR spectrum of a

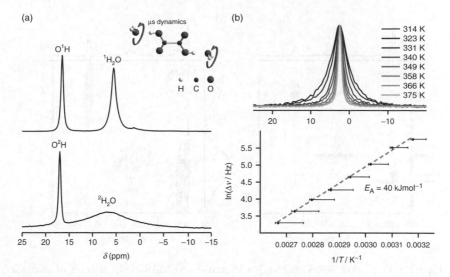

Figure 1.32 (a) ^{1}H and ^{2}H (14.1 T) MAS NMR spectra of α-oxalic acid dihydrate d6, $D_2C_2O_4 \cdot 2D_2O$, exhibiting two lineshapes. In the ^{2}H spectrum, the D_2O resonance is broadened by motional reorientation on the microsecond timescale, leading to a variation in the quadrupolar interaction. (b) Variable-temperature ^{2}H (9.4 T) MAS NMR spectra of clinohumite, $4Mg_2SiO_4 \cdot Mg(OD)_2$, and corresponding Arrhenius plot of the natural log of the linewidth, $\Delta \nu$, against $1/T$. An activation energy of 40 ± 4 kJ mol^{-1} can be extracted from the gradient of the plot. Reproduced with permission from Griffin *et al.* (2010) [53]. Copyright (2010) Royal Society of Chemistry.

hydrous silicate mineral (clinohumite, $4Mg_2SiO_4 \cdot Mg(OH)_2$), and the corresponding Arrhenius plot shows E_A to be 40 ± 4 kJ mol^{-1}.[53] While diffraction measurements indicate a disordered material (each of the two possible H sites has an occupancy of 0.5), there is no indication that there is anything other than a 'static' disorder of the hydroxyls. However, the sensitivity of NMR to dynamics enables this picture to be revised, revealing dynamic disorder, with exchange of H between the two sites occurring on the microsecond timescale (see also Section 1.7.3).

For quadrupolar nuclei, we have seen that resolution of distinct species may be difficult in simple MAS NMR spectra, and 2D experiments are often required to improve resolution. Although both the MQMAS and STMAS approaches discussed earlier result in similar spectra for crystalline materials, they exhibit significant differences when dynamics are present. For STMAS, the satellite transitions used in the experiment are broadened by the large first-order quadrupolar interaction, and any motion on the microsecond timescale results in a broadening of the

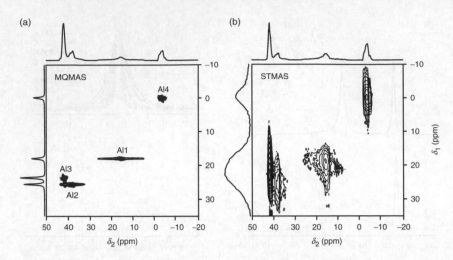

Figure 1.33 ^{27}Al (14.1 T) (a) MQMAS and (b) STMAS NMR spectra of as-made AlPO-14 prepared with an isopropylammonium template. In the MQMAS NMR spectrum, the four sharp resonances correspond to the four distinct Al species, while the corresponding resonances in the STMAS NMR spectrum are broadened by microsecond-timescale dynamics of the template within the pores. After Antonijevic *et al.* (2006) [54].

spectral resonances in an analogous manner to that in Figure 1.32.[20,21] However, MQMAS NMR spectra are not sensitive to dynamics on this timescale and the resonances remain unaffected. Differential line-broadening in STMAS and MQMAS NMR spectra can be used to detect dynamics, and changes with temperature to measure the rate constant and activation energy for the motional processes. Figure 1.33 shows ^{27}Al MQMAS and STMAS NMR spectra of as-made AlPO-14, with the MQMAS NMR spectrum showing four sharp resonances, corresponding to the four distinct Al species. However, the STMAS NMR spectrum displays motionally broadened resonances as a result of microsecond-timescale dynamics of the nearby template (isopropylam-monium) molecules.[54]

1.5.5   Challenging Nuclei and Systems

As all elements with stable isotopes in the periodic table (with the exception of Ce and Ar) have at least one NMR-active isotope, NMR would appear to be a widely applicable technique. However, while acquisition of solid-state NMR spectra is straightforward in many cases, in others it can be a considerable challenge. The intrinsic sensitivity of an NMR experiment is proportional to γ^3, making observation of so-called 'low-γ' nuclei a particular problem. For example, ^{89}Y has a receptivity ~10 000 times lower than that of ^1H, despite having 100% natural abundance.[1,2] For quadrupolar nuclei, the problems are intensified by the dependence of the (second-order) quadrupolar broadening upon $1/\gamma$, resulting in a further reduction in sensitivity. In general, relaxation processes are also typically less efficient for low-γ nuclei, adding to the difficulty and timescale of experiments.[2,20] Many conventional probes are also unable to tune to the low Larmor frequencies of low-γ nuclei, and specialist equipment or adaptation is required. (This latter point has resulted in the common definition of 'low' γ as being a nucleus with a Larmor frequency below that of ^{15}N (60.8 MHz at 14.1 T), the typical lower end of the tuning range of most probes.) One final challenge for the study of low-γ nuclei is the low rf field strengths ($\omega_1 = -\gamma B_1$) typically available; this can have an impact upon complex NMR pulse sequences, resulting in problems not only with sensitivity but also in the accuracy of the information extracted. Other sensitivity challenges are posed by nuclei with low natural abundance, *e.g.* ^{15}N (0.364%) or ^{17}O (0.037%), and by samples that either have a very small volume (*e.g.* high-pressure materials, or proteins) or contain very little of the spin under study (*e.g.* dopant elements in minerals or semiconductors).

There are a number of possible solutions to the problems outlined above. Perhaps the simplest (and yet not always the most convenient in practice) is to acquire spectra at the highest available magnetic field strength. This improves the absolute sensitivity for any spin, and has the additional effect of reducing any second-order quadrupolar broadening and narrowing lines. It is also possible to build or modify probes to have optimal performance at low frequencies, improving both sensitivity and B_1 field strength. A number of sensitivity enhancement techniques are also available, including CP (if a suitable high-γ, high-abundance nucleus is available for magnetisation transfer). CP is able to improve sensitivity for nuclei with low γ, long relaxation times and low natural abundance.[14,15] As shown in Figure 1.18, CPMG can also overcome some of the challenges associated with poor sensitivity, long experimental times and long relaxation times. For quadrupolar nuclei, if high-resolution spectra are required, the use of STMAS rather than MQMAS[20,21] will also result in increased sensitivity. For example, STMAS has been successfully applied to obtain high-resolution NMR spectra at a moderate magnetic field strength (9.4 T) for both ^{25}Mg

(at natural abundance)[55] and ^{17}O (enriched to 35%) in only 9.6 mg of the high-pressure mineral wadsleyite (β-Mg_2SiO_4).[56] In some cases, the sensitivity of isotopes with low natural abundance can be enhanced by isotopic enrichment (as either uniform enrichment or selective labelling, in order to aid spectral assignment). Enriched starting materials may be prohibitively expensive or require major modification of the synthetic procedure, making this an unattractive initial approach. However, as the SNR is improved by the square of the number of nuclei contributing to the signal, even a low level of enrichment can provide great savings in experimental time, particularly for 2D experiments.

1.5.6   Paramagnetic Materials and Metals

The presence of an unpaired electron can have a significant effect upon the NMR spectrum, as the gyromagnetic ratio of the electron is ~660 times larger than ^1H, resulting in significant electron–nucleus interactions. Although it can be difficult to obtain NMR spectra for the paramagnetic species itself, it is often possible to see the effects of the unpaired electron on more remote nuclei, and the large magnitude of the interaction provides good chemical sensitivity.[2,57]

The effect of an unpaired electron on the NMR spectrum comprises a number of contributions.[58] Perhaps the most important is the Fermi contact interaction, where the unpaired electron spin density induces a polarisation of the nuclear s orbitals. This interacts with the nuclear magnetic moment to give an isotropic shift that depends upon B_0, the nuclear gyromagnetic ratio, γ, and the hyperfine coupling constant, A. Even if the electron does not originate from the atom containing the nucleus under observation, polarisation can still be induced by the 'transferred hyperfine interaction', mediated (as in the J coupling) by the chemical bonding. The coupling is sensitive to the number, geometry and covalency of the bonding pathways linking the observed nucleus and the electron. A second contribution is the through-space dipolar coupling between the electron and the nuclear spin. This 'pseudocontact shift' has an isotropic and anisotropic component, with the latter exhibiting orientation dependence similar to the dipolar coupling. In principle, MAS can remove this anisotropy, but its large magnitude typically results in many SSBs. The pseudocontact shift is dependent on the inverse cube (*i.e.* r^{-3}) of the distance between the electron and the nucleus, but is independent of any bonding pathway between the two. In general, the fast relaxation of the electron spins results in a collapse of any coupling into a single line. However, the difference in energy of the electron spin states is appreciable in comparison to the thermal energy, $k_B T$, resulting in a population-weighted average shift of the resonance. The NMR resonances for paramagnetic materials are, therefore, often significantly shifted from those found for diamagnetic analogues, and can display a considerable sensitivity to temperature.

In addition to isotropic and anisotropic shifts, the interaction with unpaired electrons usually leads to rapid T_1 and T_2 relaxation, typically dominated by the electron–nuclear dipolar interaction. This can be advantageous, reducing the time required to acquire spectra, although any broadening of the spectral lines reduces sensitivity. Doping of diamagnetic materials with very small amounts of paramagnetic ions (*e.g.* Co^{2+}, Mn^{2+}) in order to increase the rate of T_1 relaxation for materials where this is prohibitively slow is common for ceramics, oxides and minerals. For microporous zeolites and phosphates, the presence of paramagnetic O_2 within the pores of the material can also lead to more rapid relaxation.

Overall, the effects of localised unpaired electrons in solid-state NMR can be separated into three distinct regimes: (i) the paramagnetic atom itself is rarely visible in the spectrum, owing to extreme shifts, significant broadening and rapid relaxation; (ii) nearby nuclei may be visible, but resonances will typically be broadened and shifted, and relaxation may be too rapid to allow some conventional NMR techniques to be applied, requiring different experimental approaches; and (iii) for remote nuclei, there may be negligible changes in isotropic shift, but line broadening and rapid relaxation may still be evident. Figure 1.34 shows ^{13}C NMR spectra of the copper(II)-based metal-organic framework (MOF), HKUST-1.[59] When the MAS rate is relatively slow, the spectrum displays a considerable number of SSBs, owing to the large anisotropy associated with the electron–nuclear interaction. At faster MAS rates, more efficient removal of the anisotropy results in fewer sidebands and the observation of an isotropic resonance at 227 ppm, corresponding to C3 in the linker molecule. However, two further resonances are also now clear, at −50 and 853 ppm, which were not easily observed when the rotation was slower. These resonances are much broader, reflecting a closer proximity to Cu, and can be assigned as C2 and

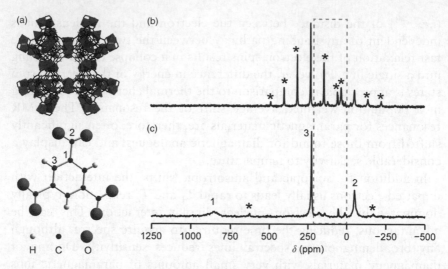

Figure 1.34 (a) Structure of HKUST-1 (a copper(II)-based MOF) and of the benzene tricarboxylate organic linker. (b,c) ^{13}C (14.1 T) MAS NMR spectra (acquired using a spin echo) of HKUST-1, with MAS rates of (b) 10 and (c) 60 kHz. The 'normal' range of ^{13}C chemical shifts is highlighted by the dashed box and SSBs are denoted by asterisks.

C1, respectively.[60] Note that all three resonances are shifted outside of the 'normal' range expected for ^{13}C chemical shifts (0–220 ppm, as shown in Figure 1.34). Furthermore, for all resonances, T_1 relaxation is very rapid (~20 ms), and despite the addition of 262 144 transients being required to obtain this natural-abundance (~1%) ^{13}C spectrum, the total experimental time is only 7.3 hours.

For conducting solids, the electron–nucleus interactions are more complicated, as the delocalised electronic band structure has the effect that every nucleus in the structure effectively interacts with many electrons, and their collective influence must be considered. This results in the 'Knight shift', first observed for Cu in 1949.[61] As the Knight shift depends upon the wavefunction at the nucleus, the effect is dominated by the s electrons. The Knight shift is very large (generally >0.1%) and always positive when the valence band is composed only of s electrons. When p or d electrons are present, the situation is more complicated, and the Knight shift may be smaller or negative. For noncubic materials, the interaction is anisotropic (described by a tensor, **K**, and its principal components, K_{xx}, K_{yy} and K_{zz}), resulting in both a shift and a broadening of the spectral resonances.

1.6   COMMONLY STUDIED NUCLEI

In order to be NMR active, a nucleus must have $I > 0$; however, many other factors influence the ease with which NMR spectra can be acquired. Of key importance are fundamental properties such as the spin quantum number (I), natural abundance (N) and gyromagnetic ratio (γ). This information can be combined in a 'receptivity' (given by $\gamma^3 N(I(I + 1))$), usually quoted relative to a specific nuclide (*e.g.* ^{1}H). For quadrupolar nuclei, the relative magnitude of the quadrupole moment, Q, also has a significant impact upon the magnitude of the quadrupolar broadening and the sensitivity of NMR spectra. Also of practical importance for experiments are the magnitude of the interactions that affect a spectrum and the typical relaxation rates; however, these properties vary significantly for different materials. Table 1.3 gives nuclear properties (for the full table, see reference [62]), and Figure 1.35 shows a log–log plot of the relative receptivities and Larmor frequencies for a range of isotopes, reflecting the general ease or otherwise of acquisition. Key aspects for some of the most commonly studied nuclei are described below.[2,20]

1.6.1   Hydrogen

Hydrogen is present in organic crystals, minerals, microporous framework materials, polymers and energy-related materials. ^{1}H ($I = 1/2$) has high natural abundance (99.99%) and very high γ, making ^{1}H NMR very sensitive, although very fast MAS or homonuclear decoupling is often required to achieve high-resolution spectra, due to large dipolar interactions. A significant background signal is often observed, arising from protons in the rotor or probehead, and additional approaches may be required to remove this, particularly if there are relatively few protons in the sample. CP from ^{1}H can significantly improve the sensitivity of NMR experiments for many other nuclei, most notably ^{13}C. ^{2}H ($I = 1$) has a low natural abundance (0.01%), moderate γ and small quadrupole moment, making natural-abundance ^{2}H NMR much less sensitive – but often higher resolution under MAS – than the corresponding ^{1}H NMR spectrum. Typical ^{2}H C_Q values are on the order of a few hundred kHz, enabling variable-temperature ^{2}H NMR to be used as a sensitive probe

Table 1.3 Properties of selected NMR-active nuclei[62].

Isotope	Spin quantum number, I	Natural abundance, %	Gyro-magnetic ratio, γ/ 10^7 rad s^{-1} T^{-1}	Larmor frequency $(14.1\ \text{T})^a$/ MHz	Quad-rupole moment, Q/ fm^2	Relative receptivity[b]
^{1}H	1/2	99.9885	26.752	600.130		1.00
^{2}H	1	0.0115	4.106	92.124	0.2860	1.11×10^{-6}
^{6}Li	1	7.59	3.937	88.316	−0.0808	6.46×10^{-4}
^{7}Li	3/2	92.41	10.397	233.233	−4.01	2.71×10^{-1}
^{11}B	3/2	80.10	8.585	192.546	4.059	1.32×10^{-1}
^{13}C	1/2	1.07	6.728	150.903		1.70×10^{-4}
^{14}N	1	99.636	1.934	43.367	2.044	1.01×10^{-3}
^{15}N	1/2	0.364	−2.713	60.834		3.80×10^{-6}
^{17}O	5/2	0.038	−3.628	81.356	−2.588	1.11×10^{-5}
^{19}F	1/2	100	25.162	564.686		8.33×10^{-1}
^{23}Na	3/2	100	7.081	158.746	10.4	9.28×10^{-2}
^{25}Mg	5/2	10.00	−1.639	36.738	19.94	2.69×10^{-4}
^{27}Al	5/2	100	6.976	156.375	14.66	2.08×10^{-1}
^{29}Si	1/2	4.685	−5.319	119.229		3.68×10^{-4}
^{31}P	1/2	100	10.839	242.938		6.66×10^{-2}
^{33}S	3/2	0.75	2.056	46.066	−6.78	1.70×10^{-6}
^{35}Cl	3/2	75.76	2.624	58.800	−8.165	3.58×10^{-3}
^{37}Cl	3/2	24.24	2.184	48.9500	−6.435	6.61×10^{-4}
^{39}K	3/2	93.258	1.250	28.004	5.85	4.75×10^{-4}
^{43}Ca	7/2	0.135	−1.803	40.389	−4.08	8.69×10^{-6}
^{45}Sc	7/2	100	6.509	145.782	−22.0	3.03×10^{-1}
^{47}Ti	5/2	7.44	−1.511	33.833	30.2	1.56×10^{-4}
^{49}Ti	7/2	5.41	−1.511	33.842	24.7	2.04×10^{-4}
^{51}V	7/2	99.750	7.046	157.852	−5.2	3.83×10^{-1}
^{59}Co	7/2	100	6.332	142.393	42	2.79×10^{-1}
^{67}Zn	5/2	4.102	1.677	37.549	15.0	1.18×10^{-4}
^{71}Ga	3/2	39.892	8.181	183.020	10.7	5.71×10^{-2}
^{77}Se	1/2	7.63	5.125	114.454		5.37×10^{-4}
^{81}Br	3/2	49.31	7.250	162.074	25.4	4.91×10^{-2}
^{87}Rb	3/2	27.83	8.786	196.365	13.35	4.94×10^{-2}
^{89}Y	1/2	100	−1.316	29.408		1.19×10^{-4}
^{93}Nb	9/2	100	6.567	146.889	−32	4.89×10^{-1}
^{109}Ag	1/2	46.161	−1.252	27.927		4.94×10^{-5}
^{119}Sn	1/2	8.59	−10.032	223.792		4.53×10^{-3}
^{125}Te	1/2	7.07	−8.511	189.340		2.28×10^{-3}
^{129}Xe	1/2	26.4006	−7.452	166.897		5.71×10^{-3}
^{131}Xe	3/2	21.2324	2.209	49.474	−11.4	5.98×10^{-4}
^{133}Cs	7/2	100	3.533	78.714	−0.343	4.84×10^{-2}
^{183}W	1/2	14.31	1.128	25.004		1.07×10^{-5}
^{207}Pb	1/2	22.1	5.577	125.551		2.01×10^{-3}

^{a}Larmor frequency at 14.1 T.
^{b}Receptivity ($\gamma^3\ N(I(I+1))$), quoted relative to ^{1}H.

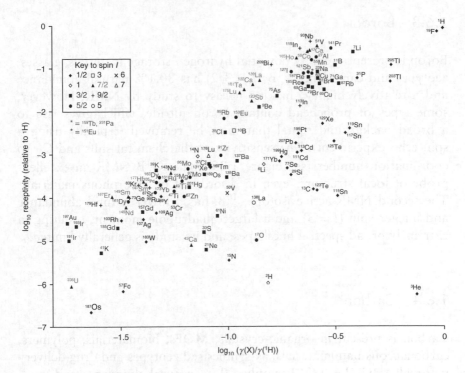

Figure 1.35 A log–log plot of receptivity (relative to ^{1}H) as a function of γ relative to that of ^{1}H for NMR-active isotopes.

of microsecond-timescale dynamics, particularly given the possibility of selective deuteration of many materials.

1.6.2 Lithium

Lithium is present in many ceramics and energy materials. ^{7}Li $(I = 3/2)$ has a 92.4% natural abundance, relatively high γ and, typically, fairly low C_Q values. The chemical shift range for lithium in diamagnetic materials is relatively small (just a few ppm), but significant shifts of the CT are observed for paramagnetic materials. ^{6}Li $(I = 1)$ has low natural abundance (7.59%), moderate γ and a very small quadrupole moment, enabling direct observation despite the lack of a CT. Lithium NMR can also be used to probe dynamics (of particular interest for battery materials) by following the linewidth or chemical shift as a function of temperature.

1.6.3   Boron

Boron is present in many potential hydrogen storage materials, glasses, zeotypes and carboranes. ^{11}B $(I = 3/2)$ has 80.1% natural abundance and a relatively high γ, making it easy to study by NMR. However, some types of probehead contain boron nitride, which gives rise to a broad background signal that must be removed (typically using a spin echo experiment). The sensitivity of the chemical shift and C_Q to coordination number (see Figure 1.25) makes ^{11}B NMR an excellent probe of local structure, even in disordered or amorphous materials. The second NMR-active isotope, ^{10}B, has a lower natural abundance and integer spin $(I = 3)$ and a larger quadrupole moment, resulting in extremely broad spectral lineshapes, and its study is generally limited.

1.6.4   Carbon

Carbon is present in organic crystals, MOFs, biomaterials, polymers, carbonaceous nanomaterials, as-synthesised zeotypes and drug-delivery materials. ^{13}C $(I = 1/2)$ has only 1.07% natural abundance and moderate γ, meaning that CP from ^{1}H is almost always required to acquire spectra on a reasonable timescale. As in the solution state, the ^{13}C chemical shift is a sensitive probe of the local chemical environment, and ^{13}C shifts in paramagnetic MOFs can act as indirect probes of the adjacent metal centres where binding of guest molecules can take place. High-power decoupling is generally required in order for many materials to remove the heteronuclear dipolar coupling to ^{1}H and improve spectral resolution.

1.6.5   Oxygen

Oxygen is a ubiquitous element in most inorganic compounds, and ^{17}O NMR should have a range of applications that includes geochemistry, materials science, biochemistry and catalysis. However, the extremely low natural abundance (0.037%) has limited its study. For routine observation, isotopic enrichment (involving both significant cost and effort) is, therefore, usually necessary, and this has hindered the development of ^{17}O solid-state NMR. An additional problem is the quadrupolar broadening ($I = 5/2$), which usually requires the use of MQMAS, DOR or another high-resolution approach. However, recent developments in methodology have enabled a much wider application, and ^{17}O has been shown to be an excellent probe of detailed local structure, with variations in both chemical shift and quadrupolar parameters linked to coordination number, covalency, nearest-neighbour coordinating atoms and local geometry.

1.6.6 Fluorine

Fluorine is present in many minerals, synthetic zeotypes and fluoropolymers. ^{19}F ($I = 1/2$) has 100% natural abundance and very high γ, making it easy to study, although as with ^{1}H, fast MAS rates may be required to overcome large CSA and homonuclear dipolar interactions. The large shift range, dipole moment and large J couplings mean that ^{19}F can act as a very sensitive probe of coordination environment, disorder and bond connectivities and lengths. A number of different chemical shift references have been used for ^{19}F in the literature, so care must be taken when comparing spectra. As $[OH]^{-}$ and F^{-} appear almost identical when using X-ray diffraction techniques, ^{19}F NMR has frequently been used to investigate cases of $[OH]^{-}/F^{-}$ disorder in materials such as minerals and zeotypes.

1.6.7 Sodium

Sodium is present in many inorganic materials, minerals, clays, ceramics, zeotypes and biomaterials. With 100% natural abundance, moderate γ and quadrupole moment, ^{23}Na ($I = 3/2$) is relatively straightforward to study using solid-state NMR, although MQMAS NMR spectra are often required to resolve distinct Na sites, owing to sodium's small chemical shift range. The sensitivity of ^{23}Na quadrupolar parameters to the local environment allows ^{23}Na NMR to provide information on polymorphism, as previously shown in Figure 1.24, and ^{23}Na NMR plays an important role in understanding local structure and disorder in many glasses.

1.6.8   Aluminium

Aluminium is present in minerals, zeotypes, MOFs, clays and glasses. ^{27}Al $(I = 5/2)$ has 100% natural abundance, moderate γ and quadrupole moment, making ^{27}Al NMR relatively straightforward. In many cases, experiments such as MQMAS are used to improve spectral resolution. As both δ_{iso} and C_Q are very sensitive to the coordination number and local geometry (as shown in Figures 1.22 and 1.25), ^{27}Al NMR experiments have been widely used to provide information on coordination number, structure and bonding, particularly in glasses and zeolites.

1.6.9   Silicon

After oxygen, silicon is the most abundant element in the Earth's crust, and is an integral component of many zeotypes, glasses, minerals and semiconductors. ^{29}Si $(I = 1/2)$ has a low natural abundance (4.7%) and typically very long T_1 relaxation times, making ^{29}Si NMR straightforward but time-consuming. The large shift range and sensitivity to local structure (*e.g.* bond angles and lengths), coordination number and next-nearest-neighbour type (as discussed in Section 1.5.1) make ^{29}Si NMR a sensitive probe of the local structure in many materials. ^{29}Si NMR is a vital tool in the study of silicate glasses, disorder in zeolites and catalysts, and has been widely applied for the study of natural minerals.

1.6.10   Phosphorus

The high γ and 100% natural abundance of ^{31}P $(I = 1/2)$ make it readily observable by solid-state NMR, leading to a range of applications in the study of glasses, zeotypes and biomaterials. ^{31}P exhibits a large chemical shift range, and the isotropic shift is very sensitive to the local geometry (*i.e.* bond lengths and angles), coordination number and condensation and NNN substitutions. Furthermore, as J and dipolar couplings to ^{31}P are often relatively large, longer-range structures and connectivites can be easily probed using a range of 1D and 2D NMR experiments.

1.6.11   Xenon

While not itself a solid, xenon gas is an informative probe of catalytically active surfaces, clathrates and other porous materials. ^{129}Xe ($I = 1/2$) has moderate natural abundance (26.4%) and γ, making ^{129}Xe NMR relatively straightforward. ^{129}Xe chemical shifts and T_1 relaxation are extremely sensitive to the local environment (*e.g.* the composition of the surface on which Xe is physisorbed) and partial pressure of Xe (*e.g.* the number of Xe atoms *per* pore of a zeolite). Furthermore, significant changes are observed in the shielding anisotropy and NMR lineshape when Xe is in contact with a surface or within a pore. ^{131}Xe ($I = 3/2$) is also NMR active, but its lower γ and natural abundance (21.2%), combined with its quadrupole moment, make it less attractive for study.

1.7   NMR OF MATERIALS

Over the years, NMR spectroscopy has helped to answer a range of structural questions in areas of chemistry, biology, materials science and geochemistry. Clearly, it is not possible in this brief overview to describe in detail its wide variety of applications. However, in this section we describe some of the main aspects associated with the study of some important types of material, provide some key examples and direct the reader to more detailed descriptions and reviews in the literature.

1.7.1   Simple Ionic Compounds and Ceramics

NMR has been widely used to study many simple compounds, including oxides, fluorides and phosphates. In many cases, the information provided can be straightforward, *e.g.* information about the number of distinct sites in the asymmetric unit, space groups or the polymorph(s) present, as shown in Figure 1.24. NMR spectroscopy is a useful tool for the identification and quantification of impurity phases, providing important information concerning the properties or processing of a bulk material. For series of related compounds, many studies into the dependence of NMR parameters upon local geometry and upon neighbouring atoms have been performed.[2]

For ordered crystalline compounds, NMR is often used alongside Bragg diffraction, in some cases to provide confirmatory evidence or support for a structural model. However, for materials where there is disorder (either in the position of atoms or in the nature of the atom occupying a particular crystallographic site), the sensitivity of NMR to the atomic-scale environment provides a vital structural probe. In recent years, first-principles calculations have been used to help to interpret and assign the complex NMR spectra that result from disordered materials, such as many important functional ceramics. For example, DFT calculations were used to assign the resonances observed in the ^{89}Y and ^{119}Sn MAS NMR spectra of $Y_2(Ti, Sn)_2O_7$ pyrochlores (of interest for the encapsulation of radioactive waste), an example of which was shown in Figure 1.28b.[63,64] As it is not generally possible to carry out calculations on truly disordered materials (as an impractically large number of atoms would be required to ensure all possible local environments were present), a simplified, systematic approach was used (shown in Figure 1.36a), in which the NNN environment of just one of the cations was altered to include all possible numbers and arrangements of Sn/Ti on the six surrounding B sites. The calculated chemical shifts, plotted in Figure 1.36b, show that for ^{89}Y, a systematic change of ~18 ppm was observed as Ti substituted onto the surrounding B sites. However, for ^{119}Sn, although there was a change in shift when Ti was substituted into Sn-rich NNN environments, as the Ti content of the B sites increased this change became smaller and the chemical shifts predicted for differing environments overlapped. These calculations were used to assign the distinct resonances in the ^{89}Y MAS NMR spectra, and to explain the broad, overlapping resonances observed for ^{119}Sn. It was shown that both sets of spectra were consistent with a random distribution of B-site cations.[63,64]

A particularly elegant example of the application of NMR spectroscopy in simple oxides is a recent study of ZrW_2O_8, in which the ^{17}O NMR spectrum contains three resonances for the four distinct oxygens.[65] At higher temperatures, the peaks broaden and coalesce into a single, Lorentzian-like line, indicating that all the oxygen sites are involved in the exchange process, as shown in Figure 1.37a. To provide more detail, variable-temperature 2D exchange spectroscopy (EXSY) experiments were performed, showing exchange between sites through the appearance of cross peaks linking the isotropic peaks on the diagonal (Figure 1.37b). The 2D spectra were able to discriminate unambiguously between two possible models for the dynamic process.

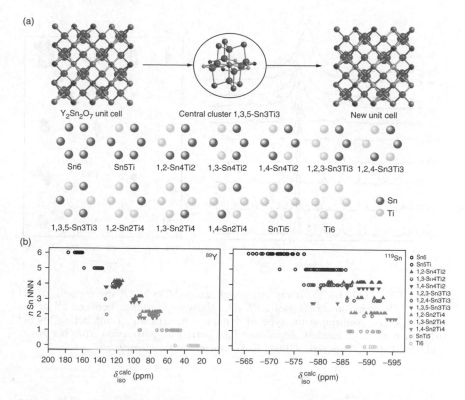

Figure 1.36 (a) Schematic showing the computational approach employed for DFT calculations of the $Y_2(Ti, Sn)_2O_7$ pyrochlore solid solution, with the modification of the local environment of just one of the cations within the unit cell to all possible arrangements of Sn/Ti on the six NNN B sites. (b) Plots showing the calculated ^{89}Y and ^{119}Sn isotropic chemical shifts as a function of the number of Sn NNN. Reproduced with permission from Reader *et al.* (2009) [63]. Copyright (2009) American Chemical Society.

For a more detailed discussion of the application of NMR to oxides and ceramics, see references [2, 41].

1.7.2 Microporous Materials

Microporous materials consist of frameworks that contain pores and channels of similar sizes to small- and medium-sized molecules, enabling chemistry to take place at the internal surfaces of the pores.[66] This has resulted in a variety of applications, including gas storage and

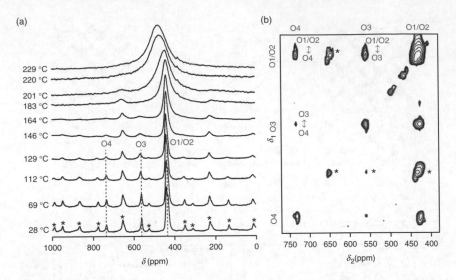

Figure 1.37 ^{17}O (11.7 T) (a) variable-temperature MAS NMR spectra and (b) 2D EXSY NMR spectrum (57 °C) of ZrW_2O_8, showing exchange between all of the oxygen species, confirming the type of dynamics present (*i.e.* a 'ratcheting' model rather than an 'S_N2' model). Reproduced (in part) with permission from Hampson *et al.* (2004) [65]. Copyright (2004) Royal Society of Chemistry.

separation, selective adsorption, catalysis and drug delivery. The main types of such materials include aluminosilicate zeolites, phosphate-based frameworks and MOFs. Solid-state NMR is a valuable tool for the study of microporous materials, as many of the components of the frameworks are NMR active, *e.g.* $^{27}Al/^{29}Si/^{17}O$ for zeolites, $^{27}Al/^{31}P/^{17}O$ for phosphates and $^{1}H/^{13}C$ for MOFs. Furthermore, NMR can be used to study the cations ($H^+/Na^+/Mg^{2+}/K^+/Ca^{2+}$) typically found in zeolites, the extra-framework charge-balancing anions ($[OH]^-/F^-$) in phosphates and many of the guest molecules or templates found within the pores of each type of material. Although framework structures can often be studied by diffraction (unless significant compositional disorder is present), the extra-framework anions, metal cations, water and guest molecules rarely possess the long-range order necessary for diffraction, or else (in many cases) are dynamic.

Zeolites are composed of corner-sharing SiO_4 and AlO_4 tetrahedral units, with any net framework charge balanced by cations (typically Na^+ or H^+) in the channels.[66] For high-silica zeolites, the most useful piece of information NMR can provide is the number of crystallographically distinct species and their occupancies. For example, ZSM-5 undergoes a

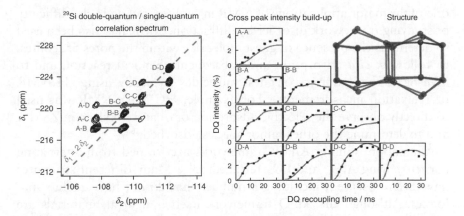

Figure 1.38 Determination of the structure of an unknown high-silica zeolite (later identified as ITQ-4) from ^{29}Si NMR. Using information on the unit cell size and space group from diffraction, the number of peaks in the ^{29}Si MAS NMR spectrum and Si–Si distances derived from the build up of cross peak signal in the double-quantum spectra, it was possible to unambiguously determine the zeolite structure in a blind test. Adapted with permission from Brouwer *et al.* (2005) [68]. Copyright (2005) American Chemical Society.

number of temperature-dependent and absorption-induced phase transitions that can be easily monitored by ^{29}Si MAS NMR.[67] Framework connectivity can be studied using 2D homonuclear correlation experiments, exploiting either dipolar or *J* coupling (Section 1.3.1.3). In recent work by Brouwer *et al.*,[68] such experiments were used to solve the complete structure of a silica zeolite. As shown in Figure 1.38 (with the unit cell size and space group taken from diffraction), information from ^{29}Si 2D double-quantum experiments (specifically the rate at which signal intensity built up for each cross peak) was used to measure Si–Si distances and ultimately (and unambiguously) determine the full zeolite structure (of materials later identified as ITQ-4 and ferrierite). Subsequent improvements to this approach have enabled structure solution without prior knowledge of the space group, utilised the CSA (in addition to the isotropic shift) to help refine the structure and, most recently, combined experimental measurements with DFT calculations within the structure refinement.

The substitution of aluminium into zeolites can be directly observed by ^{29}Si NMR, with a change in the chemical shift dependent upon the number of Al species present. From the relative intensities of the resonances, it is possible to obtain information on the Si/Al ratio and on any ordering in the material, as discussed in Section 1.5.3. This has been one of the major applications of NMR in zeolites, owing to the difficulty of studying framework disorder by diffraction. NMR has also been used to determine the presence of guest molecules within the pores of zeolites, to follow *in situ* adsorption and subsequent chemical reaction and to investigate the dynamics of guest molecules (typically using ^{2}H NMR or relaxation measurements). For example, ^{13}C MAS NMR was used to directly observe the selective adsorption of *o*- and *p*-xylene in ZSM-5 and to determine the other minor species adsorbed.[67]

Aluminophosphate (AlPO) frameworks are formed from alternating corner-sharing AlO_4 and PO_4 tetrahedra.[66] Many different structures have been observed, including a number that do not have zeolite analogues. Although the AlPO framework itself is neutral, materials are typically prepared using a structure-directing agent or template. These are usually positively charged amine bases, so charge balancing in as-made materials is achieved by the incorporation of extra-framework anions (F^- or $[OH]^-$), which bond to Al. As can be seen in Figures 1.20 and 1.22a, the ^{27}Al MAS NMR spectra clearly show the coordination number of Al in both as-made and calcined forms of AlPO-14.[34] Both ^{31}P and ^{27}Al NMR of AlPOs are used to provide information on the number and intensities of the species present, although, as ^{27}Al is quadrupolar, high-resolution experiments such as MQMAS may be required to separate the distinct resonances. As the extra-framework anions may be disordered or have a different periodicity to that of the framework itself, the structures of as-made AlPOs can be difficult to solve from powder diffraction, and NMR can provide many useful contributions. Framework connectivity can be investigated using 2D ^{27}Al/^{31}P heteronuclear or ^{31}P/^{31}P dipolar homonuclear correlation experiments.

Water plays a significant role in the structure of both calcined and as-made materials, with many of the former readily absorbing water from the atmosphere, causing a change in their structure that can be followed by ^{31}P and ^{27}Al NMR. Many as-made materials also contain water (often disordered over multiple crystallographic positions). In general, variation of the water content for as-made AlPOs is much rarer, although this has been observed *in situ* by ^{27}Al and ^{31}P NMR for AlPO-53(A), as shown in Figure 1.39.[69] This material contains two methylamine template molecules and two water molecules within each pore. Under MAS (at a rate of 30 kHz), the ^{27}Al and ^{31}P MAS NMR spectra undergo a significant change in appearance. It has been shown that this is the result of a facile dehydration process (which occurs due to the frictional heating of the sample under MAS), which can be

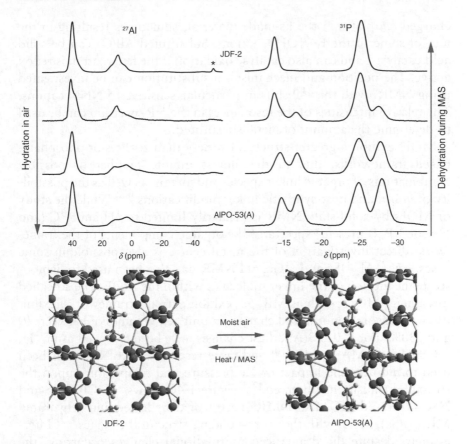

Figure 1.39 ^{27}Al and ^{31}P MAS NMR spectra (11.7 T) of AlPO-53(A), demonstrating the facile dehydration (to JDF-2) resulting from the frictional heating under MAS (30 kHz). The water is readily and reversibly absorbed when the sample is left standing in air. Adapted with permission from Ashbrook *et al.* (2009) [69]. Copyright (2009) American Chemical Society.

reversed simply by leaving the sample standing in air. The dehydrated material, later shown to be JDF-2, has a different symmetry, as shown in Figure 1.39, with half the number of distinct Al and P species. As with zeolites, NMR can be used to investigate the nature and number of guest species within the pores of the framework, and can also be used to study their dynamics (as was demonstrated for AlPO-14[34] in Figure 1.33).

AlPOs may be substituted by a wider range of heteroatoms than zeolites, including Ga, Co, Mg and Si. Substitution can occur at both Al and P sites, *e.g.* divalent Mg for Al, or Si for P.[66] In many cases, the net negative charge imparted to the framework is balanced by the positively charged templates in the as-made material, sometimes resulting in the loss of some of the F^-/[OH]$^-$ anions. Substituted AlPOs can be solid acid catalysts, and can also catalyse oxidation if the heteroatom is redox active. The position and mechanism of substitution can be investigated using NMR, with the ^{31}P shift in particular sensitive to NNN cations. The relative intensities of the resonances in the ^{31}P spectrum can be used to determine the amount of metal substituted.

MOFs exhibit a greater structural variety than zeolites or phosphate-based frameworks, due to the almost infinite number of possible combinations of organic linker species and metals, as well as the possibility of *in situ* and post-synthetic linker modifications.[66] While the study of MOFs by solid-state NMR is primarily limited to ^1H and ^{13}C (and ^{15}N or ^{31}P, if present) spectra of the organic components of the framework, direct investigation of the metal centre is also possible in some cases, *e.g.* ^{27}Al, ^{45}Sc and ^{71}Ga. ^2H NMR has also been used to demonstrate the motion of the linker molecules within the MOF. The so-called 'breathing' MOFs, such as MIL-53, exhibit extreme framework flexibility, with adsorption-induced changes in unit cell volumes of between 40 and 230%. For MIL-53(Al), these changes have been followed using ^1H, ^{13}C and ^{27}Al MAS NMR,[70] and more recently ^{129}Xe NMR has been used to monitor the impact of Xe pressure and temperature upon the structural transitions observed.[71] Figure 1.40 shows ^{13}C CP MAS and ^{45}Sc MAS and MQMAS NMR spectra of dehydrated and rehydrated MIL-53(Sc), along with the corresponding structural changes.[72] Interestingly, despite the dramatic conformational changes occurring, the ScO_6 octahedra present in MIL-53(Sc) are essentially unaffected, with only relatively small changes in C_Q and δ_{iso}; however, more significant changes are observed in the ^{13}C CP MAS NMR spectrum.

Linker modification can introduce chemical functionality in MOFs, changing the chemical nature of the pores and, therefore, the adsorption behaviour of a material.[66] As the functional groups on the linker molecules are rarely fully ordered (and in some cases more than one type of linker is present), their presence can be difficult to detect by diffraction, but is very straightforward to determine using ^{13}C CP MAS NMR, as shown in reference.[73] *In situ* modification of the linker (*i.e.* during the reaction) is relatively rare, but solid-state ^1H/^2H and ^{13}C NMR spectra of the Cu-based MOF, STAM-1, show the rapid *in situ* methylation of the trimesate linker in the presence of methanol, whereas no such process occurs in the presence of ethanol (yielding the MOF, HKUST-1, as the product).[74] While many MOFs are diamagnetic and ^{13}C chemical shifts are similar to those in solution, the presence

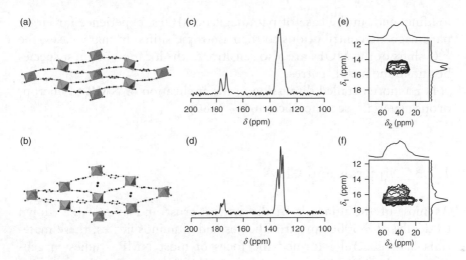

Figure 1.40 (a,b) Structural models and (c,d) ^{13}C (14.1 T) CP MAS and (e,f) ^{45}Sc (20.0 T) MAS/MQMAS NMR spectra of (a,c,e) dehydrated and (b,d,f) rehydrated MIL-53(Sc), showing the 'breathing' of the structures. Reproduced (in part) from Springuel-Huet *et al.* (2010) [71]. Copyright (2010) American Chemical Society.

of paramagnetic ions in many MOFs of interest can complicate ^{13}C NMR spectra. While relatively few ^{13}C NMR spectra have so far been reported for paramagnetic MOFs,[75] it is clear that the most significant effect of the paramagnetic ions is the large shift range (over 1500 ppm). In many cases, fast MAS (*i.e.* above 50 kHz) is required to achieve high-resolution spectra. Owing to the large positive and negative shifts observed, assignment of these spectra is nontrivial. Furthermore, the fact that the paramagnetic ions are connected in an infinite network means that assignments based on relaxation measurements and distances to metal ions are often ambiguous.

As with zeolites and phosphates, solid-state NMR is perhaps of the greatest use in MOFs for the study of guest species. As examples, ^{31}P MAS NMR has been used to confirm that the $H_3PMo_{12}O_{40}$ poly-oxometalate cluster remained intact when loaded into the pores of MIL-100(Fe),[76] and ^{13}C MAS or CP MAS NMR has been used to study a variety of drug species.[77] It is interesting to note that the major-ity of work to date has been carried out for diamagnetic guests within paramagnetic MOFs. The ^{13}C isotropic shifts of the guests (mainly drug species) are shifted slightly by pseudocontact shifts (0–15 ppm), but are generally distinguishable from the framework resonances, which are typically of greater intensity (owing to sub-stoichiometric guest loading) and, in the case of paramagnetic MOFs, experience far larger paramagnetic contributions to their isotropic shifts. In many cases, the ^{13}C shifts of the MOFs are also sensitive to the location of guest species bound to the metal centres.

For a more detailed discussion of the application of NMR to microporous materials, see references [66, 78–80].

1.7.3   Minerals and Clays

As most of the minerals in the Earth's crust and mantle (*i.e.* down to depths of $\sim$3000 km) are silicates and aluminosilicates, these materials have generally formed the focus of most NMR studies in geochemistry.[2,20,37] The principal use of NMR spectroscopy has been the investigation of Al/Si disorder in the framework aluminosilicates that account for over 90% of the crust, exploiting the sensitivity of the ^{29}Si chemical shift to NNN composition. The stability of these phases depends not only on the composition, *i.e.* upon the Al/Si ratio, but also on the local order. As described in Section 1.5.3, the number of relatively high-energy Al–O–Al linkages is often minimised (Lowenstein's rule), and the relative intensity of the resonances in the ^{29}Si MAS NMR spectrum can provide information on both composition and any short-range ordering (see Figure 1.22b).[37]

Clay minerals are (hydrous) layered aluminosilicates, containing 2D tetrahedrally and/or octahedrally coordinated silicate sheets.[20,37] The materials lack long-range order (and so are difficult to investigate by conventional diffraction), but the short-range (Al/Si) order can be easily quantified using ^{29}Si NMR, as described above. The exact distribution of Al within such layered materials determines the charge on the layers, which ultimately plays a significant role in their physical and chemical properties. The sensitivity of the ^{27}Al chemical shift to the coordination number enables NMR to provide an accurate measurement of the Al(IV)/Al(VI) ratio. Recent work has also utilised ^{17}O NMR (MAS and MQMAS) to probe the distribution of the T–O–T (T = Si or Al) linkages.

NMR spectroscopy has also been utilised for the study of anion disorder in many rock-forming minerals, where the substitution of F$^-$ for [OH]$^-$ is relatively common.[37] It has long been known that in many types of material the anion distribution is rarely random, but that F$^-$ often has a strong preference to be located closer to particular cations.

For layered silicates, both ^{1}H and ^{19}F NMR can be used to determine the local environment (*i.e.* the number and nature of the cations in close proximity), as the chemical shifts of both nuclei are sensitive to these changes. In the humite series of minerals, $n\text{Mg}_2\text{SiO}_4 \cdot \text{Mg(OH, F)}_2$, it had been suggested from diffraction measurements that F^- substitution preferentially forms $\text{F} \cdots \text{HO}$ hydrogen bonds, and minimises the number of $\text{HO} \cdots \text{OH}$ pairs (and therefore dynamics in the system;[53] see Section 1.5.4). However, the ^{19}F MAS NMR spectrum of a 50% fluorinated clinohumite ($4\text{Mg}_2\text{SiO}_4 \cdot \text{Mg(OH)F}$), shown in Figure 1.41, reveals four distinct fluorine resonances, and DFT calculations are able to assign these to specific local environments, demonstrating the presence of considerably more disorder in the material than was previously anticipated, with both F-rich and OH-rich regions.[81]

Many minerals undergo phase transitions with changes in either temperature or pressure. In the former case, the transitions can often be followed *in situ* using variable-temperature NMR, and generally result from structural rearrangement rather than the destruction or creation of chemical bonds. Transitions are indicated by changes in ^{29}Si and ^{17}O MAS NMR spectra, or by differences in relaxation rates. While changes with pressure are currently much more difficult to follow *in situ*, there can be significant changes in the NMR spectrum for

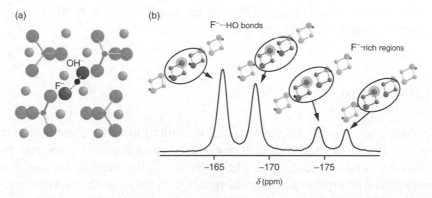

Figure 1.41 (a) A $\text{F} \cdots \text{HO}$ hydrogen bond formed by F substitution in clinohumite. (b) ^{19}F (14.1 T) MAS NMR spectrum of 50% fluorinated clinohumite $4\text{Mg}_2\text{SiO}_4 \cdot \text{Mg(OH)F}$, showing the presence of four distinct F resonances. DFT calculations were used to assign the four environments (which differ in the nature of the anions on the two closely spaced sites). The spectrum reveals the presence of considerably more disorder in the material than was previously anticipated, with both F-rich and OH-rich regions. Reproduced with permission from Griffin *et al.* (2010) [81]. Copyright (2010) American Chemical Society.

phases synthesised under pressure. An example of this was shown in Figure 1.21b for $MgSiO_3$ phases synthesised at 1 GPa/1100 °C (orthoenstatite), 19.5 GPa/1400 °C (akimotoite), 19.5 GPa/1950 °C (majorite) and 23 GPa/1650 °C (perovskite).[22] High-pressure magnesium silicates are of considerable interest as they constitute the majority of the Earth's mantle. Studying these materials by NMR can be challenging, since the high-pressure synthesis, usually in a multi-anvil apparatus, often produces only small amounts of material. For example, in the synthesis of perovskite described above, only ∼4 mg of sample was produced. This necessitates the use of sensitivity-enhancing techniques, such as CPMG, or the use of STMAS rather than MQMAS to acquire high-resolution spectra of quadrupolar nuclei. While most experiments have focused on ^{29}Si and ^{17}O NMR, ^{27}Al NMR has also been used to investigate the position(s) of Al substitution in these materials, providing insight into the mechanism by which the substitution is charge balanced.[20,37] Current and future effort may well be focused on the 'hydration' of the nominally anhydrous high-pressure silicates in the mantle. The incorporation of hydrogen into these materials has important implications for the physical and chemical properties of the inner Earth and is difficult to study by diffraction as the protons are generally disordered and often included only at ppm levels. Most recent work has utilised ^{1}H and ^{2}H NMR, but it would appear that ^{17}O NMR will play a significant role in future investigation.

For a more detailed discussion of the application of NMR to geochemistry, see references [20, 37].

1.7.4 Energy Materials

Growing environmental concerns have resulted in an increased interest in the design of materials that are able to produce and store energy efficiently and economically. The study of such materials focuses not only on the determination of local structure but also on the investigation of dynamic processes, and solid-state NMR has played a significant role in both of these areas.[82] NMR has been particularly useful in the study of lithium ion batteries, which are of interest due to their high energy density and their role in energy storage. Investigations have focused primarily on ^{6}Li and ^{7}Li NMR. Although both are quadrupolar, with relatively small chemical shift ranges, the large paramagnetic shifts (often up to 1600 ppm) that can be observed provide considerable chemical sensitivity. These shifts can often be directly related to the local geometry of the overlap between the paramagnetic centres and the Li $2s$ orbitals. Although much of the early work focused on basic structural characterisation (often by comparison to model compounds and known phases), recent work by Grey and co-workers[82,83] has transformed the use of NMR spectroscopy in this field. As shown in Figure 1.42a, the structural changes that occur upon discharge can be followed (using static ^{7}Li experiments) *in situ*, *i.e.* under realistic operating conditions during the NMR experiment.[83] For example, it has been demonstrated that for lithium ion batteries containing silicon, the first discharge occurs *via* the lithiation of Si to form isolated Si and Si–Si clusters. The clusters are then destroyed to form isolated Si ions and eventually a crystalline phase. A number of phases were thus observed that were not present in *ex situ* measurements, highlighting the need for direct, real-time observation. ^{6}Li/^{7}Li NMR can also be used to study dynamics in a material, usually through the use of variable-temperature measurements of linewidths or T_1. 2D EXSY experiments have also been employed to identify the dynamic Li species and to provide quantitative information on the mobility. Figure 1.42b shows variable-temperature ^{7}Li NMR spectra of a composite polymer electrolyte, where the line narrowing

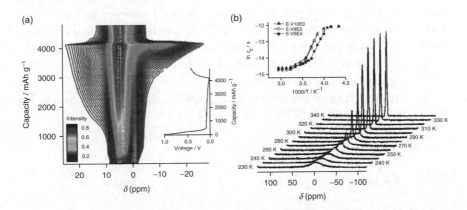

Figure 1.42 (a) Contour plot of *in situ* ^{7}Li (4.7 T) static NMR spectra and (inset) electrochemical profile of the first discharge of an actual crystalline Si *vs* Li/Li$^+$ battery (the colour bar shows the relative intensity scale for the spectra). Reproduced with permission from Key *et al.* (2009) [83]. Copyright (2009) American Chemical Society. (b) Variable-temperature ^{7}Li (7.05 T) NMR spectra for a polymer electrolyte, and the corresponding correlation time for Li dynamics. Reproduced with permission from Jeon and Kwak (2006) [84]. Copyright (2006) American Chemical Society.

can be directly related to the correlation time and the activation energy of the ionic transport.[84]

The use of oxide ion conductors in a range of devices, including solid oxide fuel cells, has also resulted in an increase in their study.[82] In many cases, and unlike the cationic dynamics described above, anionic conductivity often occurs at relatively high temperatures, e.g. 800 °C and above. There have been many investigations of materials with perovskite-, pyrochlore- and fluorite-like structures, the majority focusing on variable-temperature ^{17}O NMR to identify the species responsible for conduction and provide insight into the mechanism. Such materials can typically be enriched in ^{17}O relatively easily, owing to the dynamics of the anion lattice. An example of an investigation into oxygen dynamics was discussed in Section 1.7.1 (Figure 1.37).[65]

Another field of intense research is boron-based hydrogen storage materials, which release hydrogen gas via a complicated series of decomposition pathways. Understanding which decomposition mechanisms are active in the presence of various additives could lead to a more efficient release of hydrogen at lower temperature. As the ^{11}B chemical shift is sensitive to the number of B–H bonds present, and the C_Q is sensitive to the overall coordination number of B (Section 1.5.1), ^{11}B MAS NMR spectra can be particularly valuable in understanding the amorphous decomposition products of hydrogen release from borohydride materials. This process was recently studied by Doroodian et al.[85] for a series of methylguanidinium borohydrides, with DFT calculations on model clusters confirming the spectral assignments and highlighting the differences between decomposition of the hydrogen storage material in the presence of different catalysts.

For a more detailed discussion of the application of NMR to energy materials, see reference [82].

1.7.5   Glasses

Although they possess macroscopic properties associated with the solid state, amorphous materials, such as glasses, lack the long-range structural order exhibited by crystalline solids. This poses a number of challenges when determining the structure of these materials, which NMR – as a sensitive probe of the atomic-scale environment – is uniquely placed to address.[86] These include understanding the exact composition of the glass, determining the polymerisation of the network, determining the NNN composition and ordering, understanding the distribution of network modifiers and investigating the distribution of local geometries present. Furthermore, it is also important to understand the structural role water plays in glasses and, additionally, to understand the influence of the preparation/processing of the material upon the ultimate structure and the properties observed. The majority of NMR investigation has focused on silicate-based glasses, as a result both of the sensitivity of ^{29}Si NMR to the local environment and of the important role these materials play in nature. However, research into borate, phosphate and chalcogenide glasses is steadily increasing.

Glassy SiO_2 retains the corner-sharing tetrahedral silicate network found in many crystalline silicates, but the variation in local geometry, *i.e.* the Si–O–Si bond angle, provides an intrinsic structural disorder. This has been studied using ^{29}Si NMR spectroscopy (see Figure 1.28c),[49] and more recently ^{17}O NMR (where both δ_{iso} and C_Q can be probed), along with support from DFT calculations, demonstrating that the most probable bond angle lies between 142 and 151°.[86] Incorporation of additional elements into silica can result in more favourable physical/chemical properties, either during the processing and handling of the material or in terms of its ultimate end use. The elements incorporated tend to fall into three distinct categories: (i) network formers (*e.g.* B, Ge); (ii) network modifiers (*e.g.* Ca, Pb, Na, K), typically present as ions; and (iii) intermediates (*e.g.* Al, Mg, Zr), which can act either as modifiers or as formers, depending upon the glass composition. The incorporation of a modifier leads to depolymerisation and the formation of nonbridging oxygen (NBO) species. This process (generally followed as a function of the amount of modifier incorporated) has been investigated using ^{29}Si, NMR (as the chemical shift is very sensitive to the degree of condensation, Q^n) and ^{17}O NMR (as bridging oxygens and NBOs have very different C_Q and δ_{iso}). Figure 1.43a shows the compositional dependence of the ^{29}Si MAS NMR spectra of an Na_2O–SiO_2 glass, demonstrating that the incorporation of increasing amounts of Na_2O results in the formation of less polymerised networks. The relative intensities of the peaks obtained can be compared to those predicted using various structural models.[87]

The introduction of alumina into silicate or alkali silicate glasses can improve mechanical and thermal stability, and aluminosilicate glasses are used in a variety of applications. ^{29}Si NMR can in principle be used to probe the incorporation of Al into the framework, *i.e.* the NNN environment, although resonances can often be overlapped for the more complex glass compositions. ^{27}Al MAS NMR is often more useful, providing a simple measurement of the Al coordination number and, therefore, its role within the network. The type of structure produced generally depends upon the exact composition of the glass, specifically the Si/Al and M/Al ratios. ^{17}O NMR can also provide useful information, particularly for more complex compositions in which the resolution obtained using ^{29}Si NMR is poor. For bridging oxygens, the different X–O–X' linkages exhibit different C_Q and δ_{iso} as, of course, do NBOs.[86] In many cases, MQMAS is required to resolve the broad, overlapping resonances, as in Figure 1.43b, which shows the ^{17}O MQMAS NMR spectrum of a calcium-containing boroaluminosilicate glass. NBOs on Si and B can be clearly resolved, and Al–O–Si, Si–O–Si and B–O–Si linkages can also be identified.[88] From their various proportions (corrected for any non-uniform efficiency), it is possible to understand the ordering within the framework and whether there is any clustering or avoidance of the various network formers.

Although there has been less study of borate glasses, borosilicate glasses are widely used for the encapsulation of radioactive waste (as

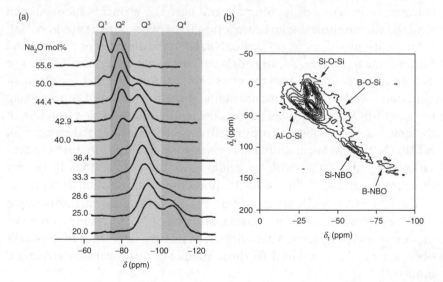

Figure 1.43 (a) Compositional dependence of ^{29}Si MAS NMR spectra of $Na_2O–SiO_2$, demonstrating the decrease in polymerisation with increasing Na content. Reproduced with permission from Maekawa *et al.* (1991) [87]. Copyright (1991) Elsevier Ltd. (b) ^{17}O (14.1 T) MQMAS NMR spectrum of calcium-containing boroaluminosilicate glass, in which NBOs on Si and B are clearly resolved, as are the different types of bridging oxygen. Reproduced with permission from Du and Stebbins (2005) [88]. Copyright (2005) Elsevier Ltd.

the incorporation of B_2O_3 improves the thermal properties of the glass, without compromising its chemical durability).[86,89] In both types of material, ^{11}B offers a sensitive probe of the B coordination environment, with both C_Q and δ_{iso} changing significantly for trigonal and tetrahedral boron. The ratio of three- and four-coordinate boron provides information on the role of network modifiers and the number of NBOs. In a recent example, ^{11}B NMR was used to demonstrate the difference in glass structure between a melt ($Na_2O-B_2O_3-SiO_2$) quenched at ambient pressure and at 5 GPa.[90] For the ambient sample, four-coordinate boron accounted for 62% of the boron in the sample, whereas in the sample quenched under pressure it accounted for 85%, indicating the increased density of the network.

For phosphate-based glasses, ^{31}P NMR is widely used, with the chemical shift sensitive to the polymerisation (Q^n speciation).[86] The 100% natural abundance of ^{31}P also enables 2D heteronuclear and homonuclear correlation experiments to be used, in order to probe through-bond and through-space connectivity in glassy materials. In addition to ^{29}Si and ^{31}P NMR of chalcogenide glasses (typical based on SiS_2, $SiSe_2$ or phosphorus–selenium compositions), ^{77}Se (spin $I = 1/2$) NMR has also been employed. Although the chemical shift is very sensitive to the local environment, this method does suffer from poor sensitivity, owing to the low natural abundance ($\sim$7.9%) of ^{77}Se and the typically large CSA, resulting in a considerable number of SSBs.[86]

For a more detailed discussion of the application of NMR to glasses, see references [86, 89, 91].

1.7.6 Polymers

Many polymeric materials are semicrystalline, with similar amounts of amorphous and crystalline regions, and can therefore be difficult to study using conventional diffraction approaches. Furthermore, dynamics (either of functional groups or of sections of the polymer chain itself) also play an important role in the microscopic behaviour of polymeric materials. NMR spectroscopy is well placed to provide both structural and dynamic information on these systems, with the majority of work using ^{13}C, ^{1}H/^{2}H and ^{19}F NMR.[92] Basic MAS (and CP MAS) experiments can be used to resolve chemically distinct species, although lines are often relatively broad due to the local disorder. Identification of the number and proportions of different species can be of particular importance to identifying the exact composition of the material in copolymers and polymer blends. The CSA is also a sensitive probe of the local structure and is often measured (indirectly in a 2D experiment) to help spectral assignment. Information on the conformation of polymer chains can also be obtained from ^{13}C NMR, as the chemical shift is sensitive to whether *trans* (t) or *gauche* (g) conformations are present. This can be seen in Figure 1.44a, in which the ^{13}C CP MAS NMR spectrum of (atatic) polypropylene is shown, with resolution of chemically different species (*i.e.* CH_3, CH and CH_2, in order of increasing δ_{iso}) and, for the CH_2 groups, different conformations.[93]

It is possible to obtain a range of more detailed structural information using more complex NMR experiments. These include 'spin counting', in which the size of a particular domain is estimated by counting the number of a particular type of nucleus in spatial proximity through the creation of multiple-quantum coherences (*i.e.* measuring the number of spins that are linked by the dipolar interaction). Relaxation measurements can also provide information on the relative amounts of crystalline (*i.e.* rigid) and amorphous (*i.e.* mobile) regions of the polymer, and these will have differing relaxation rates according to the different dynamics present.

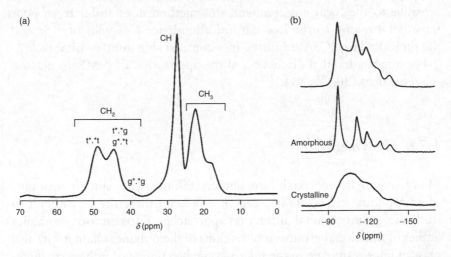

Figure 1.44   (a) ^{13}C CP MAS NMR spectrum of amorphous (atactic) polypropylene at 252 K. The abbreviation describes the confirmation around the carbon centre denoted with a '.', where t and g denote *trans* and *gauche* conformations, respectively. Reproduced with permission from Zemke *et al.* (1994) [93]. Copyright (2003) Wiley-VCH. (b) ^{19}F MAS NMR spectrum of a copolymer of vinylidene fluoride and trifluoroethylene, along with 'filtered' spectra in which only the amorphous or crystalline parts of the sample are observed. Reproduced with permission from Hazendonk *et al.* (2003) [94]. Copyright (2003) Elsevier Ltd.

It is possible to design 'filters' (*i.e.* series of pulses and/or delays) to select only mobile or rigid parts of the material, resulting in spectra containing signals only from these parts. This can be seen in Figure 1.44b, where the ^{19}F spectrum of a copolymer of vinylidene fluoride and trifluoroethylene is shown, along with 'filtered' spectra where only the amorphous and crystalline parts of the sample are observed.[94] This is useful not only for quantifying these various regions but also for improving the spectral resolution by reducing the overlap from different resonances.

The dynamics in polymers, including rotational motions (*e.g.* of CH_3 or phenyl groups) in crystalline regions, chain diffusion in both crystalline and amorphous regions and chain motion close to the glass transition temperature can be followed using a variety of NMR experiments. The simplest include the acquisition of MAS, CP MAS and filtered spectra at varying temperatures or temperature-dependent relaxation measurements. Analysis of the lineshapes in 2D exchange experiments, typically for ^2H (but also ^{13}C), is also able to provide information on the geometry of any rotational motion (as described in Section 1.5.4).[95]

For a more detailed discussion of the application of NMR to polymers, see references [92, 95].

1.8 CONCLUSION

This chapter has demonstrated just a few of the many applications of solid-state NMR to the study of inorganic materials, from solving the crystal structure of zeolites to studying the dynamics in amorphous regions of polymers and hydrated minerals. However, we hope that this shows the scope of solid-state NMR as a powerful tool for investigating structure, disorder and dynamics from an atomic-scale viewpoint.

While the range of information available from NMR is impressive, there are still areas of materials science that pose a considerable challenge. In particular, these include the study of extremely dilute species (*e.g.* surface species in porous silicas or on nanoparticles), nuclei with very broad lines or poor sensitivity (*e.g.* most of the halogens, except ^{19}F, and many important metals) and phases occurring at very high or low temperatures (particularly under MAS). At the time of writing, much effort is being directed towards tackling all of these issues. Techniques such as dynamic nuclear polarisation (transfer of magnetisation from unpaired electrons to nuclei by microwave irradiation), leading to 'surface-enhanced NMR', with vast enhancement (one or two orders of magnitude) of the signals of surface species, are being actively developed.

Broadband excitation pulse sequences (generally in combination with sensitivity-enhancing methods) are constantly being improved in order to enable rapid acquisition of very broad resonances, and laser-heated MAS probes capable of reaching temperatures of $\sim$1000 K are now commercially available. In tandem with this improved experimental capability, computational techniques (both analytical line-fitting software and quantum-chemical predictions) can only increase in speed, accuracy and capability, providing more rigorous theoretical support to the interpretation of the experimental data. Of course, it is not just hardware and software that are important to understanding a material, but also the mindset of the experimentalists who interpret the data. Recent years have seen advances in this aspect of materials science, with the field of 'NMR crystallography' seeking to combine simultaneously data from calculations, crystallography and NMR in a single, highly-detailed model of a material, rather than publishing several datasets for the same material, but with conflicting interpretations.

As solution-phase NMR spectroscopy has grown to become an integral part of all molecular chemistry, it seems that solid-state NMR will in the near future come to play a similarly crucial role in materials science, providing a probe of all aspects of materials, from their structures and motional behaviour to their properties and applications.

REFERENCES

[1] D. C. Apperley, R. K. Harris and P. Hodgkinson, *Solid State NMR Basic Principles and Practice*, Momentum Press, New York, 2012.

[2] K. J. D. MacKenzie and M. E. Smith, *Multinuclear Solid-State NMR of Inorganic Materials*, Pergamon Press, Oxford, 2002.

[3] E. L. Hahn, *Phys. Rev.*, 80, 580 (1950).

[4] D. M. Grant, 'Chemical Shift Tensors' in *Encyclopedia of Magnetic Resonance*, Eds-in-chief R. K. Harris and R. E. Wasylishen, John Wiley: Chichester. Published online 15 Mar 2007.

[5] R. E. Wasylishen, 'Dipolar and Indirect Coupling Tensors in Solids' in *Encyclopedia of Magnetic Resonance*, Eds-in-chief R. K. Harris and R. E. Wasylishen, John Wiley: Chichester. Published online 15 Mar 2007.

[6] A. J. Vega, 'Quadrupolar Nuclei in Solids' in *Encyclopedia of Magnetic Resonance*, Eds-in-chief R. K. Harris and R. E. Wasylishen, John Wiley: Chichester. Published online 15 Mar 2010.

[7] H. Koller, G. Engelhardt, A. P. M. Kentgens and J. Sauer, *J. Phys. Chem.*, 98, 1544 (1994).

[8] E. R. Andrew, A. Bradbury and R. G. Eades, *Nature*, 182, 223 (1958).

[9] E. R. Andrew, 'Magic Angle Spinning' in *Encyclopedia of Magnetic Resonance*, Eds-in-chief R. K. Harris and R. E. Wasylishen, John Wiley: Chichester. Published online 15 Mar 2007.

[10] A. Samoson, 'Magic-Angle Spinning Extensions' in Encyclopedia of Magnetic Resonance, Eds-in-chief R. K. Harris and R. E. Wasylishen, John Wiley: Chichester. Published online 15 Mar 2007.

[11] J. M. Griffin, L. Clark, V. R. Seymour, D. W. Aldous, D. M. Dawson, D. Iuga, R. E. Morris and S. E. Ashbrook, *Chem. Sci.*, **3**, 2293 (2012).

[12] P. Hodgkinson, *Prog. Nucl. Magn. Reson. Spectrosc.*, **46**, 197 (2005).

[13] P. K. Madhu, *Solid State Nucl. Magn. Reson.*, **35**, 2 (2009).

[14] A. Pines, J. S. Waugh and M. G. Gibby, *J. Chem. Phys.*, **56**, 1776 (1972).

[15] D. P. Burum, 'Cross Polarization in Solids' in Encyclopedia of Magnetic Resonance, Eds-in-chief R. K. Harris and R. E. Wasylishen, John Wiley: Chichester. Published online 15 Mar 2007.

[16] S. R. Hartmann and E. L. Hahn, *Phys. Rev.*, **128**, 2042 (1962).

[17] R. R. Ernst, *Chimica*, **29**, 179 (1975).

[18] G. A. Morris and J. W. Emsley, 'Multidimensional NMR: an Introduction' in Encyclopedia of Magnetic Resonance, Eds-in-chief R. K. Harris and R. E. Wasylishen, John Wiley: Chichester. Published online 15 Sept 2011.

[19] A. Lesage, C. Auger, S. Caldarelli and L. Emsley, *J. Am. Chem. Soc.*, **119**, 7867 (1997).

[20] R. E. Wasylishen, S. E. Ashbrook and S. Wimperis, *NMR of Quadrupolar Nuclei in Solid Materials*, John Wiley & Sons, Ltd, Chichester, 2012.

[21] S. E. Ashbrook and S. Wimperis, *Prog. Nucl. Magn. Reson. Spectrosc.*, **45**, 53 (2004).

[22] S. E. Ashbrook, A. J. Berry, D. J. Frost, A. Gregorovic, C. J. Pickard, J. E. Readman and S. Wimperis, *J. Am. Chem. Soc.*, **129**, 13213 (2007).

[23] A. Samoson, E. Lippmaa and A. Pines, *Mol. Phys.*, **65**, 1013 (1988).

[24] G. Engelhardt, A. P. M. Kentgens, H. Koller and A. Samoson, *Solid State Nucl. Magn. Reson.*, **15**, 171 (1999).

[25] A. Llor and J. Virlet, *Chem. Phys. Lett.*, **152**, 248 (1988).

[26] L. Frydman and J. S. Harwood, *J. Am. Chem. Soc.*, **117**, 5367 (1995).

[27] Z. Gan, *J. Am. Chem. Soc.*, **122**, 3242 (2000).

[28] A. W. MacGregor, L. A. O'Dell, R. W. Schurko, *J. Magn. Reson.*, **208**, 103 (2011).

[29] S. Meiboom and D. Gill, *Rev. Sci. Instrum.*, **29**, 688 (1958).

[30] R. M. Martin, *Electronic Structure: Basic Theory and Practical Methods*, Cambridge University Press, Cambridge 2004.

[31] J. R. Yates and C. J. Pickard. 'Computations of Magnetic Resonance Parameters for Crystalline Systems: Principles' in Encyclopedia of Magnetic Resonance, Eds-in-chief R. K. Harris and R. E. Wasylishen, John Wiley: Chichester. Published online 15 Mar 2008.

[32] C. J. Pickard and F. Mauri, *Phys. Rev. B*, **63**, 245101 (2001).

[33] J. R. Yates, *Magn. Reson. Chem.*, **48**, S23 (2010).

[34] S. E. Ashbrook, M. Cutajar, C. J. Pickard, R. I. Walton and S. Wimperis, *Phys. Chem. Chem. Phys.*, **10**, 5754 (2008).

[35] D. H. Brouwer, in *NMR Crystallography*, R. K. Harris, R. E. Wasylishen and M. J. Duer (Eds), John Wiley & Sons, Ltd, Chichester, 2009.

[36] A. Othman, J. S. O. Evans, I. Radosavljevic Evans, R. K. Harris and P. Hodgkinson, *J. Pharm. Sci.*, **96**, 1380 (1997).

[37] B. L. Phillips, in *NMR Crystallography*, R. K. Harris, R. E. Wasylishen and M. J. Duer (Eds), John Wiley & Sons, Ltd, Chichester, 2009.

[38] M. S. Ironside, R. S. Stein and M. J. Duer, *J. Magn. Reson.*, **188**, 49 (2008).
[39] M. R. Mitchell, D. Carnevale, R. Orr, K. R. Whittle and S. E. Ashbrook, *J. Phys. Chem. C*, **116**, 4273 (2012).
[40] K. E. Johnston, C. C. Tang, J. E. Parker, K. S. Knight, P. Lightfoot and S. E. Ashbrook, *J. Am. Chem. Soc.*, **132**, 8732 (2010).
[41] S. E. Ashbrook and M. E. Smith, *Chem. Soc. Rev.*, **35**, 718 (2006).
[42] T. J. Bastow, M. A. Gibson and C. T. Forwood, *Solid State Nucl. Magn. Reson.*, **12**, 201 (1998).
[43] C. Gervais, M. Profeta, F. Babonneau, C. J. Pickard and F. Mauri, *J. Phys. Chem. B*, **108**, 13249 (2004).
[44] M. H. Levitt, *J. Chem. Phys.*, **128**, 052205 (2008).
[45] J. Schaefer. 'REDOR and TEDOR' in *Encyclopedia of Magnetic Resonance*, Eds-in-chief R. K. Harris and R. E. Wasylishen, John Wiley: Chichester. Published online 15 Mar 2007.
[46] R. D. Gougeon, E. B. Brouwer, P. R. Bodart, L. Delmotte, C. Marichal, J. M. Chézeau and R. K. Harris, *J. Phys. Chem. B*, **105**, 12249 (2001).
[47] S. E. Lister, A. Soleilhavoup, R. L. Withers, P. Hodgkinson and J. S. O. Evans, *Inorg. Chem.*, **49**, 2290 (2010).
[48] W. Lowenstein, *Am. Mineral.*, **39**, 92 (1953).
[49] R. Oestrike, W. H. Yang, R. J. Kirkpatrick, R. L. Hervig, A. Navrotsky and B. Montez, *Geochim. et Cosmochim. Acta*, **51**, 2199 (1987).
[50] M. Castro, V. R. Seymour, D. Carnevale, J. M. Griffin, S. E. Ashbrook, P. A. Wright, D. C. Apperley, J. E. Parker, S. P. Thompson, A. Fecant and N. Bats, *J. Phys. Chem. C*, **114**, 12698 (2010).
[51] C. Martineau, V. Bouchevreau, Z. J. Tian, S. J. Lohmeier, P. Behrens and F. Taulelle, *Chem. Mater.*, **23**, 4799, (2011).
[52] S. E. Ashbrook, S. Antonijevic, A. J. Berry and S. Wimperis, *Chem. Phys. Lett.*, **364**, 634 (2002).
[53] J. M. Griffin, A. J. Miller, A. J. Berry, S. Wimperis and S. E. Ashbrook, *Phys. Chem. Chem. Phys.*, **12**, 2989 (2010).
[54] S. Antonijevic, S. E. Ashbrook, S. Biedasek, R. I. Walton, S. Wimperis and H. X. Yang, *J. Am. Chem. Soc.*, **128**, 8054 (2006).
[55] N. G. Dowell, S. E. Ashbrook and S. Wimperis, *J. Phys. Chem. B*, **108**, 13292 (2004).
[56] S. E. Ashbrook, A. J. Berry, W. O. Hibberson, S. Steuernagel and S. Wimperis, *J. Am. Chem. Soc.*, **125**, 11824 (2003).
[57] A. Nayeem and J. Yesinowski, *J. Chem. Phys.*, **89**, 4600 (1988).
[58] M. Kaupp and F. H. Köhler, *Coord. Chem. Rev.*, **253**, 2376 (2009).
[59] S. S. Y. Chui, S. M. F. Lo, J. P. H. Charmant, A. G. Orpen and I. D. Williams, *Science*, **283**, 1148 (1999).
[60] D. M. Dawson, L. E. Jamieson, M. I. H. Mohideen, A. C. McKinlay, I. A. Smellie, R. Cadou, N. S. Keddie, R. E. Morris and S. E. Ashbrook, *Phys. Chem. Chem. Phys.*, **15**, 919 (2013).
[61] C. H. Townes, C. Herring and W. D. Knight, *Phys. Rev.*, **77**, 852 (1950).
[62] R. K. Harris, E. D. Becker, S. M. Cabral de Menezes, R. Goodfellow and P. Granger, *Pure Appl. Chem.*, **73**, 1795 (2001).
[63] S. W. Reader, M. R. Mitchell, K. E. Johnson, C. J. Pickard, K. R. Whittle and S. E. Ashbrook, *J. Phys. Chem. C*, **113**, 18874 (2009).

[64] M. R. Mitchell, S. W. Reader, K. E. Johnson, C. J. Pickard, K. R. Whittle and S. E. Ashbrook, *Phys. Chem. Chem. Phys.*, **13**, 488 (2011).

[65] M. R. Hampson, P. Hodgkinson, J. S. O. Evans, R. K. Harris, I. J. King, S. Allen and F. Fayon, *Chem. Commun.*, **4**, 392 (2004).

[66] P. A. Wright, *Microporous Framework Solids*, RSC Publishing, Cambridge, 2008.

[67] C. A. Fyfe, H. Grondey, Y. Feng and G. T. Kokotailo, *J. Am. Chem. Soc.*, **112**, 8812 (1990).

[68] D. H. Brouwer, R. J. Darton, R. E. Morris and M. H. Levitt, *J. Am. Chem. Soc.*, **127**, 10365 (2005).

[69] S. E. Ashbrook, M. Cutajar, J. M. Griffin, Z. A. D. Lethbridge, R. I. Walton and S. Wimperis, *J. Phys. Chem. C*, **113**, 10780 (2009).

[70] T. Loiseau, C. Serre, C. Huguenard, G. Fink, F. Taulelle, M. Henry, T. Bataille and G. Férey, *Chem. Eur. J.*, **10**, 1373 (2004).

[71] M. A. Springuel-Huet, A. Nossov, Z. Adem, F. Guenneau, C. Volkringer, T. Loiseau, G. Férey and A. Gédéon, *J. Am. Chem. Soc.*, **132**, 11 599 (2010).

[72] J. P. S. Mowat, V. R. Seymour, J. M. Griffin, S. P. Thompson, A. M. Z. Slawin, D. Fairen-Jimenez, T. Düren, S. E. Ashbrook and P. A. Wright, *Dalton Trans.*, **41**, 3937 (2012).

[73] J. P. S. Mowat, S. R. Miller, J. M. Griffin, V. R. Seymour, S. E. Ashbrook, S. P. Thompson, D. Fairen-Jimenez, A. M. Banu, T. Düren and P. A. Wright, *Inorg. Chem.*, **50**, 10844 (2011).

[74] M. I. H. Mohideen, B. Xiao, P. S. Wheatley, A. C. McKinlay, Y. Li, A. M. Z. Slawin, D. W. Aldous, N. F. Cessford, T. Düren, X. P. Zhao, R. Gill, K. M. Thomas, J. M. Griffin, S. E. Ashbrook and R. E. Morris, *Nature Chem.*, **3**, 304 (2011).

[75] G. de Combarieu, M. Morcrette, F. Millange, N. Guillou, J. Cabana, C. P. Grey, I. Margiolaki, G. Férey and J.-M. Tarascon, *Chem. Mater.*, **21**, 1602 (2009).

[76] R. Canioni, C. Roch-Marchal, F. Sécheresse, P. Horcajada, C. Serre, M. Hardi-Dan, G. Férey, J.-M. Grenèche, F. Lefebvre, J.-S. Chang, Y.-K. Hwang, O. Lebedev, S. Turner and G. Van Tendeloo, *J. Mater. Chem.*, **21**, 1226 (2011).

[77] P. Horcajada, C. Serre, M. Vallet-Regí, M. Sebban, F. Taulelle and G. Férey, *Angew. Chem., Int. Ed.*, **45**, 5974 (2006).

[78] J. Klinowski, *Prog. Nucl. Magn. Reson. Spectrosc.*, **16**, 237 (1984).

[79] J. Klinowski, *Anal. Chim. Acta*, **283**, 929 (1993).

[80] M. Houas, C. Martineau and F. Taulelle, '*Quadrupolar NMR of Nanoporous Materials*' in *Encyclopedia of Magnetic Resonance*, Eds-in-chief R. K. Harris and R. E. Wasylishen, John Wiley: Chichester. Published online 15 Dec 2011.

[81] J. M. Griffin, J. R. Yates, A. J. Berry, S. Wimperis and S. E. Ashbrook, *J. Am. Chem. Soc.*, **132**, 15 651 (2010).

[82] F. Blanc, L. Spencer and G. R. Goward, '*Quadrupolar NMR of Ionic Conductors, Batteries, and Other Energy-related Materials*' in *Encyclopedia of Magnetic Resonance*, Eds-in-chief R. K. Harris and R. E. Wasylishen, John Wiley: Chichester. Published online 15 Dec 2011.

[83] B. Key, R. Bhattacharyya, M. Morcrette, V. Seznéc, J. M. Tarascon and C. P. Grey, *J. Am. Chem. Soc.*, **131**, 9239 (2009).

[84] J. D. Jeon and S. Y. Kwak, *Macromolecules*, **39**, 8027 (2006).

[85] A. Doroodian, J. E. Dengler, A. Genest, N. Rösch and B. Reigler, *Angew. Chem., Int. Ed.*, **49**, 1871 (2010).

[86] H. Eckert, '*Amorphous Materials*' in *Encyclopedia of Magnetic Resonance*, Eds-in-chief R. K. Harris and R. E. Wasylishen, John Wiley: Chichester. Published online 15 Mar 2007.

[87] H. Maekawa, T. Maekawa, K. Kamura and T. Yowokawa, *J. Non-Cryst. Solids*, **127**, 53 (1991).

[88] L. S. Du and J. F. Stebbins, *J. Non-Cryst. Solids*, **351**, 3508 (2005).

[89] S. Kroeker, '*Nuclear Waste Glasses: Insights from Solid-State NMR*' in *Encyclopedia of Magnetic Resonance*, Eds-in-chief R. K. Harris and R. E. Wasylishen, John Wiley: Chichester. Published online 15 Dec 2011.

[90] L. S. Du, J. R. Allwardt, B. C. Schmidt and J. F. Stebbins, *J. Non-Cryst. Solids*, **337**, 196 (2004).

[91] A. Ananthanarayanan, G. Tricot, G. P. Kothiyal and L. Montagne, *Crit. Rev. Solid-State and Mater. Sci.*, **36**, 229 (2011).

[92] H. W. Spiess, '*Polymer Dynamics and Order from Multidimensional Solid State NMR*' in *Encyclopedia of Magnetic Resonance*, Eds-in-chief R. K. Harris and R. E. Wasylishen, John Wiley: Chichester. Published online 15 Mar 2007.

[93] K. Zemke, K. Schmidt-Rohr and H. W. Spiess, *Acta Polym.*, **45**, 148 (1994).

[94] P. Hazendonk, R. K. Harris, S. Ando and P. Avalle, *J. Magn. Reson.*, **162**, 206 (2003).

[95] E. R. deAzevedo, T, J. Bonagamba and D. Reichert, *Prog. Nucl. Magn. Reson. Spectrosc.*, **47**, 137 (2005).

2

X-ray Absorption and Emission Spectroscopy

Pieter Glatzel[a] and Amélie Juhin[b]

[a] *European Synchrotron Radiation Facility, Grenoble, France*
[b] *Institut de Minéralogie et de Physique des Milieux Condensés (IMPMC), CNRS, Pierre-and-Marie-Curie University, Paris, France*

2.1 INTRODUCTION: WHAT IS PHOTON SPECTROSCOPY?

Photon spectroscopy exploits the interaction between electromagnetic radiation and matter. The radiation perturbs the system and creates excited states. Spectroscopy records the energics of the excited states and the probabilities of creating them, *i.e.* the spectral intensities. This allows the spectroscopist to describe the atomic and electronic structure of the system, ideally supported by calculations based on quantum mechanics.

The system under study is a certain configuration of electrons and nuclei. The spectroscopist is often interested in specific properties of the system, and the merits of spectroscopy are greatly improved when the selectivity is increased. The elements of the Periodic Table differ in the binding energies of their electrons, and this can be used to develop an element-selective spectroscopy. This difference in binding energy is most pronounced for electrons that are not in the valence shell, *i.e.* **core**

Local Structural Characterisation, First Edition. Edited by Duncan W. Bruce, Dermot O'Hare and Richard I. Walton.
© 2014 John Wiley & Sons, Ltd. Published 2014 by John Wiley & Sons, Ltd.

or **inner-shell** electrons. The energy required to excite these electrons into unoccupied levels above the Fermi energy is at least several tens of electron volts (eV) and may be as high as 10^5 eV. The photon energy needed is thus in the vacuum ultraviolet (VUV) or X-ray range.

When only one atom of the system is considered, we use the atomic cross-section to describe the probability of exciting an electron of this atom from a certain electronic level. When dealing with more than one atom, the meaningful quantity is the absorption coefficient, which is typical for the entire system. It must thus include the full composition, *i.e.* concentration, of all elements, the density of the system and the sample thickness along the X-ray beam path. The excitation of an inner-shell electron results in a clear signature in the absorption coefficient of the material under study, *i.e.* the absorption edge. The absorption process leaves the system in an excited state, which persists for a certain time before decaying upon emission of an electron (Auger decay) or a photon (fluorescence decay). In this chapter, we are not concerned with detection of electrons; we will focus instead on photon-in and photon-out spectroscopy. The energy of the emitted photon again depends on the electron binding energies and consequently can also be used for element-selective spectroscopy.

Figure 2.1 shows the absorption and emission lines of $CoFe_2O_4$, using tabulated values for the cross-sections, fluorescence yields and transition

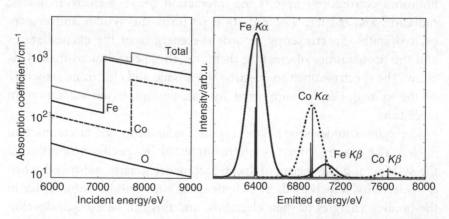

Figure 2.1 Absorption edges (left) and fluorescence lines (right) of Co and Fe in $CoFe_2O_4$. The elemental and total absorption coefficients are shown. Note the logarithmic scale in the absorption coefficient. The fluorescence lines are shown for Fe and Co, assuming an instrumental (Gaussian) energy bandwidth of 200 eV (lines) and 1 eV (sticks) full width at half maximum (FWHM). The fine structure of the $K\alpha$ lines becomes visible when the instrumental broadening is reduced.

energies. At this point we have two element-selective signatures in the absorption and emission process. In the next step we need to explore to what extent these techniques can provide information on the environment of the atoms, *i.e.* whether the techniques can also be chemically sensitive.

Spectroscopy involves scanning of the photon energy. In most cases, this is the energy, ω_{in}, of the photons that are incident on the sample. A detector monitors the intensity preceding (I_0) and transmitted through (I_1) the sample as a function of incident energy, and this is used to obtain an absorption spectrum. This technique follows the definition of the absorption coefficient, μ, by Bouguer (also called the Beer–Lambert law):

$$I_1(x) = I_0 e^{-\mu(\omega_{in})x} \tag{2.1}$$

where x is the thickness along the incident beam path. In many cases it is experimentally very challenging to use the transmitted X-rays to obtain μ, *e.g.* if the sample is very thick (so no beam is transmitted) or very thin (so the absorption is weak). In such cases, one can choose to detect a **secondary process**, *i.e.* a process whose probability of occurring depends on the absorption process. A secondary process might be a fluorescence line that is emitted as a consequence of the core hole creation during the absorption process. The experimentalist can chose a secondary process and record its intensity as a function of incident energy. This can be done in various fashions, some of which we will discuss here. It is of paramount importance to note that measuring the absorption intensity using secondary process detection is *always* an approximation to the absorption coefficient μ.

The experimental principle is shown in Figure 2.2. Experimental spectra, when recorded with sufficient energy resolution, show a fine structure containing chemical information. A fluorescence-detected absorption spectrum is often recorded with a small energy bandwidth for the incident beam ($\Delta E_{in} \sim 1$ eV) and a large energy window (no energy resolution or $\Delta E_{out} \sim 200$ eV) for the emission detection. However, chemical information may not only be contained in the X-ray absorption but also in the emission; this becomes accessible if the energy bandwidth for the emission detection is significantly reduced. This is realised in X-ray emission spectroscopy (XES), which consists in recording the emitted X-rays with a small energy bandwidth that is on the order of the core hole lifetime.

The inner-shell spectroscopist is concerned with the electronic and atomic structure around the element of interest. The experimental energy resolution, *i.e.* the spectral energy bandwidth, should be such that the

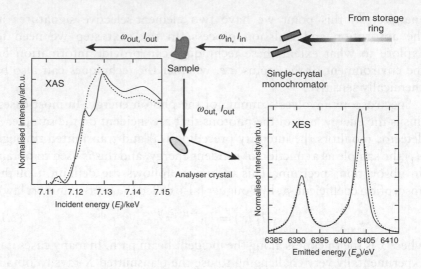

Figure 2.2 Top: experimental set-up to record an X-ray absorption spectrum in transmission mode or using secondary process detection. The absorption coefficient as a function of incident energy (bottom left) is obtained either by applying the Beer–Lambert law using the incoming and transmitted photon intensity or by recording the emitted photon intensity, which may give a good approximation to the absorption coefficient. An X-ray emission spectrum (bottom right) reveals the fine structure of the $K\alpha$ fluorescence lines. The spectra show examples for high-spin Fe ($Fe(phen)_2(NCS)_2$, dashed) and low-spin Fe ($[Fe(bpy)_3]Cl_2$, solid).

spectral features show detectable changes according to the atomic and electronic structure. The energies for bond formation and breaking, electron–electron interactions and crystal field splitting can be on the order of a few eV. This has to be seen in the context of the spectral broadening due to the lifetime of the photoexcited state, which, following the Heisenberg uncertainty principle,[a] causes a broadening of the spectral features.[1] The core levels of $3d$ transition metals have lifetime broadenings on the order of a few eV. This is therefore the desired spectral energy bandwidth in inner-shell spectroscopy. We note that in some cases it is possible to observe chemical changes with lower-energy resolution. Furthermore, using resonant scattering techniques it is possible to circumvent the limitation of the core hole lifetime broadenings (Section 2.4.4).

Spectroscopy does not only measure the energies of excited states but also the probability of reaching these states from the ground state,

---

[a] The lifetime of a core hole τ is linked to the uncertainty in its energy Γ *via* the Heisenberg uncertainty relation: $\Gamma\tau \geq \hbar/2 \sim 10^{-16}$ eV s^{-1}. A lifetime of 1 fs implies a lifetime broadening of ~ 0.1 eV.

which translates into spectral intensity. The spectral intensity is closely linked to the symmetry of the atomic and electronic arrangement and the population of electronic levels.

This chapter presents X-ray absorption and emission spectroscopy (XAS-XES). The subject is vast, but many aspects are already covered in other excellent texts, which allows the present authors to be topic-selective and to fill the gaps with appropriate references. We first give a basic introduction to the theory of the interaction of X-rays with atoms, which allows us to discuss fundamental mechanisms and derive from them different spectroscopic techniques. This approach allows showing of the relations between the techniques and their common origins. We then discuss the information that can be obtained on the system under study by the various techniques, by giving selected examples. We hope that this strategy results in a text that allows the reader to understand the basics of inner-shell spectroscopy while presenting some selected new trends. We point out that the International Union of Crystallography (IUCr) and the International X-ray Absorption Society (IXAS) maintain Web sites with dictionaries and tutorial material (www.iucr.org and www.ixasportal.net).

We will focus in our examples on hard X-ray spectroscopy, covering the range of the $3d$ transition metal K edges and $5d$ transition metal L edges. Energies below 1000 eV are referred to as 'soft' and cover the $3d$ transition-metal L edges. The energy range between 1000 and 5000 eV is referred to by some authors as the 'tender' range.

2.2 ELECTRONIC STRUCTURE AND SPECTROSCOPY

2.2.1 Total Energy Diagram

The quantum mechanical state of a sample whose properties we are interested in is described by the Dirac equation (the relativistic version of the Schrödinger equation), *i.e.* by obtaining a Hamiltonian and a wavefunction, which in principle fully explains all properties. It is impossible to solve the Dirac equation (or the Schrödinger equation) and the task of the theoretically orientated scientist is to find out how to best approximate the full equation, *i.e.* to examine what interactions are most relevant. We are not concerned for now with computational difficulties and simply assume that the system is in a ground state $|g\rangle$ that considers all electrons and nuclei in the system. The ground state

$|g\rangle$ is, however, not the only solution to the Dirac equation: there are infinitely more with higher energies, which are excited states. We can therefore draw a very simple scheme in which the vertical axis is the total energy of the system and simply take note of the fact that excited states exist at certain energies (Figure 2.3).

We now assume that we have successfully calculated, or approximated, all excited states within a certain energy range and we would like to find out whether our calculations are correct. We want to create the excited states and record how much energy we need to put into the system in order to do so. A common way of injecting energy into a system is to use photons and have the system interact with them. This is what we term 'spectroscopy'. Depending on the energy of the photon, we may call the technique visible light, VUV or X-ray spectroscopy.

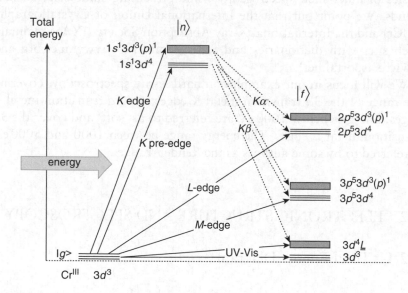

Figure 2.3 Total energy diagram of a system containing chromium in formal oxidation state III. The ground state $|g\rangle$ and excited states $|n\rangle$ and $|f\rangle$ energy levels are approximated using atomic configurations and a single-particle picture of the electronic transitions (see Section 2.3). Closed shells are omitted for clarity. Black lines indicate that each configuration actually corresponds to a collection of several many-body states (thus building the many-body picture). Rectangles symbolise a band state, where the notation (p) represents mixed orbitals between the atomic $4p$ level and the bands of the solid. The notation $\underline{L}$ corresponds to a hole created on a ligand orbital. Solid-line arrows indicate the absorption edge that allows the corresponding excited state to be reached directly, *i.e.* in one step. Dotted-line arrows indicate the emission line that allows excited states of lower energies to be reached in a second step, *i.e.* after a core hole has been primarily created.

The fundamental principle of photon spectroscopy is thus the connection of two states of a system by interaction with a photon. This is done by evaluating the interaction in transition matrix elements $\langle s_1 | \widehat{O} | s_2 \rangle$, where $|s_1\rangle$ and $|s_2\rangle$ are two states of the system and $\widehat{O}$ is a transition operator that considers the properties of the photon. The squared transition matrix element, either summed over all states or multiplied by the density of allowed states, provides the probability *per* unit time that a transition between $|s_1\rangle$ and $|s_2\rangle$ will occur (for a mathematically exact treatment, see *e.g.* reference [2]). This rather simple method (based on numerous approximations for the time-dependent perturbation) of evaluating transition probabilities is often called 'Fermi's golden rule', although it was in fact derived by Paul Dirac.

An important observation is that the energies of the excited states of the system are of course independent of the spectroscopy that is used to study them. Thus, different spectroscopic techniques may be used to study the same excited state. Furthermore, the transition matrix elements may not always give finite values for the transition probability. In fact, in most cases the matrix element will be zero. Whether or not the probability is finite can often be evaluated using simple considerations with symmetry arguments. These considerations result in **selection rules** that can be exploited for the point group symmetry of the system under investigation.[3]

We can now extend the energy scheme and connect some excited states with the ground state *via* absorption matrix elements. Probing of the magnitude of such matrix elements, which involve inner-shell electrons, is done by XAS. One can also imagine connecting one excited state with another at a lower energy, which will ultimately decay into the ground state (cascade decay). The decay of an excited state can occur radiatively or nonradiatively. The former is generally referred to as 'X-ray emission' (or fluorescence) and the latter as 'Auger decay'. Recording all radiative (nonradiative) decay channels is referred to as total fluorescence (electron) yield (TFY or TEY) detection. Turning back to our requirement for element selectivity, we note that both the absorption and the emission occur at energies that are characteristic for an element. We can thus use either XAS or XES, or even both, in order to increase the sensitivity. This is known as 'resonant inelastic X-ray scattering' (RIXS). The principle of these two spectroscopies is explained in Section 2.4.

Spectroscopy may be used as a 'fingerprint' technique to characterise an unknown sample. One first needs to measure a representative number of reference samples in order to understand the behaviour of the spectra as a function of a known modification in the sample.[4−6] The unknown system can then be characterised using this library of spectra. This is a valid approach, and often the only feasible one. It is more ambitious but also possibly more gratifying to understand the nature of the ground state and the excited states, and this can be done at varying levels of sophistication using different theoretical approaches (Section 2.3). The most convenient and easy way to label ground or excited states is to use atomic configurations (*i.e.* assuming spherical symmetry), as in Figure 2.3. However, even for an isolated ion, this notation represents a drastic simplification, which often does not allow for a correct representation of the electronic states. Indeed, an atomic configuration corresponds in most cases to several electronic states that differ in energy and other properties (*e.g.* orbital symmetry or spin) not accounted for in this simplified picture. In Section 2.3, we discuss the different pictures that one can use to describe electronic states, as well as the different theoretical approaches by which to calculate their properties.

2.2.2   Interaction of X-rays with Matter

An incoming photon enters the sample with frequency ω_{in}, wavevector k_{in} and polarisation ε_{in} and leaves it with ω_{out}, k_{out} and ε_{out} (see Figure 2.4). The change of frequency (*i.e.* energy), wavevector and polarisation gives insight into the electronic properties of the sample. In 1925, Hans Kramers and Werner Heisenberg published an expression which treats the interaction of the vector potential **A** (which defines the properties of the photon) and atomic electrons.[7] The famous Kramers–Heisenberg equation describes the probability of detecting a

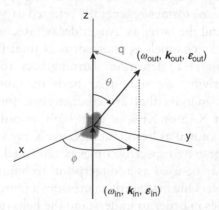

Figure 2.4   Geometry of photon scattering. For $\phi = 0$, ε_{in} points along x.

photon with energy ω_{out} at a point in space that is defined relative to the interaction point and the incoming photons by the angles θ and ϕ. It is a double differential scattering cross-section (DDSCS) with respect to emitted energy and solid angle. The outgoing photon may be referred to as 'scattered', 'transmitted', 'emitted' or 'fluorescence', but the theoretical framework remains the same. The original Kramers–Heisenberg equation was subsequently refined and additional interaction terms were identified to account for weak effects, *e.g.* the interaction of the photon magnetic field with the electron spin. Consequently, the literature, while agreeing on the fundamental physical process, does not provide a common definition of what terms should be called a 'Kramers–Heisenberg equation'.

2.2.2.1 The Scattering Cross-Section

The interaction of vector field **A** (which describes the electromagnetic radiation) with matter is treated within perturbation theory as retaining only the first- and the second-order terms.[2,8] The relevant interactions in the Hamiltonian that describes the perturbation are represented by the operators $\mathbf{A}^2$ and $\mathbf{p} \cdot \mathbf{A}$, where $\mathbf{p}$ is the electron momentum operator.[9,10] We neglect interactions of the electron spin with **A**, but, as we will show below, this does not exclude the possibility of studying the spin state of the system. The operators connect the ground state $|g\rangle$ of the system with excited states $|e\rangle$ in matrix elements. Several terms containing such matrix elements are obtained. In this section, we discuss only those terms that are relevant to the presented spectroscopies. We discuss these terms separately, with the implicit assumption that the experimental conditions are such that one term dominates over the others. Furthermore, the different terms may interfere with one another. For the majority of practical applications with which we are concerned, these approximations hold.

The term containing $\mathbf{A}^2$ contributes to the first-order perturbation term of the interaction Hamiltonian. The spectral intensity is proportional to the following expression:

$$F^{TH}(\omega_{in}, \omega_{out})$$
$$= \frac{\omega_{out}}{\omega_{in}} |\varepsilon_{in} \cdot \varepsilon_{out}^*|^2 \sum_f |\langle f|\widehat{O}|g\rangle|^2 \delta(E_f - E_g - \hbar(\omega_{in} - \omega_{out}))$$

$$= \frac{\omega_{\text{out}}}{\omega_{\text{in}}} \, |\boldsymbol{\varepsilon}_{\text{in}} \cdot \boldsymbol{\varepsilon}^*_{\text{out}}|^2 \, S(q, \omega_{\text{in}} - \omega_{\text{out}}) \qquad (2.2)$$

with $\widehat{O} = \sum\limits_i e^{iq \cdot r_i}$ where r_i is the position vector of the electron.

The incoming and outgoing photon polarizations are ε_{in} and ε_{out} (cf. Figure 2.4) and E_{g}, E_{n} and E_{f} denote the ground, intermediate and final state energies (cf. Figure 2.3), respectively. The δ function requires the energy transfer $\hbar(\omega_{\text{in}} - \omega_{\text{out}})$ to be equal to the difference between ground- and final-state energy. The operator connecting the ground state $|g\rangle$ and final state $|f\rangle$ of the system contains the momentum transfer $q = k_{\text{in}} - k_{\text{out}}$ in the exponential function. Equation 2.2 thus describes a 'transfer' or 'loss' spectroscopy. The incident photon transfers part of its momentum and energy to the system. It gives rise to **elastic** ($\omega_{\text{in}} = \omega_{\text{out}}$, zero energy loss) and **inelastic** ($\omega_{\text{in}} \neq \omega_{\text{out}}$) scattering. The expression $S(q, \omega_{\text{in}} - \omega_{\text{out}})$ is the dynamic structure factor that provides information on elementary excitations caused by charge fluctuation. Such excitations may be similar to an X-ray absorption spectrum (see below) but can also be collective, such as plasmon excitations.[8]

Experimental evidence for X-ray events that arise from Equation 2.2 was recorded well before the interaction of photons with atomic electrons was described in a general quantum mechanical formalism. These observations were explained using classical or semiclassical theory and named after their discoverers, e.g. Raman, Compton and Thomson. They are special cases of the general interaction term. We emphasise that it is important to consider the whole of Equation 2.2 one contribution to the scattering (or spectral) intensity, in order to keep in mind that Raman, Compton and Thomson scattering have the same origin. This point becomes clear when considering the angular dependence of the scattering cross-section that arises from the polarisation vectors. Synchrotron radiation is linearly polarised (with $\boldsymbol{\varepsilon}_{\text{in}}$ along x, cf. Figure 2.4) in standard experiments, and the angular dependence of Equation 2.2 is:[2]

$$\sin^2 \varphi + \cos^2 \vartheta \, \cos^2 \varphi \qquad (2.3)$$

This angular dependence thus holds for Raman, Compton and Thomson scattering.

The nomenclature found in the literature has been subject to historical developments and is thus not always strict. Compton and Raman scattering are always inelastic. Some authors refer to the entire Equation 2.2 as 'Thomson scattering', while others exclude $S(q, \omega_1 - \omega_2)$ in the definition of Thomson scattering and some restrict it to the case of elastic scattering. Historically, the latter appears to be more correct as Thomson derived an expression for elastic scattering.

The interaction term containing $\mathbf{p} \cdot \mathbf{A}$ contributes to the second-order term, which is proportional to:

$$F^{KH}(\omega_{in}, \omega_{out}) = \frac{\omega_{out}}{\omega_{in}} \sum_f \left| \sum_n \frac{\langle f|\widehat{O}'^{\dagger}|n\rangle \langle n|\widehat{O}|g\rangle}{E_n - E_g - \hbar\omega_{in} - i\Gamma_n} \right|^2$$

$$\times \delta(E_f - E_g - \hbar(\omega_{in} - \omega_{out})) \tag{2.4}$$

with $\widehat{O}'^{\dagger} = \sum_{j'}(\boldsymbol{\varepsilon}^*_{out} \cdot \boldsymbol{p}_{j'})e^{-ik_{out}r_{j'}}$ and $\widehat{O} = \sum_j(\boldsymbol{\varepsilon}_{in} \cdot \boldsymbol{p}_j)e^{ik_{in}r_j}$.

We consider only one ground state, $|g\rangle$. The sum includes all intermediate $|n\rangle$ and final states $|f\rangle$, as well as all electrons in the scattering system (j, j'). The denominator is a complex expression of a Lorentzian profile in which the linewidth, Γ_n, due to the finite lifetime of the intermediate state appears in the imaginary part. As in Equation 2.2, the δ function requires the energy transfer to equal an energy difference in the system. The lifetime τ_f of the final state can be considered by replacing the δ function with a second Lorentzian function. The difference between Equation 2.2 and Equation 2.4 is the intermediate state, $|n\rangle$, which exists for time τ_n. During this time, the incoming photon has been absorbed (annihilated) and no photon has yet been emitted (created). This photonless period is absent in Equation 2.2.

In Equation 2.4, the operator that governs the transition matrix elements does not contain the momentum transfer (as in Equation 2.2), but does contain the photon polarisation and wavevectors, as well as the atomic electron momentum operators. We will analyse the transition operators below. Equation 2.4 alone is sometimes referred to as the 'Kramers–Heisenberg scattering cross-section'. We will use this term below, even though it is historically not entirely correct. The resonance term is, however, a key observation by Kramers and Heisenberg, and this nomenclature honours their achievements.

2.2.2.2 X-ray Transmission: The Linear Absorption Coefficient

Inspection of Figure 2.4 shows that tuning angle θ to $0°$ represents a beam that is transmitted through the sample. It may seem counterintuitive at first to refer to a beam that passes straight through a sample as 'scattered', but it is consequent and physically correct. We thus obtain a coherent forward scattering amplitude at energy $k_{in} = k_{out}$, $\omega_{in} = \omega_{out}$ (i.e. $|k_{in}| = |k_{out}|$) and $\boldsymbol{\varepsilon}_{out} = \boldsymbol{\varepsilon}_{in}$ that can be evaluated using all interaction terms containing A^2 and $p \cdot A$. 'Coherent' means that the phases of the incoming and outgoing waves have a defined relation. A very clear picture is given by Collins and Bombardi:[11] the only way to reduce the intensity of a beam (*i.e.* absorption) is by a coherent superposition of waves, a fraction of them interfering destructively.

An analysis of the forward scattering amplitude thus only requires measurement of the change of X-ray intensity before and after the sample. The Beer–Lambert law (Equation 2.1) relates the intensity transmitted through the sample normalised to the incident intensity to μ, the linear absorption coefficient, which depends on the material. We consider a homogeneous sample containing only one element and write $\mu = N \, \sigma_{\text{tot}}$, where σ_{tot} is the total cross-section of one atom and N is the density of atoms. Inhomogeneous samples require a more detailed treatment of μ.[12]

Tuning the incident energy, ω_{in}, to a resonance energy, E_n, of the system will cause Equation 2.4 to dominate the interaction. In this case, *'resonance'* has a rather general meaning, because E_n may refer to any eigenvalue of the Hamiltonian of the system. We are interested in an element-specific response and thus wish to excite a core electron. Therefore, we require that the excited state with eigenvalue E_n has a hole in a core level. Often, isolated excitations near an absorption edge, *i.e.* transitions to assumed localised electronic states (see Section 2.4.4), are referred to as *'resonant'* excitations. Equation 2.4 is more general and also considers excitations into delocalised electronic states well above an absorption edge. Consequently, *'non-resonant'* X-ray emission (*i.e.* the fluorescence lines that are measured following excitation well above an absorption edge), *'resonant'* scattering and element-selective X-ray absorption (*i.e.* absorption due to the photoelectric cross-section, as opposed to *e.g.* Compton scattering) measured in transmission mode arise from the same interaction term, *i.e.* Equation 2.4.

Ground state, $|g\rangle$, and final state, $|f\rangle$, are identical, and we find that the coherent forward scattering amplitude is proportional to $\sum_n |\langle n|\widehat{O}|g\rangle|^2$, *i.e.* the square of the absorption transition matrix element. In the energy range that is mostly used for X-ray absorption studies, this expression is the dominant term, corresponding to the photoelectric cross-section, σ_{PE}. The optical theorem expresses the imaginary part of the coherent forward scattering amplitude as proportional to the total cross-section, $\sigma_{\text{tot}} \sim \sigma_{\text{PE}}$, which is defined in the Beer–Lambert law. We thus have a connection between the classical Beer–Lambert law and a quantum-mechanical description of the interaction of X-rays with matter.

Comparing Equation 2.4 with Equation 2.2, we find that the matrix elements are governed by different transition operators. The photon momentum transfer in Equation 2.2 is replaced by the photon wavevector in Equation 2.4 and the exponential function is multiplied by the scalar product between polarisation vector and electron momentum operator. As a consequence, different electronic states may be probed in the two expressions. This is an important observation because, by using different experimental configurations (by varying scattering angle, photon polarisation and energies), it is possible to emphasise interactions that arise from either Equation 2.2 or Equation 2.4 and thus to probe different aspects (e.g. different orbital angular momentum) of the electronic structure.

2.2.2.3 Full Expressions of the Kramers–Heisenberg Cross-Section

Above, we have qualitatively explained how the case of XAS measured in transmission and the case of resonant scattering can be related to the general expression of X-ray scattering. We now wish to give the complete expressions of the absorption and resonant scattering cross-sections, i.e. including all factors needed for a quantitative evaluation. The various theoretical approaches by which to calculate the cross-sections are discussed in Section 2.3.

The full Kramers–Heisenberg scattering cross-section is expressed as:[2,13]

$$\sigma_{KH}\left(\omega_{in},\omega_{out}\right) = \frac{r_e^2}{m^2}\frac{\omega_{out}}{\omega_{in}}\sum_f\left|\sum_n\frac{\langle f|\widehat{O}'^{\dagger}|n\rangle\langle n|\widehat{O}|g\rangle}{E_n - E_g - \hbar\omega_{in} - i\Gamma_n}\right|^2$$

$$\times \frac{\frac{\Gamma_f}{\pi}}{(E_f - E_g - \hbar(\omega_{in} - \omega_{out}))^2 + \Gamma_f^2} \quad (2.5)$$

where r_e is the classical radius of the electron and the δ function in Equation 2.4 has been replaced by a Lorentz function to account for the finite lifetime of the final state.

X-ray transmission (forward scattering) leads to the X-ray absorption cross-section as a special case of Equation 2.5. The XAS cross-section can be written as:[14]

$$\sigma_{XAS}\left(\omega_{in}\right) = 4\pi^2\alpha\sum_n|\langle n|\widehat{O}|g\rangle|^2 \frac{(E_n - E_g)\frac{\Gamma_n}{\pi}}{(E_n - E_g - \hbar\omega_{in})^2 + \Gamma_n^2} \quad (2.6)$$

where $\alpha = \frac{e^2}{4\pi\varepsilon_0 \hbar c} \approx \frac{1}{137}$ is the fine structure constant. We are considering elastic scattering and do not observe the decay process. The intermediate state in the Kramers–Heisenberg equation thus becomes the final state of the absorption process.

The XAS cross-section as written in Equation 2.6 is a sum of discrete transitions including all final states. Note that in the case where the initial state $|g\rangle$ is degenerate, Equation 2.6 can be transformed into an equivalent expression, where the sum now runs over both ground and final states, and where the matrix element $|\langle n|\widehat{O}|g\rangle|^2$ is weighted by $1/d_g$, where d_g is the degeneracy of the ground state.

An important difference between the XAS and the general Kramers–Heisenberg formula is that in the former, the matrix elements are squared. The phase of the transition is therefore irrelevant for X-ray absorption, while it may become important in other scattering processes. In the Kramers–Heisenberg equation, the sum of the products of matrix elements are squared, possibly leading to interference effects (Section 2.4.4).

2.2.2.4 The Transition Operator

The transition probability $|\langle s_1|\widehat{O}|s_2\rangle|^2$ between two states of the system can be further evaluated by using commutation relations. We obtain:[14]

$$\widehat{O} = (\boldsymbol{\varepsilon} \cdot p)e^{ikr} \approx \boldsymbol{\varepsilon} \cdot \boldsymbol{r} + \frac{i}{2} (\boldsymbol{\varepsilon} \cdot \boldsymbol{r}) (\boldsymbol{k} \cdot \boldsymbol{r}) \qquad (2.7)$$

The expansion of $|\langle s_2|\widehat{O}|s_1\rangle|^2$ yields: (i) a term, $|\langle s_2|\boldsymbol{\varepsilon} \cdot \boldsymbol{r}|s_1\rangle|^2$, which represents electric dipole transitions (E1); (ii) a term, $|\langle s_2 |(\boldsymbol{\varepsilon} \cdot \boldsymbol{r}) (\boldsymbol{k} \cdot \boldsymbol{r})|s_1\rangle|^2$, which represents electric quadrupole transitions (E2); and (iii) two cross-terms between E1 and E2. The electric quadrupole transition probability is approximately $(\alpha Z_{\text{eff}})^2 \approx \left(\frac{Z_{\text{eff}}}{137}\right)^2$ times that of the electric dipole.[b] Quadrupole transitions are therefore considerably weaker than dipole transitions.

The cross-terms coming from the interference of E1 and E2 amplitudes are exactly zero either if the system is centrosymmetric or if it is nonmagnetic and linear polarised X-rays are used.[15] We should point out that magnetic dipole transitions also arise from the expansion of

---

[b] Z_{eff} is the effective atomic number.

e^{ikr}, but they are negligible in the X-ray range (see reference [14] for a detailed presentation of the different terms).[c]

According to the abovementioned considerations, only the electric dipole E1 and electric quadrupole E2 transitions are considered to contribute significantly to the X-ray absorption spectrum. If the E1–E2 interference terms are zero, the XAS cross-section can be expressed as the sum of the E1 cross-section and the E2 cross-section:

$$\sigma_{XAS} = \sigma_{XAS}^{E1} + \sigma_{XAS}^{E2} \tag{2.8}$$

Selection rules use the symmetry of the electronic states to rule out transitions with zero probability ('forbidden' transitions). They must be evaluated for a good quantum number. Spin and orbital angular momentum are only good when spin–orbit interaction can be neglected. Otherwise, the coupled angular momentum must be considered. The selection rules depend on the operator, *e.g.* dipole or quadrupole. They may be applied to either one-electron transitions (l, s, j) or transitions between multi-electronic states (see Section 2.3) where the symmetry is assigned to the full configuration, considering all electrons (L, S, J). The literature often simply states that a transition is forbidden, meaning that the dipole operator yields zero matrix elements. Spectral intensity may still arise from higher-order, *e.g.* quadrupole, transition operators.

Electric dipole one-electron transitions which neglect spin–orbit interaction correspond to a change in angular momentum of ± 1 between the core orbital and the excited electron: $\Delta \ell = \ell_n - \ell_g = \pm\,1$, while electric quadrupole transitions correspond to a change in angular momentum of 0 or $\pm 2 :$ $\Delta \ell = \ell_n - \ell_g = 0, \pm\,2$. In the case of an excitation from a $1s$ core level (*i.e.* XAS at the K edge), only orbitals with p-orbital angular momentum can be reached by E1 transitions, while only d orbitals can be reached by E2 transitions. At the K edge of $3d$ elements, E2 transitions between $1s$ and $3d$ orbitals may be visible in the region of the pre-edge. Similarly, $2p$ to $4f$ transitions have been observed for rare earths.[16,17]

2.2.2.5   *Polarisation Dependence of the Cross-Section*

For some materials and in certain conditions, the Kramers–Heisenberg (and thus XAS) cross-sections may depend on the polarisation vector, $\boldsymbol{\varepsilon}$.

---

[c] The expansion of e^{ikr} to the next order gives rise to an electric octupole term that interferes with the electric dipole amplitude. Electric dipole–octupole transitions are usually neglected.

The corresponding variation of the cross-section is called **pleochroism** ('many colours'), or sometimes dichroism. The term 'dichroic' initially designated samples which showed two colours when observed at different angles under visible light. The same phenomenon is observed in the X-ray energy range. Using a single crystal with crystallographic orientation or a sample with some preferential arrangement (*e.g.* a surface, a layered material *etc.*) one measures the dependence of the cross-section for different incident or emitted polarisation vectors with respect to the sample and, if relevant, by changing the direction of the wavevector. The theoretical framework used to interpret the polarisation dependence of the cross-section is established for XAS[14] and nonresonant XES,[18,19] where either the incident or the emitted photon polarisation must be considered, but not both (some scattering processes require consideration of both: a textbook example is given by Kotani *et al.*[20]). We will limit our introduction to the case of XAS.

Magnetic dichroism can be observed **in the presence of a magnetic field**. The Zeeman effect lifts the energy degeneracy of the atomic levels of the absorber. Photons with different polarisation vectors induce transitions to states with different symmetries, leading to a different absorption coefficient. Such an effect is called **X-ray magnetic circular dichroism (XMCD)** or **X-ray magnetic linear dichroism (XMLD)**, depending on whether the dichroism is observed using incident X-rays that are circularly or linearly polarised.[d]

XMCD is the difference measured in absorption for left circularly polarised and right circularly polarised photons, in the presence of a fixed external magnetic field (Figure 2.5). XMLD is the difference measured in absorption with linearly polarised photons using a magnetic field direction either parallel or perpendicular to the incoming X-rays. Both techniques have been used for 25 years as magnetic spectroscopies, elegantly combining the element selectivity of XAS and the sensitivity to magnetism. The XMLD technique is used to study antiferromagnetic materials, while XMCD is employed for ferro- and ferrimagnets. Note that for antiferromagnets, XMCD gives no intensity. In Section 2.4.4.3 we introduce a newly developed hard X-ray magnetospectroscopy, RIXS-MCD, which is based on the combination

---

[d] Adopting the usual choice in X-ray spectroscopies, X-rays propagate along (Oz), the direction of the incident wavevector $\mathbf{k}_{in}$. Note that we always consider vectors to be normalised. For linearly polarised X-rays, the coordinates of ε_{in} are real numbers, *e.g.* $\varepsilon_{in} = (1, 0, 0)$ (polarisation along Ox), $\varepsilon_{in} = (0, 1, 0)$ (polarisation along Oy), $\varepsilon_{in} = (1/\sqrt{2}, 1/\sqrt{2}, 0)$ (polarisation at 45° to Ox). For circularly polarised X-rays, the coordinates of ε_{in} are written as complex numbers: $\varepsilon_{in} = 1/\sqrt{2}(1, i, 0)$ and $\varepsilon_{in} = 1/\sqrt{2}(1, -i, 0)$ for left and right polarised light, respectively.

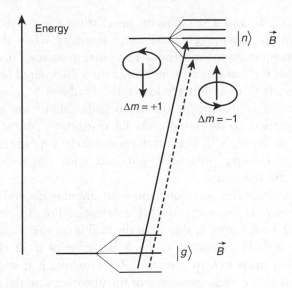

Figure 2.5 Total energy (multi-electron) scheme for the MCD effect. The degeneracy of the ground (and excited) state is lifted by applying a magnetic field. Different helicities of the circular polarised light lead to different final states, giving rise to spectral differences.

of XMCD with RIXS, and is well adapted to the study of ferro- and ferrimagnets under extreme conditions.

Natural dichroism can be observed for certain compounds in the absence of a magnetic field. If an effect is measured using circularly polarised X-rays, it is called 'X-ray natural circular dichroism' (XNCD). If it is observed using linearly polarised X-rays, this is called 'X-ray natural linear dichroism' (XNLD). XNCD is the difference measured in absorption for left and right circularly polarised light. It can only be detected in the case of chiral centres such as enantiomers, and is caused by the interference between electric dipole and quadrupole transitions. XNLD is the difference measured in absorption for two different directions of the incident linear polarisation vector with respect to the sample. A prior requirement for a detectable XNLD is to perform the measurements on a single crystal or a sample with some macroscopic (~X-ray beam-sized) alignment. The following are then considered: (i) the nature of the electronic transitions probed and (ii) the point symmetry group of the system.[e] Indeed, the electric dipole cross-section shows an angular dependence only for crystals with symmetry lower

---

[e] We underline that the symmetry of the crystal is relevant, not the local symmetry around the absorbing atom.

than O_h, T_d, O, T_h and T.[14] This means that for cubic crystals, electric dipole transitions always give the same intensity when the direction of the polarisation vector is varied. The corresponding cross-section is 'isotropic': it is the same as that measured on a disordered (*e.g.* powder) sample. Crystals with symmetry lower than cubic are anisotropic: the E1 cross-section is dependent on the orientation of the polarisation vector with respect to the sample. As E1 transitions contribute mainly to the XAS spectrum, XNLD is often encountered experimentally when working with linearly polarised light and with single crystals with symmetry lower than cubic.[f]

Electric quadrupole transitions show an angular dependence for all crystal symmetries, *i.e.* even for cubic crystals. For 3d elements, the XNLD of E2 transitions is thus mainly visible at the K pre-edge. An example for XNLD measured at the K pre-edge of a 3d element in a cubic crystal is presented in Section 2.4.1.4, where it is shown how it can be related to the local symmetry of the absorber and the crystal field properties.

2.3   CALCULATION OF INNER-SHELL SPECTRA

The problem of the calculation of a spectrum involves several aspects. We first need to determine the ground-state electronic structure of the system of interest (Sections 2.3.1–2.3.4). We then consider the interaction of the system with an electromagnetic field and determine the excited states that can be reached (Sections 2.3.5 and 2.3.6). The calculation of inner-shell spectra is a daunting task. A theoretical approach may be very suitable for calculating the electronic structure but limited in considering the interaction with the electromagnetic field. We have made no assumptions concerning the forms of the initial, intermediate and final states. Approximations must be made to calculate the absorption (or the Kramers–Heisenberg) cross-section, and the spectroscopist must therefore choose which theoretical approach is most suitable for the problem at hand.

The Dirac equation, with all relativistic effects and including all possible interactions in the Hamiltonian of relevance at the intra-atomic scale (such as intra-atomic electron–electron repulsions) as well as the

---

[f] The complete angular dependence of a crystal can be determined using an appropriate combination of 'fundamental' spectra, the number of which is determined by the crystal symmetry.

surrounding electrons and nuclei (*e.g.* interatomic electronic repulsions), must be solved to provide an exact description of ground and excited states. This will yield a relativistic, many-body, extended (*i.e.* beyond an isolated ion) description of the electronic states. In practice, it is not possible to perform such a calculation. In most cases, we instead solve the Schrödinger equation and introduce relativistic effects such as spin–orbit interactions. The Born–Oppenheimer approximation is normally used, thus separating the electronic properties of the system from the dynamics of the nuclei.[g] In order to describe the electronic part of the wavefunction, several theoretical frameworks can be used to describe and then calculate electronic states: the **single-particle extended** picture (density functional theory (DFT)-based approaches, Section 2.3.1), the **many-body atomic picture** (ligand field multiplet theory, LFMT, Section 2.3.2) and the **many-body extended** picture ('beyond DFT methods', Section 2.3.4).

2.3.1 The Single-Particle Extended Picture of Electronic States

DFT based methods enable a description of electronic states using a single-particle extended picture. In principle, DFT should only work for ground-state calculations; in practice, it often also works quite well for excited states probed in core-level spectroscopies.

In DFT, the quantity to determine is no longer the ground-state wavefunction of N electrons, but the electron charge density of the system. The correct charge density minimises the total energy of the system. In order to determine it, the system of N electrons is replaced by a fictitious, non-interacting system of independent electrons, which has the same electronic density as the real system. The Schrödinger equation is transformed into a system of equations (the Kohn–Sham equations) with an effective Hamiltonian and wavefunctions that are functions of only *one* space variable. DFT is therefore called a 'single-particle' approach, although some many-particle (many-body) interactions are contained in the effective one-particle Hamiltonian. These many-body effects are hidden in the exchange and correlation term of the electronic potential. The exact analytical expression of this electronic exchange and correlation term is not known. Hence, DFT can only be applied in an

---

[g] Note that vibronic coupling can lead to sizeable effects in core-level spectra. A vibronically coupled expression of the cross-section is therefore needed to explain the observed temperature dependence of the spectra. See references [21, 22], and references therein.

approximate form, *i.e.* using an approximation to model the exchange and correlation between electrons (*e.g.* local density approximation, LDA, or generalised gradient approximation, GGA).

The wavefunction solution to the independent particle problem is a single Slater determinant, which is constructed from the single-particle wavefunction solutions of the Kohn–Sham system of equations. This means that DFT is unable to calculate electronic states that are linear combination of Slater determinants. This is an important point because it limits the ability to treat a many-body response of a system to an interaction with an electromagnetic field.

To illustrate this, we take as an example two electrons. Coupling $s = 1/2$ to $s = 1/2$ yields $S = 1$ or 0. This corresponds to four $|S, M_s>$ wavefunctions: $|S = 1, M_s = 1>$, $|S = 1, M_s = -1>$, $|S = 1, M_s = 0>$ and $|S = 0, M_s = 0>$. These functions need to meet the property of being antisymmetric under exchange, and it can be shown that only $|S = 1, M_s = 1>$ and $|S = 1, M_s = -1>$ can be expressed individually as a single Slater determinant. Suppose the electrons are placed in spatial orbitals φ_m and $\varphi_{m'}$; the triplet $|S = 1, M_s = 1>$ wavefunction is $\frac{1}{\sqrt{2}}(\varphi_m(1)\varphi_{m'}(2) - \varphi_{m'}(1)\varphi_m(2))|\uparrow\uparrow>$; that is, a single Slater determinant (m indicates the one-electron magnetic quantum number). Such an electronic state can be calculated in DFT. However, the $|S = 1, M_s = 0>$ and $|S = 0, M_s = 0>$ states can only be expressed as combinations of two Slater determinants and thus cannot be calculated in DFT.

One can group the many different DFT-based methods according to their characteristics:

- *Cluster or periodic:* some methods solve the Kohn–Sham equations for a cluster centred around the absorbing atom (direct or real-space methods); others start from a unit cell of the crystal (or multiple unit cells, known as a 'supercell') in order to take advantage of the 3D periodicity (reciprocal space methods).
- *Self-consistent or not:* the calculation of the charge density using the Kohn–Sham equations can be carried out without or (preferably) with self-consistency, *i.e.* using an iterative cycle in which two successive steps are mixed until a convergence criterion is reached.
- *Type of basis functions used to expand the orbital solutions of the Kohn–Sham equations:* either localised (linear combination of atomic orbitals, LCAO; linear muffin-tin orbitals, LMTO) or delocalised (plane waves, PW; full-potential linearised augmented plane waves, FLAPW) functions.

- *Approximation of the shape of the electronic potential:* in LMTO or multiple scattering theory, the potential is approximated to be spherically symmetric in the atoms and constant between them (muffin tin). In full-potential methods, such as FLAPW or projected augmented wavefunction (PAW)-pseudopotentials,[h] no approximation is made, which is generally preferable, even though it makes the calculations heavier.

2.3.2 The Many-Body Atomic Picture of Electronic States

The electrons of the atom of interest (*i.e.* the X-ray absorbing atom) are explicitly taken into account in this approach. This description is restricted to the absorbing atom; the other atoms of the solid, and therefore their electrons, are not considered in this approach. Their effect may be included empirically.

2.3.2.1 Simple Introduction to the Many-Body Atomic Picture

We consider a Cr^{3+} ion in octahedral (O_h) environment. The atomic electronic configuration is ($1s^2 2s^2 2p^6 3s^2 3p^6 3d^3$). Although electrons of atoms surrounding the chromium are not included, we can consider the effect of the octahedral crystal field. The degeneracy of the Cr $3d$ levels is lifted and the d orbitals are split into two groups, the e_g orbitals pointing towards the ligands and the t_{2g} orbitals pointing between the ligands. The number of possible electronic states is given by the number of allowed arrangements of three electrons in 10 spin orbitals. This gives $^{10}C_3 = 120$ microstates, *i.e.* 120 possibilities, potentially spread over several eV. The energy separation between these states is due to a combined effect of the crystal field splitting, electronic repulsions, and spin–orbit coupling (SOC), *i.e.* 'multiplet effects'. Note that electrons occupying closed shells do not actually contribute to the energy splitting of the electronic levels; they contribute only to the average energy of the configuration.

All these states can be further grouped in so-called 'spectroscopic terms' (or term symbols) according to their energy and their spin and orbital moments. The diagrams established by Tanabe and Sugano, available in several references,[23] give the relative energy positions of the spectroscopic terms for a $3d^n$ transition metal ion in O_h symmetry,

---

[h] The PAW method uses plane waves to expand the Kohn–Sham orbitals.

neglecting $3d$ SOC. For some symmetries lower than O_h, such as trigonal or tetragonal, similar diagrams are available in reference [24] for given values of distortion parameters and SOC.

The relative energies of the electronic states depend on the crystal field parameters and the Racah parameters, which are related to the electronic repulsions. At room temperature, the ground state for $3d^3(Cr^{3+})$ is $^4A_{2g}$ ($S = 3/2$) in O_h symmetry. If we include SOC and the Zeeman term in the Hamiltonian and/or consider lower symmetries, the description and determination of the spectroscopic terms become very complex. LFMT[25-27] takes these effects into account and has been realised in computer codes.[28]

2.3.2.2   Key Ideas of the Ligand Field Multiplet Theory

Atomic multiplet theory, crystal field multiplet theory, LFMT and charge transfer multiplet theory (sometimes collectively referred to as the 'multiplet theory') are based on concepts developed in atomic physics and make large use of group theory.[29,30]

One solves the Schrödinger equation for the ion with its N electrons in a given configuration: $H_{ion}|g\rangle = E|g\rangle$, where H_{ion} is the Hamiltonian of the system for the chosen configuration and E and $|g\rangle$ are the eigenvalue and eigenstate, respectively. The different eigenstates $|g\rangle$ are functions of N electrons, hence they are called many-body (or multi-electronic) states.

The Hamiltonian is expressed as: $H_{ion} = H_{kin} + H_{eN} + H_{ee} + H_{SO} + H_Z + H_{crystal}$, where H_{kin} is the kinetic energy of the electrons, H_{eN} is the Coulomb attraction between electrons and nucleus, H_{ee} is the electron–electron Coulomb interaction, H_{SO} is the SOC interaction, $H_{crystal}$ is the crystal field Hamiltonian (which takes into account the local environment of the absorbing atom) and H_Z is the Zeeman Hamiltonian (which presents the interaction of an external magnetic field with the atomic magnetic moment).

The different terms in the Hamiltonian are expressed as k^{th}-rank spherical tensors, $T_q^{(k)}$, in order to take advantage of the analogy to the spherical harmonics, $Y_m^{(l)}$.[i] These tensors thus transform under symmetry operations as spherical harmonics. The many-body eigenstates corresponding to the electronic configuration (*cf.* Figure 2.3) are determined by the diagonalisation of the matrix constituted by the $\langle\psi|T_q^{(k)}|\psi'\rangle$ matrix elements, where $|\psi\rangle$ and $|\psi'\rangle$ are many-body basis functions,

---

[i] k and q are analogous to l and m, respectively.

which are chosen as irreducible representations of the symmetry group of the Hamiltonian.

The kinetic energy of the electrons and the Coulomb interaction of the electrons with the nucleus are fixed for a given atomic configuration and contribute only to the average energy of the configuration. Often, the spectroscopist is mainly interested in relative energies and therefore concentrates on those terms that determine the energy difference between excited states. In a first step, we take advantage of the spherical symmetry of the interelectronic interactions and of the SOC to calculate the $\langle \psi | T_q^{(k)} | \psi' \rangle$ matrix elements for an isolated ion in spherical symmetry. This is **atomic multiplet theory**. In the case of a spherical symmetry, the 'good' basis functions, $|\psi\rangle$, are $|\alpha(L, S)JM\rangle$ functions, *i.e.* the eigenfunctions of L^2, S^2, J^2 and J_z operators (where α represents all other quantum numbers needed to completely identify these basis functions). The values of L, S, J and M are obtained by angular momentum coupling of all electrons. The eigenstates, $|g\rangle$, solving the Schrödinger equation are thus linear combinations of these basis functions, $|\psi\rangle$.

When reasoning using the true point group symmetry around the absorbing atom, *i.e.* when adding the crystal field term to the Hamiltonian, the previous functions are no longer adapted. What **crystal field multiplet theory** does is to use group theory arguments and subduction theory to transfer ('branch') the results obtained in spherical symmetry to the point group symmetry of the absorbing atom. In this way, we obtain the matrix elements needed for diagonalisation, and all $|g\rangle$ states are calculated in the true (point group) symmetry of the absorber.

The calculation in spherical symmetry uses as input values some radial integrals (repulsion Slater integrals, SOC constants) which are, in a very first step, calculated self-consistently using a Hartree–Fock model. In point group symmetry, where the absorber is considered along with its environment, these values are reduced empirically ("scaled") in order to take into account the effect of the chemical bond which delocalises the electrons. This reduction factor is an adjustable parameter, and so are the crystal field parameters. A more refined way to treat the chemical bond is to consider orbital mixing and use **charge-transfer multiplet theory**, which adds several atomic configurations to build the initial states, as well as the final states.[30] The additional configurations mimic charge transfer from a ligand to the absorbing metal ion.

Thole developed the first charge transfer multiplet code in the 1980s, using the theoretical framework of Butler and Cowan,[31,32] thereby making multiplet theory accessible to a large community. The code has since been extended, but the physics has not fundamentally changed since

1991.[29,30,33] Some codes by other authors do not require specifying the local cluster point group symmetry; instead, the positions and charges of surrounding atoms are required as input, considerably increasing the flexibility of the code to adapt to any symmetry.[34-36]

2.3.3   Comparison of Theoretical Approaches

The chemical environment in LFMT is treated semi-empirically. The calculations require the adjustment of parameters, *e.g.* the crystal field splitting, by comparison of the theoretical spectrum with the experimental. This is the price that must be paid for considering explicitly all the electrons of the atom of interest. The effect of the chemical bond can be considered by reducing the electron–electron repulsion integrals or by including several configurations to simulate metal–ligand orbital mixing. Some of these parameters can be obtained experimentally from optical absorption spectroscopy or electron paramagnetic resonance. The advantage of LFMT is that all intra-atomic electronic interactions can in principal be taken into account, including the **electron–hole interaction**, which often gives rise to an additional multiplet structure. This is a crucial point because inner-shell spectroscopy always involves the creation of a core hole. The core hole effect is one of the major challenges in the theory of inner-shell spectroscopy. Spectra in which multiplets are important to understanding the spectral shape often require an analysis using LFMT. A prominent example is the L edges of $3d$ transition metals.

Single-particle methods, in contrast, are called *ab initio* as there is no fitting (or empirical) parameter in the solution of the Kohn–Sham equations. There are some convergence parameters, which are optimised but not *in the strictest sense* fitted. DFT can only calculate states represented by single Slater determinants, while LFMT can account for the multi-Slater determinant character of an electronic state, which is often important to the inclusion of multi-electron excitations.

Electronic exchange and correlation in standard DFT approaches (such as DFT-LDA, DFT-GGA) are treated in a much more approximated fashion than in LFMT. Including at least a cluster of atoms surrounding the absorbing atom makes it impossible to take into account all the electrons involved with current computing power. As the exact formulation of exchange and correlation is unknown, this implies that correlations are described using a mean field approach. In most DFT implementations, valence electrons (*i.e.* those participating in chemical bonds) and core electrons are generally treated differently, *e.g.* in the PAW-pseudopotential approach, only the valence electrons are considered explicitly, but the possibility of reconstructing all-electron wavefunctions using a linear operator is retained.[37,38] The core–hole effect is much more difficult to address in single-particle methods than in LFMT. In standard DFT approaches, the electron–hole interaction is treated in a static way by inclusion of a core hole on the absorber and use of a supercell (or a large cluster) of atoms. To ensure neutrality of the system, a background charge is added when solving the Kohn–Sham equations. In most cases, this does not yield an accurate description of the core–hole effect. Improvement requires going beyond DFT approaches (see Section 2.3.8).

Note that a direct comparison between LFMT- and DFT-based calculations is often difficult due to their fundamentally different ways of describing the electronic structure. Bridges can be built by considering certain properties of the system (*e.g.* spin state).

2.3.4 The Many-Body Extended Picture of Electronic States

Recently developed approaches aim at circumventing the shortcomings of both DFT and LFMT. They try to combine the 'extended scale' aspect with a comprehensive description of many-body interactions. While remaining within the framework of DFT, an example of a rather manageable improvement is to add a static Hubbard parameter, U, to account for onsite electronic repulsion.[39]

There are other approaches that go well beyond DFT. These can be grouped into two main types: the quantum chemistry approaches (mainly wavefunction-based methods) and the Green's functions methods. Quantum chemistry approaches (DFT configuration interaction (CI), multiconfigurational self-consistent field method, coupled cluster theory, quantum Monte Carlo *etc.*) are many-body extended approaches (see reference [40], and references therein). In DFT-CI, for example, the total wavefunction of the system is determined as a combination of Slater determinants (CI). The number of configurations is limited by computing power and the challenge is to decide what configurations to include. These approaches are applied for small clusters and molecules.

Green's functions-based methods, such as GW and dynamic mean field theory (DMFT), are used to calculate the electronic structure of strongly correlated materials. The GW method calculates a screened

Coulomb interaction (W) following a DFT calculation of the charge density.[41] It requires a non-local, energy-dependent self-energy operator. DMFT consists in mapping a lattice problem onto a single-site quantum impurity problem, using a local screened Coulomb interaction (U).[42] In both GW and DMFT approaches, correlations (many-body effects) are significantly better described than in standard DFT. We add that the Bethe–Salpeter equation (BSE), which is also based on the Green's formalism, is a two-particle theory, which improves significantly the description of electron–hole interaction.[43]

2.3.5   Single-Particle Calculation of the Absorption Cross-Section

The underlying assumption behind using a single-particle approach to calculating transition probabilities is that the multi-electron matrix element can be approximated by one-electron transitions. As an example, we can consider matrix elements of the form $\langle \varphi_c^{-1} \varphi_u | \widehat{O} | 0 \rangle$, where φ_c and φ_u are orbitals for a core and an unoccupied orbital, respectively. The ground state is denominated $|0>$ and the excited state has one core electron less and one electron more in a previously unoccupied orbital.

A single-particle approximation often works well, but can also dramatically break down. The L edges of $3d$ transition metals (except $3d^0$ and $3d^9$ configurations) show strong multi-electron effects, and a single-particle approach fails. It is possible to include part of the multi-electron response to the photoexcitation by an energy dependent factor, *e.g.* the amplitude-reduction factor S_0^2.[44]

The single-particle extended approach provides the possibility of describing the transitions using DFT. An initial natural idea is to try to connect the absorption single-particle cross-section to the empty density of states (DOS). This relationship can be achieved in some way, and it provides an intuitive explanation of what is measured in an X-ray absorption or emission experiment. We will discuss below to what extent this DOS-based picture holds. Then we will describe the different single-particle frameworks that allow the full calculation of the cross-section, *i.e.* with the sole assumption that the initial and final states are single-particle states. Finally, we will describe in which cases the single-particle extended (DFT) approach is well adapted.

2.3.5.1   The Empty Density of States and X-ray Spectroscopy

It is possible to approximate the single-electron XAS cross-section as the product of an empty local partial density of states (LPDOS), ρ, times a squared radial matrix element, M^2:[45]

$$\sigma(E) \approx M(E)^2 \rho(E) \tag{2.9}$$

where E is the energy of the final state with respect to the Fermi energy. The empty DOS of a system describes the number of empty electronic states *per* interval of energy. 'Local' means that the DOS that is considered in Equation 2.9 is restricted to the absorbing atom. It is obtained by a projection of the total DOS of the system onto a basis of localised orbitals centred on the absorber. 'Partial' means that the projection is done using basis orbitals with principal, angular and azimuthal quantum numbers, thus defining an LPDOS of p type (p_x, p_y, p_z), d type *etc*. It is important to underline that rigorously speaking, the LPDOS considered in Equation 2.9 is the one projected on the absorber with a core hole. The matrix element M corresponds to a radial integral. Equation 2.9 is a general expression that holds independently of the chosen method (PAW method, multiple scattering theory *etc*.). The analytical expressions of M and ρ, however, depend on the method.[j] When comparing the validity of a chosen single-particle approach to the experimental results or another theoretical model, it is therefore necessary to use the cross-section $\sigma(E)$ rather than $\rho(E)$.[k]

Equation 2.9 implies that at each final state energy, E, the cross-section is an image of the empty LPDOS on the absorbing atom. This expression has the practical advantage that a direct connection between the absorption spectrum and the empty DOS in the presence of a core hole on the absorber is achieved. This is a convenient image through which to interpret the data. As a matter of fact, historically the very first interpretation of XAS spectra was based on this relation, implicitly assuming that M is independent of the energy. For example, the spectrum measured at the O K edge in α-SiO$_2$ nicely corresponds to the empty p DOS projected on the excited O atom calculated with a pseudopotential

---

[j] See for example the derivation of the expressions of M and ρ in the case of a muffin-tin potential, at D. Cabaret's page: http://www.impmc.upmc.fr/~cabaret/memoireHDR_Cabaret.pdf (last accessed 31 May 2013).
[k] The proportionality holds only if the radial matrix element, M, depends on the energy only in a very moderate way. Otherwise, the direct correspondence between the cross-section and the LPDOS becomes much less obvious, because M strongly varies as a function of the energy.

single-particle code (Figure 2.6). In this case, the cross-section is a direct image of the LPDOS on the excited atom, *i.e.* including a core hole.

We would point out that there are some favourable cases in which the cross-section looks very similar to the LPDOS *without* a core hole, because core hole effects are small. This is the case, for example, for the O K edge spectra in rutile TiO_2, VO_2 and CrO_2, which look like the oxygen p-projected DOS without a core hole.[30] Note, however, that for α-SiO_2 in Figures 2.6 and 2.7, the core hole effect is much larger and thus must be included in the LPDOS calculation.

The radial matrix element M and the LPDOS ρ are quantities which depend on the method chosen to calculate them. Methods based on PAW

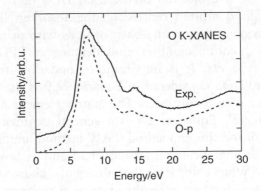

Figure 2.6   Experimental O K edge X-ray absorption spectrum measured in α-SiO_2 *vs* p-projected DOS on the excited O atom (including a $1s$ core hole) calculated using a pseudopotential method (D. Cabaret, personal communication). The energy is relative to the Fermi level.

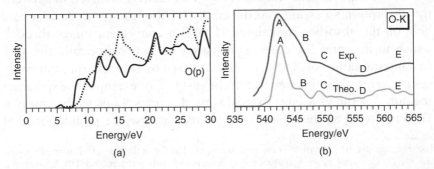

Figure 2.7   (a) LPDOS on the excited O atom without (dashed line) and with (solid line) core hole; the energy is relative to the Fermi-level energy. (b) Experimental *vs* theoretical O K edge spectrum in α-SiO_2 calculated using the LCAO method. Reproduced with permission from Mo *et al.* (2001) [46]. Copyright (2001) American Institute of Physics.

pseudopotential formalism calculate radial matrix elements which show generally little energy dependence. Therefore, as in Figure 2.6, LPDOS and cross-section are almost proportional to each other. However, when using for example the LCAO method on the same system, the LPDOS on O (Figure 2.7a) looks significantly different from the cross-section calculated with the same method (Figure 2.7b),[46] even though core hole effects are included in both calculations. The discrepancy comes from the fact that, with this method, radial matrix elements usually show strong energy dependence, and therefore the calculated LPDOS and cross-section are very different. This implies that an interpretation of the XAS spectrum based only on the calculated projected DOS on the absorber would be misleading.

We have shown that the interpretation of XAS based on the LPDOS requires a good knowledge of the method employed. Therefore, to avoid such difficulties, a more sophisticated interpretation is required. This consists in calculating directly the full single-particle cross-section, instead of using the approximation in Equation 2.9.

2.3.5.2 Full Single-Particle Calculations of the Cross-Section

There are many theoretical codes which use the single-particle approach to calculate the XAS cross-section. We do not have the space to describe them all here, but a list of the most popular mono-electronic codes for core spectroscopies is available at www.xafs.org. We wish here to convey the key ideas of the method, so that the different scientific cases used as examples will be accessible to the general reader. Details on theoretical modelling of the cross-section can be found in references [8, 9, 30, 44].

Single-particle calculations of the XAS cross-section are obviously based on a single-particle description of electronic states, *i.e.* DFT-based methods. The calculation of the cross-section is performed in two steps. The first consists in calculating the single-particle wavefunctions φ_c and φ_u. The calculation of φ_c does not raise any difficulty, since it is a core orbital. It can be done for an isolated (neutral) atom. The final-state single-particle wavefunctions φ_u are obtained by solving the Kohn–Sham equations for a system that includes a core hole on the absorbing atom. The second step of the calculation consists in calculating the matrix element for a given transition operator, $\widehat{O}$, and then, after convolution with a Lorentzian function, obtaining the cross-section. Spin-polarised cross-sections can be readily calculated by considering minority and majority spins.

2.3.6   Many-Body Atomic Calculation of the Cross-Section

Many-body atomic calculations of the Kramers–Heisenberg cross-section are currently mainly based on the LFMT. One solves the Schrödinger equation for an isolated ion with N electrons in the ground- and excited-state ($|g\rangle$ and $|n\rangle$ (or $|f\rangle$ in the case of X-ray emission)) electron configuration. All states are calculated in the true symmetry of the absorber, *i.e.* the chemical environment is considered by branching from spherical to local symmetry, which includes the absorber and its ligands (without actually having ligands). The many-electron transition amplitudes are evaluated for all ground states that are populated at a given temperature (assuming a Boltzmann distribution) and all intermediate and final states that can be formed for the given atomic electron configuration.

The XAS and RIXS cross-sections are then evaluated using Equations 2.6 and 2.5, respectively, including a convolution that takes into account the instrumental resolution. An important point is that LFMT readily allows evaluation of the unsquared matrix elements, *i.e.* the phase of the multi-electron transition matrix elements is known, which is important for possible interference effects in Equation 2.5 (see Section 2.4.4). A user-friendly interface based on the multiplet code, CTM4XAS, is available for the calculation of core spectroscopies.[28]

2.3.7   Which Approach Works Best for Inner-Shell Spectroscopy?

There is unfortunately no simple answer to this question: it depends on the element, the absorption edge and the spectroscopy. Consequently, the spectroscopist should use several theoretical models to see which works best. This is very time-consuming. We will try to help in this section with some guiding considerations.

The (assumed) localisation of the final-state wavefunction may be one criterion. Electrons that are excited into orbitals spread over several atoms (delocalised) can be expected to show small intra-atomic interactions, *i.e.* interactions with the valence electrons and the core hole of the absorbing atom. When multi-electronic effects, *i.e.* the electronic repulsions between the valence electrons and their interactions with the core hole, are small, single-particle approaches often give satisfactory results. This may be the case for the K main edge of $3d$ elements and the $L_{2,3}$

edges of heavy elements (*e.g.* rare earths or $5d$ transition metals), as well as the K edges of ligands (such as C, O, N, S). For example, at the K edge of $3d$ transition metals, the transitions involved in the edge region are dipole-allowed $1s \rightarrow p$, where the empty p states are delocalised with a broad energy distribution. Hence the electronic repulsion in the final state is moderate, as is the attraction created by the (screened) $1s$ core hole.

Electrons in localised orbitals will strongly interact with each other and with the core hole. When multi-electronic effects become significant, a description which considers only the absorber atom may be more successful. In such cases, LFM or CTM theory is generally better adapted. This is typically the case at the $L_{2,3}$ edges of $3d$ elements and at the $M_{4,5}$ edges of $4f$ elements. For the $L_{2,3}$ edge, the transitions involved in the edge region are electric dipole-allowed $2p \rightarrow 3d$. As the $3d$ orbitals are rather localised in energy and space, and close in energy to the $2p$ hole, strong multiplet effects are observed. The K pre-edge of $3d$ elements is an intermediate case in which both single-particle and many-body approaches may work or fail.

It may also happen that the different energy ranges of a spectrum are best described by different theoretical approaches. This is often the case when a weak pre-edge feature is observed before the strong main edge.[16,47,48] Pre-edges often arise from excitations into localised orbitals, while the main edge has a more delocalised character.

Whether the final state is localised or not may be deduced from the shape of the absorption edge.[49] When the edge is dominated by a step function, the final state can be assumed to be delocalised. If it exhibits distinct peaks that decrease sharply after the maximum, the small energy band may be the result of a localised wavefunction (but not necessarily).

The reader may find this confusing and unsatisfactory, and calculation of inner-shell spectra is indeed far from being a routine task. The combination of many-body calculations considering intra-atomic interactions with an extended treatment of the system is an important objective in the theory of inner-shell spectroscopy. Recent years have brought important progress in this field, and some approaches that go beyond what we have described so far are briefly explained in the next section.

2.3.8   Beyond Standard DFT Methods

Standard DFT approaches have the drawback of treating electron-electron repulsion in an approximate way, *i.e.* in a mean-field way.

This deficiency can be partly met within the framework of DFT, *e.g.* by adding an onsite Coulomb term, U.[39] In single-particle methods, the treatment of the core hole–electron interaction is a much more difficult problem to deal with. Indeed, these effects are treated in a static way by including a core hole on the absorber and using a supercell of atoms (or a large cluster). However, the attraction created by the core hole on the electronic states is often overestimated. This leads to spectral features which are not located at the right position in the spectrum.[50,51] A better modelling of the core hole–electron interaction requires going beyond the DFT framework. Several methods have undergone a significant development in the past 15 years, such as time-dependent DFT (TDDFT) and the many-body Green's function-based methods (*e.g.* BSE, GW; see review by Onida *et al.*[52]). These approaches give a better description of many-body effects relevant in core spectroscopies, such as the electron–hole interaction (BSE),[43] the electron self-energy (GW)[53] and the screened Coulomb interaction (GW). Unlike DFT, TDDFT is suitable for the calculation of excited states.[54]

All these methods that go beyond standard DFT open new opportunities to draw a fully *ab initio* many-body picture of electronic excitations and meet a growing community of users.[43,52,54–60]

2.4   EXPERIMENTAL TECHNIQUES

With Equations 2.5 and 2.6 and some theoretical background, we have all the tools we need to discuss X-ray spectroscopy. By choosing suitable experimental conditions, it is possible to emphasise a particular term of the DDSCS and thus probe a certain aspect of the electronic structure. It is very important to note that the choice of scattering angles, photon polarisation, incident energy and energy transfer (*i.e.* the difference between incident energy and emitted energy) determines which expression dominates, *i.e.* which interaction mechanism gives rise to the spectral intensity. Each expression may probe different aspects of the electronic structure. Consequently, an experimental set-up with many degrees of freedom with respect to energy and angle analysis can provide rich information on the electronic structure.

Often, experimental constraints limit the possible geometries and thus what scattering mechanism can be used. There can be several ways of probing the same physical quantity. The creative experimentalist who is knowledgeable about scattering mechanisms can access the relevant information by appropriately adjusting the experimental conditions. The most prominent example is secondary process detection of the absorption cross-section using *e.g.* electron and fluorescence yield techniques, in cases where it is not possible to measure the transmitted beam. Another example is the detection of absorption edges in low-Z elements using hard X-ray Raman spectroscopy (XRS). In this section, we describe the experimental conditions required to probe certain aspects of the scattering process.

2.4.1   X-ray Absorption Spectroscopy

An X-ray absorption spectrum is generally divided into the X-ray absorption near-edge structure (XANES) and the extended X-ray absorption fine structure (EXAFS). This distinction is motivated by the assumption that XANES arises mainly from excitations into orbitals that can be associated with the absorbing atom ('bound' states), while the EXAFS occurs against a large background due to excitations into the continuum. It is useful in this context to note that, theoretically, a clear distinction is neither possible nor desirable. Where the XANES range stops and EXAFS starts can therefore not be clearly defined. By convention, we consider that the XANES region starts a few eV below the edge jump and ends a few tens of eV ($\sim$60 eV) above the edge (Figure 2.8), where the EXAFS region begins. In practice, EXAFS is referred to as the spectral range analysed using the EXAFS equation (see Section 2.4.1.5).

Often, edges show weak spectral features that occur at energies lower than the main rising edge. These arise from transitions into the lowest unoccupied orbitals and owe their low spectral intensity to the fact that they are not dipole allowed. Examples are the K pre-edges in $3d$ transition metals and L_3 pre-edges in rare earths.

2.4.1.1   *X-ray Absorption Near-Edge Structure*

XANES (or near-edge X-ray absorption fine structure, NEXAFS) is a standard tool for chemical analysis. The technique is applied to all elements of the Periodic Table, and the available literature is extensive. One objective of XANES is to probe the lowest-lying unoccupied orbitals in order to learn about the chemical state of the absorbing atom. This is not always possible, due to the selection rules. The $3d$ transition metals,

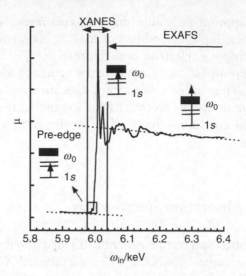

Figure 2.8   XAS spectrum measured at the Cr K edge for $MgAl_2O_4 : Cr^{3+}$, showing the different regions mentioned in the text (pre-edge, XANES, EXAFS) and the corresponding one-electron transitions. The threshold energy is $\hbar\omega_0$.

for example, are most suitably probed at the L $(2p)$ edge because the dipole selection rules allow $2p$ to $3d$ transitions. The L edge lies in the soft X-ray range and it is experimentally often more favourable to perform an experiment using hard X-rays. The K edge promotes a $1s$ electron into orbitals with p-orbital symmetry (*i.e.* electric E1 transitions). Fortunately, modern intense X-ray sources allow the study of weak transitions, *i.e.* E2 transitions are also accessible. One can therefore attempt to probe $1s$ to $3d$ E2 transitions using hard X-rays, and this is indeed done in practice.

Verification of the validity of calculated spectra is always limited by the spectral broadening. Interactions that are smaller than the core hole lifetime broadening can usually not be observed unless they split the ground state into levels that give rise to sufficiently different X-ray absorption spectra (*e.g.* XMCD). Systems in which local interactions are much smaller than the spectral broadening can often be modelled using an extended single-particle approach. This may hold for low-Z elements (*e.g.* C, O, N, Si), but also for the $2p$ to $5d$ excitations in rare earths and $5d$ transition elements. An additional complication arises from multi-electron excitations, *i.e.* transitions that are not considered within a single-electron picture. These may be strong in the XANES region. A prominent example is charge transfer excitations that result in so-called screened and unscreened XAS final states.[61]

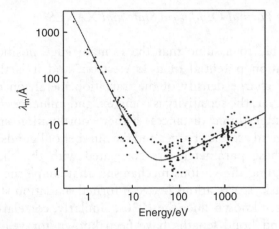

Figure 2.9 Mean free path as a function of the energy of the photoelectron. Reproduced with permission from Seah *et al.* (1979) [62]. Copyright (1979) John Wiley & Sons Ltd.

The *K* main edge in 3*d* transition metal ions is about 10–20 eV above the Fermi level and arises mainly from intense electric dipole $1s \rightarrow p$ transitions.[1] The photoelectron has low kinetic energy and thus a large mean free path (Figure 2.9). It therefore travels far and interacts with neighbouring atoms up to high coordination spheres. The excited states probed correspond to excitations into *np* orbitals ($n \geq 4$), highly mixed and delocalised over the solid. The interpretation of this spectral region is often challenging, since it is sensitive to both the electronic and the crystallographic structure around the absorber.

The different approaches to the interpretation of the XANES region can be classified into two groups: the fingerprint approach and calculations. The fingerprint approach is based on the comparison of the measured spectrum with spectra measured for reference compounds of known structure, eventually including some simple mathematical operations such as linear combinations and principal component analysis. This approach often enables simple information on the structural and electronic environment of the absorber to be obtained. The wealth of information contained in XANES spectra can potentially be fully exploited by simulating the spectrum using a theoretical model that employs mono- or multi-electronic methods. It is, however, a great challenge to account fully for all local and extended effects in the theory, and push-button modelling of *e.g.* XANES of 3*d* transition metals remains a challenge.

---

[1] For metals, the main absorption K edge starts at the energy of the Fermi level.

2.4.1.2   The Formal Oxidation State and XANES

It is reasonable to assume that the K main edge position is related to the ionisation potential of a $1s$ electron, and it is thus expected to reflect the charge density of the transition metal ion *via* screening effects. However, the sensitivity is complex, and influences from nearest-neighbours interatomic distances, higher coordination spheres, bond angles, the ligand atomic charges and the number of ligands may become visible. All these parameters are correlated with the charge density, and disentangling the various mechanisms that shape the XANES is a formidable task. Correlations between formal oxidation state and edge shifts have been known since the 1930s. Similarly, correlations between edge shifts and bond lengths have been known for years. Scattering theory predicts a direct influence of edge shifts on bond lengths. An approximate empirical $\Delta E \approx 1/r^2$ edge shift rule has been proposed by Natoli, where r is a bond distance.[63] According to this rule, expansion of bond distances implies a contraction of the XANES features. The local charge density may also influence the energy position of the main edge. de Vries *et al.* performed a detailed study of the effects of the charge and the Madelung potential on the edge position. They found that both effects have a strong influence.[64]

Experimentally, it is not possible to change the atomic configuration without changing the electrons and *vice versa*. Theory can do this, and it has been found that if the atoms are held in fixed positions, transferring charge from one to another does not cause large edge shifts.[12] Transferring charge and allowing the atoms to relax in order to minimise the energy will cause them to move, in which case the edge position will shift.

Inferring the oxidation state from the edge position is thus not straightforward.[65] If the atoms are free to move and follow the charge distribution, the charge density may be derived from the edge position. If on the other hand there are steric or crystalline constraints to the motion of the atoms, it may be unreliable.

Figure 2.10 shows the experimental K absorption edges of some model Mn compounds, comparing oxides with coordination complexes for oxidation states II, III and IV. Determining the edge position is diffi-cult because of the complex spectral shape. Furthermore, experimental artefacts can distort the spectral shape (Section 2.5.3.3.1).

The spectroscopic results can be compared to calculations of the electronic structure. Codes based on the single particle approach, such as FEFF,[44] can determine the electron density. The calculated electron

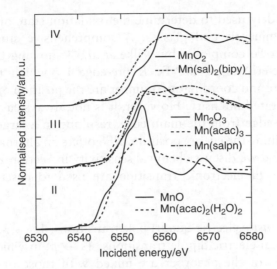

Figure 2.10 *K* absorption edges (XANES) of some Mn oxides, and coordination complexes for oxidation states II, III and IV. Reproduced with permission from Glatzel *et al.* (2009) [65]. Copyright (2009) Institute of Physics.

density can be further treated in order to assign charge to an atom. A population analysis (*e.g.* Mulliken population analysis) distributes the charge density over atomic orbitals. A bond valence analysis is the quantum chemical equivalent to the oxidation state.[66] The classification becomes more complex than for simple oxidation states, however. Orbital mixing leads to non-integer charge fluctuations and the energy position of the spectral feature flows continuously for different systems. This is often nicely confirmed in the calculations. It has, for example, been long known for a while that oxygen takes part in the charge transfer and thus cannot always be assigned the charge 2−.[67]

2.4.1.3 *The Pre-Edge Region*

Alternatively, one can analyse the *K* absorption pre-edges.[48] These also show a complex structure, due to orbital mixing and electron–electron interaction effects, but they are believed to be more directly influenced by the absorber charge density. The spectral intensities and line splitting reflect the energy ordering and symmetry of the unoccupied orbitals, which are believed to be mainly localised on the absorbing metal atom but often mix with ligand orbitals. In Earth science, fingerprint analysis is widely used to determine the oxidation state of $3d$ elements in complex minerals and glasses. A comprehensive study has been performed for Fe complexes by Wilke *et al.*,[4] showing that the most useful characteristics of the Fe K pre-edge by which to determine oxidation state and coordination number are the position of the centroid and the integrated intensity. However, it is not always easy to separate the weak pre-edge from the main edge, resulting in a large uncertainty when determining the energy position. XES offers an alternative to X-ray absorption, and we discuss chemical sensitivity in Section 2.4.3.1.

Three types of electronic transition are used to describe pre-edge features:

   (i) Local *onsite* electric quadrupole transitions, $1s \rightarrow 3d$.
  (ii) Nonlocal electric dipole transitions, $1s \rightarrow p$, in which the empty p levels of the absorber are mixed with those of the nearest-neighbour metal atoms, *via* the p orbitals of the ligands. This leads to **non-local** (*off-site*) features located a few eV above the localised *in-site* transitions;[48,50,51,68,69] this is the case for example for the Ti K edge in TiO_2 and the Co K edge in $LiCoO_2$.
 (iii) Local electric dipole transitions, $1s \rightarrow p$. These are only possible when the absorber is in a crystallographic site without local inversion symmetry: local $p - d$ mixing becomes allowed on the absorbing atom, and transitions gain an electric dipole contribution. This enhances significantly the intensity of the pre-edge.

The pre-edge features of transition metals are therefore closely related to the coordination number, oxidation state and spin state of the absorbing atom, the point symmetry of the absorbing atom site and (related to the latter) orbital mixing.

2.4.1.4   X-ray Natural Linear Dichroism

When using the fingerprint approach to analyse XANES, the use of powder samples, which yield the so-called **isotropic** spectrum, is certainly sufficient. Measurements on single crystals, however, fix the direction of the crystal axes with respect to the photon vectors (k, ε) and thus add a constraint on the transition matrix elements. The variation of the XANES spectrum is measured when the direction of the incident wavevector and linear polarisation vector are varied with respect to the single crystal (*cf.* Figure 2.11a). In practice, the single crystal is rotated with respect to the beam. Such experiments probe the angular dependence of the cross-section, or the XNLD.

It is well known that the spectra of noncubic crystals show just such a sizeable XNLD effect.[14,70] For cubic crystals, the situation is slightly different. Electric dipole transitions are isotropic (*i.e.* they do not show any XNLD), while electric quadrupole transitions are anisotropic. Therefore, the $1s \rightarrow 3d$ transitions visible in the pre-edge, even in cubic crystals, show an angular dependence.[51,71] Great advantage can be taken of this effect, such that the contributions of electric dipole and electric quadrupole transitions can be disentangled. This allows a more accurate insight into the $3d$ orbitals, which are responsible for many interesting properties, such as coloration, magnetism, reactivity in organometallics *etc.* One possibility offered by XNLD is thus to connect the atomic picture of the $3d$ orbitals as seen from X-rays with macroscopic properties.

This approach has been applied to the case of substitutional Cr^{3+} in $MgAl_2O_4$. The presence of Cr^{3+} in a slightly trigonally distorted octahedral site provides a red colour to the crystal. Both DFT and LFM calculations have been performed to interpret quantitatively the isotropic and dichroic spectra.[15,72] From this combined approach, it follows that:

(i) Electric dipole transitions do not contribute to the K pre-edge, in agreement with the inversion symmetry at the Cr site, which prohibits $p - d$ mixing.

(ii) If pre-edge features are more intense than 4% of the edge jump, pure quadrupole transitions alone cannot explain the origin of the pre-edge structures for Cr^{3+}. This result, which is consistent with other studies on Fe^{3+} and Fe^{2+} in glasses and minerals, can probably be extended to the other $3d$ ions.

(iii) The isotropic spectrum is not significantly sensitive to the trigonal distortion of the Cr site, but the dichroic signal is much more so (Figure 2.11b). Thus crystal field parameters and quantitative information on site distortion can be determined from XNLD at the K pre-edge in single crystals, using LFMT calculations. This method can be particularly interesting for systems that cannot be investigated by optical absorption spectroscopy (opaque compounds or compounds without any d electrons).

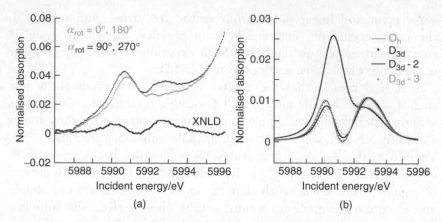

Figure 2.11 (a) Experimental Cr *K* pre-edge spectra of $MgAl_2O_4$, for four different orientations of the single crystal with respect to the incident X-rays. The XNLD is the difference signal. (b) Influence of the trigonal distortion on the theoretical isotropic and dichroic spectra, for increasing values of the distortion parameters. Reproduced with permission from Juhin *et al.* (2008) [15]. Copyright (2008) American Physical Society.

2.4.1.5 Extended X-ray Absorption Fine Structure

The field of X-ray spectroscopy was revolutionised when it was established that the intensity oscillations above the edge region reflect the local atomic coordination (for a review of the history, see reference [73]). The theory explaining the oscillations is now well established and EXAFS spectroscopy has become a routine tool; an introduction can be found in references [12, 74]. IXAS provides comprehensive tutorial material on its website (www.ixasportal.net).

In the EXAFS region, the kinetic energy of the photoelectron increases, and its mean free path is thus smaller than it is in the XANES region (*cf.* Figure 2.9). The dominating process is the outgoing photoelectron, represented by a wave that is scattered by the potentials of the neighbouring atoms. The EXAFS oscillations can be viewed as being a result of interference effects between a scattered and a backscattered electron wave. The scattering time is determined by the mean free path of the photoelectron and the core hole lifetime, *i.e.* the lifetime of the excited state. Constructive and destructive interference causes an oscillation in the absorption coefficient as a function of incident energy. The absorption coefficient, μ, is composed of a smoothly decreasing function of energy (post-edge dashed line in Figure 2.8) and an oscillatory part.

The normalised oscillatory part is defined as:

$$\chi(\omega_{\text{in}}) = \frac{\mu(\omega_{\text{in}}) - \mu_0(\omega_{\text{in}})}{\Delta\mu_0} \qquad (2.10)$$

where $\mu_0(\omega_{\text{in}})$ is the smoothly varying atomic-like background absorption (possibly including contributions from other edges) and $\Delta\mu_0$ is a normalisation factor that is often approximated by the magnitude of the jump in absorption at the edge energy (*i.e.* the difference between the post-edge and pre-edge dashed lines determined at the threshold energy in Figure 2.8).

The frequencies of the normalised EXAFS oscillations provide information on the distances between the absorber element (the element of interest) and its neighbours. The amplitudes of the normalised EXAFS oscillations can provide information on the number and kind of neighbour atoms, *i.e.* it may be possible to identify what elements are in the successive coordination shells.

The EXAFS signal is analysed using the EXAFS equation, which can be obtained using empirical arguments but can also be strictly derived based on the transition matrix elements and DFT. Using a spherical wave formalism, the EXAFS equation is expressed as:

$$\chi(k) = -\frac{1}{k} \sum_{i=1}^{N} A_i\ (k) \sin(2kR_i + \Phi_i(k)) \qquad (2.11)$$

where the energy has been converted into the photoelectron wavevector, k, defined as:

$$k = \sqrt{\frac{2m(\omega_{\text{in}} - \omega_0)}{\hbar^2}} \qquad (2.12)$$

In Equation 2.11, the index i refers to the ith atomic shell, R_i is the distance between the absorbing atom and the neighbouring atom in the ith shell and $\Phi_i(k)$ is a phase function related to both the absorbing and the backscattering atoms. $A_i(k)$ is defined as:

$$A_i(k) = \frac{N_i\ S_0^2}{R_i^2}\ |f_i(k)| \exp(-2\sigma_i^2 k^2)\ \exp\left(-\frac{2R_i}{\lambda\,(k)}\right) \qquad (2.13)$$

where N_i is the coordination number in the ith shell, $f_i(k)$ is the backscattering amplitude function of the atoms in this shell, σ_i is the standard deviation associated to the (relative!) R_i distances, S_0^2 is an overall many-body amplitude reduction factor and $\lambda(k)$ is the energy-dependent mean-free path.

The extraction of structural information from experimental EXAFS data requires several steps of data reduction (including background subtraction, normalisation and Fourier transform) and of data analysis: for each coordination shell, the aim is to determine R_i, N_i and σ_i using theoretical phase functions, $\Phi_i(k)$, and amplitude functions, $f_i(k)$, calculated using a structural model, and using values of S_0^2 and $\lambda(k)$ previously determined on a reference sample with known structure. A list of several programs that allow EXAFS data reduction and analysis is available at the IXAS Web site (www.ixasportal.net).

Note that while the analysis of the frequencies of the EXAFS oscillations is quite robust, an accurate treatment of the amplitudes is often challenging. This is in agreement with the general observation in experimental spectroscopy that energy positions can be measured much more accurately than intensities. In the case of EXAFS, the problems mainly arise from the sample composition (Section 2.5), but also from the fact that the EXAFS oscillations have to be separated from the photoelectric absorption of a free ion, which can be viewed as a very high background signal.

2.4.2 X-ray Raman Scattering Spectroscopy

The scattering (or spectral) intensity in Equation 2.2 is found at the energy transfer $\omega_{in} - \omega_{out}$. X-ray Raman scattering thus detects a spectrum as a function of the energy that was deposited in the sample (Figure 2.12). The energy loss is conveniently tuned to an absorption edge of the element of interest. A prominent example is the O K edge at 543 eV, measured *e.g.* with $\omega_{in} = 6000$ eV and $\omega_{out} = 5457$ eV. We thus have a hard X-ray probe of the oxygen K edge that is conventionally measured using soft X-rays. Another example is the K edge of Li, which is used in batteries. Because of the factor ω_{out}/ω_{in}, the X-ray Raman cross-section decreases with increasing energy loss.

Being able to measure absorption edges of low-Z elements such as O and C using hard X-rays is an invaluable tool for the experimentalist. A review with emphasis on high-pressure work has been given by Rueff and Shukla.[10] Measurements on liquids (*e.g.* water) using soft X-rays are possible but very challenging, and are easily compromised by experimental artefacts. The hard X-ray probe readily allows for measurements under ambient conditions, and artefacts are more easily controlled and evaluated.[75] The X-ray Raman scattering technique requires high

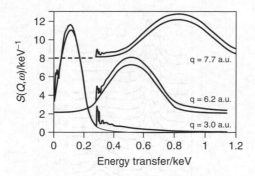

Figure 2.12 X-ray Raman spectra in diamond. The broad Compton feature moves to a larger energy transfer with increasing momentum transfer, q, while the C K edge stays at a fixed energy transfer of about 284 eV. Reproduced with permission from Huotari *et al.* (2012) [77]. Copyright (2012) IUCr.

incident photon fluxes and efficient detection of the scattered X-rays. This is increasingly being provided at synchrotron radiation sources worldwide, and XRS is used by a growing community. Experimental facilities have been improved such that even XRS tomography is now possible.[76]

X-ray Raman scattering has another interesting property. The experimentalist defines an energy transfer according to the edge of the element of interest. The momentum transfer, q, is determined by the scattering angle, θ, and the incoming and outgoing energies. The transition operator in Equation 2.2 can be approximated as:

$$e^{iqr} \approx 1 + iqr + \frac{(iqr)^2}{2} \qquad (2.14)$$

The first term can be neglected. For small values of q, we thus find that the spectra are dominated by dipole transitions. Increasing the momentum transfer will shift the spectral weight to quadrupole or octupole transitions. XRS thus provides a powerful tool by which to study excited states with different orbital angular momentums (see Figure 2.13). In a single-particle DOS picture, this means that we can probe the unoccupied p- and d-DOS in one experiment.

2.4.3 Non-Resonant X-ray Emission (X-ray Fluorescence)

Equation 2.6 dominates the detected signal if the incident energy, ω_{in}, corresponds to the energy difference between an excited state and the

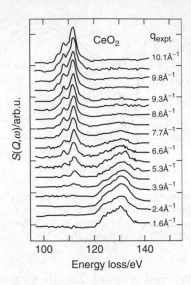

Figure 2.13 Variation of the transition matrix elements as a function of the modulus of the momentum transfer, q. Energy loss is synonymous with energy transfer. Reproduced with permission from Gordon *et al.* (2008) [78]. Copyright (2008) European Physical Society.

ground state of the system. The excited states are not limited to 'resonant' excitations into bound orbitals but also include photoionised states in which the excited state is described by a photoelectron in the continuum and the remaining ion. Thus, Equation 2.5 describes all photoexcitation processes, *i.e.* element-selective interactions, which are not considered in the X-ray Raman processes of Equation 2.2. This is an important observation because the use of the term 'resonant' in the literature generally only concerns experiments in which the incident energy is tuned to the vicinity of an absorption edge.

Distinction between resonant and nonresonant scattering is thus not straightforward because both are described by the same scattering term. We may ask whether the scattering process is coherent or not and use this as a distinction. In case of noncoherent scattering, the matrix elements are first squared and then summed over the intermediate states. Equation 2.5 then becomes:

$$\sigma_{KH} \left(\omega_{in}, \omega_{out} \right) \approx \frac{r_e^2}{m^2} \frac{\omega_{out}}{\omega_{in}} \sum_f \sum_n \frac{|\langle f|\widehat{O}'^\dagger|n\rangle|^2 |\langle n|\widehat{O}|g\rangle|^2}{(E_n - E_g - \hbar\omega_{in}) + \Gamma_n^2}$$

$$\times \frac{\frac{\Gamma_f}{\pi}}{(E_f - E_g - \hbar \left(\omega_{in} - \omega_{out} \right))^2 + \Gamma_f^2} \qquad (2.15)$$

The transition operators can be approximated by electric-dipole or electric-quadrupole transition operators in most cases. This expression neglects interference effects between different intermediate states and is simply the product of absorption and emission spectral intensities. This approximation can be assumed to be valid for photoionisation processes with an incident energy well above an absorption edge. It often also works surprisingly well for excitations near an absorption edge. We discuss this in Section 2.4.4.

The fluorescence lines may significantly change their shape when ω_{in} is just above the Fermi level. This is often ascribed to resonance effects. We discuss below the fact that this change of spectral intensity is mainly caused by a change of the distribution of the density of electronic states and not a change of the scattering mechanism or coherence (interference) effects (Section 2.4.4). We would like to make the important point that resonant and non-resonant X-ray emission and X-ray absorption all arise from the same interaction term.

Non-resonant X-ray emission or X-ray fluorescence can thus be considered as the Kramers–Heisenberg X-ray scattering following excitation into non-bound states. This shows that there can be no strict distinction between resonant and nonresonant excitations, because the transition between bound and non-bound levels is not clearly defined. One might attempt a definition using the degree of localisation of the unoccupied states but there is no strict physical argument unless it is within an approximation of the scattering process, *e.g.* one-electron transitions in muffin-tin potentials. If the electron leaves the material, *i.e.* the incident energy is sufficient to excite an electron above the work function, we are clearly in the non-bound regime.

2.4.3.1   *Chemical Sensitivity in Non-Resonant XES*

The valence electrons determine the chemical properties of the atom, and it is of great importance to understand the valence electronic structure. The valence electrons can be probed using X-ray emission either indirectly (core-to-core, c.t.c.) or directly (valence-to-core, v.t.c.) by monitoring the fluorescence emitted from the transition of the outer core or the valence electrons, respectively, to the core holes (Figure 2.14). The mechanisms that make c.t.c.-XES chemically sensitive may arise from two effects:

(i) *Screening effects:* the energy of the core levels adjusts to the modified potential caused by the change of the electron density in the valence orbitals. This shift of the core level is detected in c.t.c.-XES.

(ii) *Multiplet effects:* the spin and orbital angular momentum of the unfilled core level (the core hole) and the valence electrons strongly couple, giving rise to energy shifts and a multiplet structure, which depend on the valence shell electron configuration.

The chemical sensitivity of the $K\alpha$ lines in sulfur, for example, arises mainly from screening effects.[79] Two different regions can be distinguished in the $K\beta$ emission spectrum of $3d$ transition metal compounds, and these can be referred to as $K\beta$ main and $K\beta$ v.t.c. or satellite lines. It has been shown that $K\beta$ main ($3p \rightarrow 1s$) emission spectra have a pronounced sensitivity to the chemical environment of the $3d$ element that arises from multiplet effects. The $K\beta$ satellite lines arise from transitions from valence orbitals with metal p-orbital momentum character. We discuss these two cases below.

2.4.3.2 $K\beta$ Main Lines in $3d$ Transition Metals

The $K\beta$ main spectral features are separated into the strong $K\beta_{1,3}$ line and a broad $K\beta'$ feature at lower emission energies (Figure 2.15).[80,81] The splitting between $K\beta_{1,3}$ and $K\beta'$ is mainly due to the $3p$–$3d$ exchange

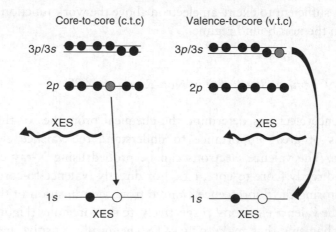

Figure 2.14 One-electron diagrams of different X-ray emission processes in sulfur.

interaction and thus reflects the spin density located on the transition metal ion. The centroid (the first moment) values of $K\beta$ spectra have been used to relate to the oxidation states of the $3d$ transition metal oxides.[82] Another data reduction technique applied to $K\beta$ spectra is called integrated absolute difference (IAD).[83] This works by taking the absolute value of the difference between the sample spectrum and a reference spectrum, summing it to give the IAD value and then relating this to the relative change in difference spin density at the metal site. IAD values have been used to correlate the relative change in difference spin density as a function of applied pressure or of molecular fraction of the doped element, and also in the linear dichroism of tetragonally distorted $LaSrMnO_4$.[10,19,84]

Figure 2.15 shows the $K\beta$ main lines in some polycrystalline Mn compounds alongside their IAD analyses, for which MnO was taken as the reference. The IAD values are plotted against the Mn nominal spin state. MnO, Mn_2O_3 and MnO_2 were used for a linear fit, as they are all compounds with Mn coordinated to six oxygens. This can be viewed as a relative calibration of IAD values which lumps together all possible effects, including changes in oxidation states, bond lengths, bond angles

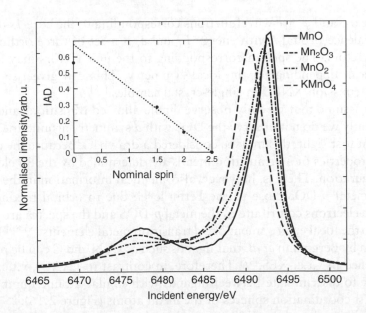

Figure 2.15 $K\beta$ main X-ray emission spectra normalised to the spectral area of manganese oxides and $KMn^{VII}O_4$. Inset: IAD values, with MnO as reference, as a function of formal spin state with linear fit that considers the oxides.

and long-range order effects. The IAD value for $KMnO_4$ suggests a spin state greater than zero, which may be explained by assuming that not all spin density is removed from the Mn ion. It may, however, also indicate a beginning reduction of the Mn site due to the incident X-ray beam, as this is known to occur in $KMnO_4$ on a fast timescale. A more rigorous analysis based on a theoretical model which includes the core hole, multi-electron excitations and the atomic environment would be desirable. Owing to recent important advances in the theory of inner-shell spectroscopy, this may be possible in the near future.

2.4.3.3 Valence-To-Core X-ray Emission Spectroscopy

The v.t.c.-XES lines arise from transitions from occupied orbitals a few eV below the Fermi level. The orbital symmetry that is probed depends on the symmetry of the core hole. In a simple one-electron picture, the transitions can be described for K lines in $3d$ transition metals using:

$$\sum_{j=1}^{3} |\langle \varphi_{1s} | \boldsymbol{\varepsilon}_j \cdot \boldsymbol{r} | \varphi_i \rangle|^2 \tag{2.16}$$

where φ_{1s} and φ_i are wavefunctions corresponding to the core $1s$ orbital and valence orbital i with energy E_i and $\boldsymbol{\varepsilon}_j$ is a set of three orthogonal unit vectors. The spectra corresponding to the unit vectors may not be identical, depending on the local symmetry. This can give rise to an X-ray emission XNLD in single-crystal samples.[18]

We assume that we only observe dipole-allowed transitions and consequently we do not observe the DOS with d symmetry around the metal ion. At first sight, this may be considered a drawback because the chemical properties of $3d$ transition metals are determined by the $3d$ electron configuration. There is, however, also chemical information in the occupied metal p-DOS close to the Fermi level: due to orbital mixing, the ligand electrons contribute to the metal p-DOS and the spectra are sensitive to the local environment of $3d$ transition metal elements.[85,86] It was shown by Bergmann et al. that ligands such as O, N and C can be readily identified in v.t.c.-XES.[87] Therefore, in contrast to EXAFS, v.t.c.-XES is able to discriminate between ligands with similar atomic numbers in the first coordination spheres of the metal atoms (Figure 2.16).[87–89]

Besides ionic solids, $K\beta$ v.t.c. spectra have also been used to study the valence electronic structure of molecular complexes.[91,92] In all cases reported so far in the literature, modelling of the experimental

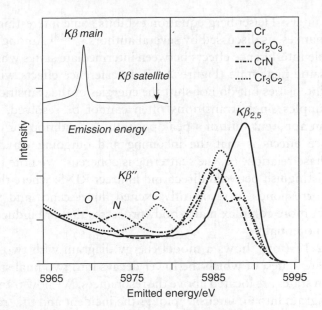

Figure 2.16 V.t.c. (*Kβ* satellite) emission lines in Cr systems with different ligands. The inset shows the full *Kβ* range. Reproduced with permission from Safonov *et al.* (2006) [90]. Copyright (2006) American Chemical Society.

data was possible using ground-state DFT calculations and enjoyed considerable success in identifying the ligand environment. The fact that v.t.c.-XES spectra can be interpreted when considering only one-electron transitions within ground-state DFT greatly helps the experimentalist to extract information on the electronic structure from the data. We show in the following section that in some cases this also applies to resonantly excited v.t.c.-XES.

2.4.4 Resonant Inelastic X-ray Scattering

We have already mentioned that all element-selective emission lines arise from the Kramers–Heisenberg term and that the distinction between resonant and nonresonant excitation is not clearly defined. We can still provide some tentative criteria by which to describe resonant scattering. The DOS changes dramatically near an absorption edge, and the shape of the emission line also changes. Excitations that are well separated in energy often arise from excitations into orbitals that are mainly localised on the absorber atom. Such excitations are often referred to as 'resonances'.

The Kramers–Heisenberg equation exhibits some interesting features near resonances, as discussed by several authors.[8,9,13] Among these are the possible interference effects between intermediate states which decay into the same final state (Figure 2.17). Interference effects will change the peak intensities but do not shift the energies of the transitions. Since in real samples single transitions often cannot be resolved, this may result in an apparent shift of a peak position. A condition for observing interference effects is that the incoming and outgoing waves have a defined phase relation, *i.e.* the scattering is coherent. We note that some authors distinguish between direct and indirect RIXS, where the former roughly corresponds to resonantly excited fluorescence and the latter arises from more complex many-body processes that are induced by the core hole potential.[9]

Figure 2.17 (top) shows a model energy diagram with two resonant intermediate states, of which the lower decays into two final states. The continuum states are located above the resonances. We now translate this energy diagram into an intensity plot *vs* the incident and the transferred (final-state) energy and call this the RIXS plane (Figure 2.18).

Comparisons between different spectra and a quantitative analysis are usually made for line plots. We therefore show some possibilities for extracting line plots from the full RIXS plane. When the incident energy is plotted *vs* the energy transfer, both lifetime broadenings extend perpendicularly to each other in the RIXS plane (*cf.* Equation 2.15). The

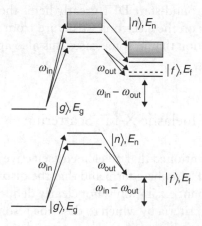

Figure 2.17 Top: total energy scheme for a model system with two sharp resonances at lower energies and one broad continuum excitation. The intermediate-state resonance at lower energy decays into two final states. Bottom: two intermediate states that decay into the same final state may interfere.

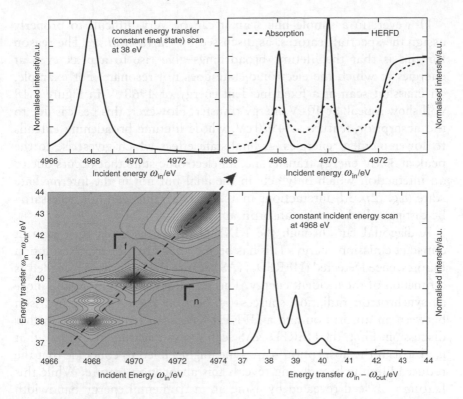

Figure 2.18 Bottom left: RIXS plane based on the model energy scheme in Figure 2.17 (top). The lifetime broadenings are indicated. The dashed line shows a cut through the RIXS plane, which results in a high-energy resolution fluorescence-detected (HERFD) XAS scan (top right, compared to an absorption scan). Cuts through the RIXS plane are shown for constant energy transfer (top left) and constant incident energy (bottom right). Reproduced with permission from Glatzel *et al.* (2009) [93]. Copyright (2009) EDP Sciences.

broader intermediate-state lifetime stretches along the incident energy axis. Scans of the emission energy (constant incident energy scans) at incident energies that select an isolated resonance yield spectra that are broadened by the final-state lifetime broadening, which is usually smaller than that for the intermediate state. **It is thus possible to record spectra that are sharper than the intermediate-state lifetime broadening.** This does not hold for excitations into a broad band or the continuum. Such excitations can be viewed as an infinite number of resonances infinitely close in energy, each broadened by Γ_n. Inspection of the RIXS plane shows that the emission line is then larger than Γ_f; in the case of a continuum intermediate state, the broadening is $\Gamma_n + \Gamma_f$ (convolution of two Lorentz functions).

However, in a simple line scan it can be very difficult to properly assign the spectral features, as discussed by Carra *et al.*[94] The reason for this is that the lifetime broadenings give rise to a peak even at energies at which the electronic state does not resonate. For example, an emission scan at a fixed incident energy of 4968 eV in Figure 2.18 will show a peak at 40 eV energy transfer. However, this peak is due to the absorption feature at 4970 eV, whose lifetime broadening extends to lower incident energies. A final-state effect which gives rise to the peak at 39 eV energy transfer, *i.e.* an electronic state that occurs due to an interaction which only acts in the final but not in the intermediate state (*e.g.* $(2p, 3d)$ interactions in $1s2p$ RIXS), thus can not necessarily be distinguished from an absorption feature.

A diagonal cut through the RIXS plane corresponds to a scan at constant emission energy. This has been coined High Energy Resolution Fluorescence Detected (HERFD) XAS because the intensity is recorded as a function of the incident energy. This technique is used more and more at synchrotron radiation sources (*cf.* Section 2.4.4.4). A comparison between an absorption and a HERFD XAS scan requires some further discussion. First, the HERFD XAS scan shows a feature at 4969 eV that is absent in the absorption scan. One might be led to conclude that the reduced lifetime broadening reveals this absorption feature. While this feature is indeed revealed by using an instrumental energy bandwidth below the lifetime broadening, it is, however, not an absorption feature. Rather, it arises from a final-state effect, as a comparison of the RIXS plane with the energy diagram in Figure 2.17 shows. Loeffen *et al.* showed that a deconvoluted absorption spectrum is different from a constant-emission energy scan.[95]

RIXS is frequently used to analyse K pre-edge features in $3d$ transition metals, and we can illustrate the pitfalls of the analysis using $1s2p$ RIXS in TiO_2 (a $1s$ to $3d$ absorption is followed by a $2p$ to $1s$ decay). Figure 2.19 shows the challenges that arise when analysing RIXS data as line scans. The scattered intensity in the line plots is normalised to unity in the maximum intensity. The peak at the lowest energy transfer remains at a constant energy transfer for low incident energies and moves with incident energy above 4970 eV. A feature at 464 eV energy transfer is observed at 4975 eV and below 4968.5 eV incident energy. At the lowest incident energy of 4965 eV, three strong spectral features are observed.

Interpretation of such line plots is challenging, and an analysis might easily lead to wrong assignments. The full RIXS plane (bottom panel in Figure 2.19) clearly shows the $K\alpha_1$ $(2p_{3/2})$ and $K\alpha_2$ $(2p_{1/2})$ diagonal streaks. The feature at 464 eV energy transfer is easily assigned to a $2p_{3/2}$

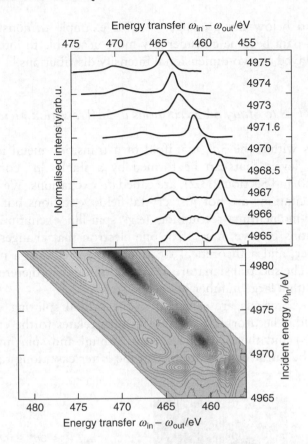

Figure 2.19 $1s2p$ RIXS at the K pre-edge of anatase TiO_2. The top panel shows the scattering intensity as line plots vs the energy transfer (final-state energy). Each scan is normalised to unity in the maximum intensity. The incident energy at which the scan was recorded is given in eV. The bottom panel shows the full RIXS plane, which is composed of RIXS line plots but is not properly normalised to the incident photon flux. The top-panel and bottom-panel energy-transfer axes coincide. Reproduced with permission from Glatzel *et al.* (2013) [97]. Copyright (2013) Elsevier Ltd.

final state at 4975 eV, but to a $2p_{1/2}$ final state below 4968.5 eV incident energy. The contours show clearly how the lifetime broadenings (Γ_n extends vertically is this plot) extend the signature of resonances. The Lorentzian shape of the broadening causes the intensities of different resonances to become similar, far away from the resonance energies. As a result, three features have similar intensities at 4965 eV incident energy even though two resonate at 4968.5 eV and one (at 462 eV energy transfer) at 4972 eV incident energy. Often, spectral features caused by main-edge resonances have strong signatures at incident energies around and below the pre-edge.[96] This example demonstrates well how RIXS data become considerably more accessible to interpretation when displayed as two-dimensional intensity distributions.[97]

2.4.4.1 RIXS to Study dd Excitations in 3d Transition Metals

Excitations within the $3d^n$ manifold of a transition-metal ion, *i.e.* all the energy levels that can be formed by a single $3d^n$ configuration (Figure 2.20 and Section 2.3.2), are called *dd* excitations. We emphasise that *dd* excitations are not just crystal-field excitations but also refer to angular-momentum recoupling (*e.g.* spin-flip excitations). Often, *dd* excitations involve more than one electron that changes its quantum number, and a theoretical treatment often requires a many-body approach. The spectral signature of *dd* excitations can therefore be very complex with a large number of excited states.

The *dd* excitations are sensitive to the orbital splitting within the $3d$ manifold. The *d*-orbital splitting in turn relates to the crystal-field parameter(s) and the electron–electron orbital and spin interactions. These can be intra- or interatomic. In the latter case, long-range order

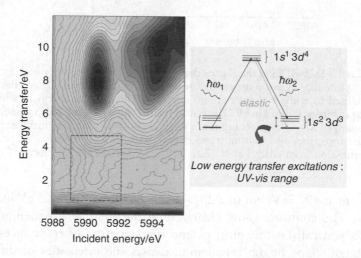

Figure 2.20 Right: scheme of the RIXS process allowing access to *dd* excitations at the *K* pre-edge of a Cr^{3+} ion. Left: valence band RIXS measured at the Cr *K* pre-edge of a powder sample of $MgCr_2O_4$ at room temperature. The incident energy range corresponds to the Cr *K* pre-edge. In the region below 5 eV, *dd* excitations can be observed. The spectral features at higher energy transfer arise from ligand-metal charge transfer excitations.

effects may be observed.[9] The energy range for dd excitations usually lies between 0 and 5 eV but may reach higher energies in some cases. On the high-energy side, charge transfer excitations which may overlap with dd excitations are often found. On the low-energy side are phonon excitations. The energy range of dd excitations is easily accessible using optical absorption techniques. While optical spectroscopy has considerably better energy resolution than X-ray techniques, it suffers from two drawbacks. First, it is not possible to distinguish metal dd from ligand excitations, due to the lack of element selectivity. Second, dd transitions are symmetry forbidden according to the Laporte rule in systems with inversion symmetry, which can be relaxed dynamically by vibronic coupling (and eventually local $p - d$ mixing on the metal site). These vibrational effects are crucial but hardly understood. As a matter of fact, the main quantitative information derived from optical absorption spectra is the energy positions of the bands. Intensities, lineshapes and linewidths are indeed difficult to interpret, and so far no theory has been able to predict them correctly.

Electron energy-loss spectroscopy (EELS) and RIXS, mainly in the soft X-ray region (L and M absorption edges of $3d$ ions), have successfully allowed the study of dd excitations in many $3d$ compounds.[98−101] RIXS is element-selective and allows for a more quantitative analysis, due to the possibility of theoretically modelling the spectra with a good agreement with experiments. In soft X-ray RIXS, dd excitations are intense, as both absorption and emission are electric dipole: in the case of a $3d^3$ ion, for example, the transitions are $2p^6 \, 3d^3 \rightarrow 2p^5 \, 3d^4 \rightarrow 2p^6 \, 3d^3$. The strong $2p$-$3d$ interaction causes significant intra-atomic multiplet effects, which make the spectrum rich but complicated to interpret. An alternative is to use hard X-rays at the K pre-edge, probing electric quadrupole transitions in absorption and emission: $1s^2 \, 3d^3 \rightarrow 1s^1 \, 3d^4 \rightarrow 1s^2 \, 3d^3$. Mixing between the metal $3d$ orbitals with ligand orbitals may induce some dipole character and thus increase the spectral intensity (but make the theoretical analysis more challenging). Finally, dd excitations have also been observed using XRS.[102]

Using hard X-rays to detect dd transitions is experimentally more challenging, since dd excitations are less intense than soft X-ray RIXS. Experimental confirmation was achieved on the benchmark system NiO, when all dd excitations were measured with hard X-rays and in high resolution by Huotari *et al.* in 2008.[101]

Figure 2.20 shows the RIXS plane measured at the Cr K pre-edge in a polycrystalline sample of $MgCr_2O_4$ at room temperature. The incident energy range corresponds to the region of the Cr K pre-edge. The valence band RIXS is plotted for energy transfer between 0 and ~ 12 eV. The elastic peak at 0 eV can be used for an absolute energy calibration of the energy transfer axis. Note that the most intense features at 8–10 eV correspond to ligand–metal charge transfer excitations, which are equivalent to resonantly excited v.t.c. fluorescence lines (*cf.* Figure 2.3). The region in which *dd* excitations can be observed corresponds to energy transfer in the ultraviolet (UV)–visible range, *i.e.* below 5 eV.

Figure 2.21 zooms in on the *dd* excitations. Two crystal-field excitations are visible at ~ 5990.6 eV incident energy and at energy transfers of 2.2 and 3.1 eV. This spectrum is compared to the one calculated in the framework of crystal field multiplet theory for a Cr^{3+} ion in O_h symmetry using the configurations shown in Figure 2.20 (although the Cr site is trigonally distorted, D_{3d}, in $MgCr_2O_4$, a first interpretation based on the O_h symmetry is sufficient). The good agreement obtained with experiment, with respect to the energy splitting and relative intensity, confirms that the two features are spin-allowed *dd* transitions (the elastic peak is not reproduced in the calculation). In O_h symmetry, the first spin-allowed transition corresponds to $^4A_{2g} \to {}^4T_{2g}$ and has a transition energy close to $10Dq$.[26] The second spin-allowed transition $^4A_{2g} \to {}^4T_{1g}$ has a transition energy close to $15\,Dq + \frac{15}{2}B - \frac{1}{2}\sqrt{(9B - 10Dq)^2 + 144B^2}$, where B is the first Racah parameter.

Ghiringhelli *et al.* showed good agreement between RIXS *dd* excitations and the energy levels in a Tanabe-Sugano diagram.[98] It is important to note that a multi-electron treatment of the scattering process is crucial to accounting for all observed transitions.

Using the observed peak positions, an estimation of the value of the crystal field splitting and Racah parameter B can therefore easily be achieved. This is particularly interesting for materials which cannot be investigated using optical absorption spectroscopy (opaque materials or empty d shells). This makes RIXS a very appealing technique for such purposes.

We turn back to Figure 2.20 to point out another interesting aspect of RIXS: the *dd* features appear at two resonant energies, but the peak at 3.1 eV disappears when the incident energy is changed from 5990.6 to 5994.2 eV. The transitions probed in the excitation processes are very different and thus different final states are reached. Based on the interpretation made for a Cr-dilute sample of $MgAl_2O_4 : Cr^{3+}$[15], the Cr $t_{2g}^{\downarrow}$ empty orbitals are probed when the incident energy is set to 5990.6 eV. The peak visible at around 5994.5 eV is likely 'non-local',

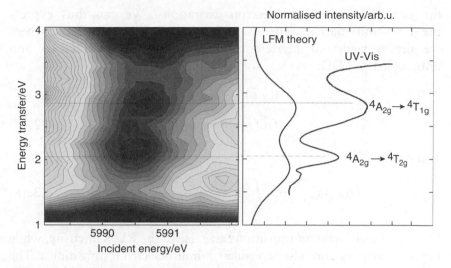

Figure 2.21 Left: close-up of the *dd* excitations of the experimental RIXS plane in Figure 2.22. Right: line scans showing crystal field multiplet theory calculations performed in O_h symmetry and UV–visible measurements. The symmetries of the transitions are indicated.

i.e. a transition due to intersite mixing involving the absorbing Cr atom, the bridging oxygen and its Cr nearest neighbours. Indeed, it is not visible in the dilute $MgAl_2O_4 : Cr^{3+}$ sample and grows together with the Cr content in the oxide. Such transitions have been observed in other transition-metal oxides.[103] Thus, the nature of the intermediate state in the RIXS process determines what *dd* excitations can be probed.

The localised character of the intermediate state (such as $1s^1 3d^4$) is not a requirement for the experimental observation of *dd* excitations. Indeed, *dd* excitations can also be measured by exciting the K edge ($1s \rightarrow 4p$) instead of the K pre-edge ($1s \rightarrow 3d$). This has recently been done by Hancock *et al.* in CuB_2O_4[104] and by van Veenendaal *et al.* in NiO,[100] and the authors also show that *dd* excitations at the K edge are caused by the Coulomb interaction between the excited $4p$ electron and the $3d$ valence electrons.

2.4.4.2 Simplified Calculations of RIXS

It is often desirable to use approximations to the Kramer–Heisenberg equation in order to employ standard theoretical codes to interpret the spectra. The most radical assumption neglects the interaction with the core hole and multi-electron excitations. We can thus express the transition matrix elements starting from Equation 2.15, where we already neglected interference effects, using the occupied (ρ) and unoccupied (ρ') DOS:

$$|\langle n|\widehat{O}|g\rangle|^2 \rightarrow \rho'(\varepsilon_u)$$

$$|\langle f|\widehat{O}'^{\dagger}|n\rangle|^2 \rightarrow \rho(\varepsilon_o) \tag{2.17}$$

and obtain:[105]

$$F(\omega_{in}, \omega_{out}) \propto \int_{\varepsilon} d\varepsilon \; \frac{\rho(\varepsilon) \; \rho'(\varepsilon + \omega_{in} - \omega_{out})}{(\varepsilon - \omega_{out})^2 + \frac{\Gamma_n^2}{4}} \tag{2.18}$$

Here, single-electron dd transitions give rise to the RIXS intensity, while the previous section also considered multi-electron transitions. The advantage of this simplified approach to RIXS is that DFT calculations can be used to understand the spectra that are often easily performed. Use of such a drastic approximation has to be made with care. This approach can, for example, not reproduce most dd excitations in $3d$ transition metals. It does, however, reproduce ligand–metal charge-transfer excitations, which are in fact resonantly excited v.t.c. X-ray emission lines.

The dipole selection rules can be considered by projecting the DOS onto a metal ion orbital angular momentum (LPDOS), as discussed in Section 2.3. It is interesting to explore to what extent the simple Equation 2.18 can model experimental data. RIXS data at the $L_{2,3}$ edges of $5d$ transition metals have been successfully simulated using DFT and Equation 2.18 within an experimental energy bandwidth down to 1.5 eV.[106,107]

We show in Figure 2.22 the valence band RIXS data of $K_2[PtCl_4]$ and $K_2[PtCl_6]$ with simulations using DFT. The agreement is good, except for a double feature at 7 eV energy transfer for $K_2[PtCl_4]$; this is predicted by theory, but the experiment only shows a single peak. We note that only the calculated incident energy has been shifted to coincide with experiment, whereas the energy transfer axis has not been modified. The RIXS planes show nicely the intermediate state (L_3) lifetime broadening that extends horizontally in the plane (*cf.* Figure 2.18). The spectral features along the energy transfer axis are broadened mainly by the experimental bandwidth in these ionic complexes.

The approximations leading to Equation 2.18 break down when the core hole potential significantly modifies the probed electronic structure, the interaction of the core hole with the valence electrons is large

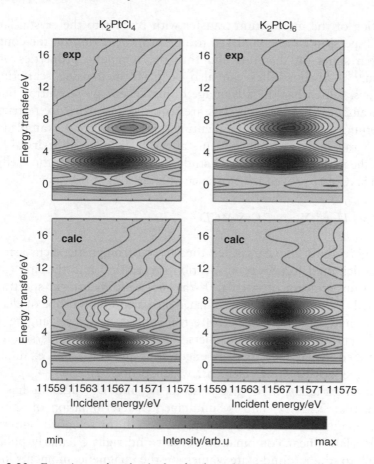

Figure 2.22 Experimental and calculated valence band RIXS planes at the L_3 edge of Pt in $K_2[PtCl_4]$ and $K_2[PtCl_6]$. The elastic peak has been added in the calculated spectra to facilitate comparison with experiment. Reproduced with permission from Glatzel *et al.* (2013) [97]. Copyright (2013) Elsevier Ltd.

compared to the spectral line broadening, the photoexcited electron strongly interacts with the valence electrons (and the core hole), multi-electron excitations become significant or interference effects modify the spectral intensities. In such cases, a more sophisticated theoretical model is necessary. Calculations that can fully treat the core hole, electron–electron interactions and atomic environment are currently not possible, and the theoretically orientated scientist must decide within what approximation the calculations should be performed.

RIXS is often used to study collective excitations in systems with long-range order. The energy of such excitations may depend on the direction of the momentum transfer with respect to the crystal lattice. This dispersion can be analysed to obtain information on the coupling between spins on the ions and thus the magnetic properties of the system.[108] Such studies can also be performed using scattering of neutrons, which has the advantage of higher energy resolution but the disadvantages that large crystals are required and there are restrictions concerning the energy and momentum range.[9] A recent example in the hard X-ray range is measurements at the Ir L_3 edge in Sr_2IrO_4[109] to study the magnon dispersion. A study using soft X-rays at the L edge of Cu in Sr_2CuO_3 proved the existence of long-sought orbitons.[110]

2.4.4.3   Hard X-ray RIXS-MCD

It is more than 25 years since the effect of magnetic dichroism was first anticipated for X-ray absorption spectra[111] and the first experimental observation of circular X-ray magnetic dichroism spectra was reported.[112] Since then, it has turned into a common probe of element-specific magnetisation in ferro(ferri)magnetic systems, ranging from multilayers to molecular magnets. XMCD is well understood and interpreted when measured at edges split by SOC. For example, in the $L_{2,3}$ edges of a $3d^n$ element, which probes $2p^6 3d^n \rightarrow 2p^5 3d^{n+1}$ transitions, the $2p$ SOC gives rise to L_3 and L_2 edges. It has been shown that XMCD enables simultaneous determination of spin and orbital magnetic moments upon application of the so-called 'sum rules', which relate linear combinations of left and right circularly polarised spectra to the ground-state values of the magnetic moments of the absorber.[113] These rules are very useful for the extraction of quantitative information from experimental spectra without the need of numerical calculations.

When applied to $3d$ transition metals and lanthanides, the main drawback of the technique is that the $L_{2,3}$ and $M_{4,5}$ absorption edges, respectively, lie in the soft X-ray range. Most soft X-ray XMCD measurements are performed using total electron yield, because significant self-absorption effects are observed when using fluorescence yield (*cf.* Section 2.5.3.3.1). Thus, L-edge XMCD is mainly sensitive to the sample surface and is not compatible with demanding sample environments such as high-pressure, liquid and gas cells, which limits the range of materials that can be investigated and excludes *de facto* buried magnetic phases, multilayered samples and ferrofluids. For these systems, the penetrating properties of hard X-rays are necessary, but at the K edge the XMCD signal is weak (a few tenths of a per cent of the edge jump) and attributed to the p-projected orbital magnetisation density of unoccupied states, which is difficult to interpret quantitatively. Also, due to the absence of spin–orbit split edges, the separation of spin and orbit contributions is not permitted.

The element-specific studies of bulk magnetism under extreme conditions have been largely limited to very weak K-edge magnetic dichroism and to $K\beta$ emission spectroscopy.[114,115] The latter is sensitive only to the spin and orbital kinetic moments (S and L) and not to the magnetic moments (m_s and m_l). As such, it does not provide quantitative information on the ordering of interatomic magnetic interactions. We discuss here 1s2p RIXS combined with MCD, which not only overcomes some of the experimental difficulties for studies of magnetic properties but also demonstrates the origin of the MCD effect. We build on our discussion of the K absorption pre-edges and the MCD effect in previous sections.

The 1s2p RIXS probes the evolution of $K\alpha$ emission ($2p \rightarrow 1s$) upon excitation at the K edge ($1s \rightarrow p$). In the K-edge absorption spectra of most transition-metal compounds, pre-edge features are visible that arise from weak quadrupole $1s \rightarrow 3d$ excitations, possibly combined with some additional intensity due to dipole $1s \rightarrow 4p$ transitions. Therefore, the K pre-edge of a $3d$ transition-metal is predominantly sensitive to the $3d$ density of unoccupied states, and 1s2p RIXS probes the same final state as the $L_{2,3}$ absorption edges, $2p^5 3d^{n+1}$, but using a hard X-ray *photon in–photon out* probe.[116] Combined with the idea of XMCD, *i.e.* using circularly polarised X-rays and an external magnetic field, it becomes a promising technique by which to study the $3d$ magnetic structure of transition metals.

2.4.4.3.1 1s2p RIXS-MCD at the Fe K Pre-Edge in Bulk Magnetite

Figure 2.23 shows the experimental RIXS-MCD plane on magnetite, $[Fe^{III}]_{tetra}[Fe^{II}Fe^{III}]_{octa}O_4$, at the Fe K pre-edge, plotted as the difference between the spectra measured for left and right circular polarised light. Comparison with the RIXS plane averaged over the two photon helicities shows that only the resonant features give rise to the MCD. The features caused by nonresonant fluorescence (visible as diagonal structures in the RIXS plane) do not show any detectable MCD effect. The spectra show a characteristic dispersion along incident photon energy, due to 1s hole lifetime broadening, and along the energy transfer, due to final-state effects and $2p$ hole lifetime broadening. The experimental RIXS-MCD plane reveals two groups of final states, which correspond, respectively, to $K\alpha_1$ and $K\alpha_2$ emission lines. These are each composed mainly of two resonances with different spin polarisations, and the signs of their

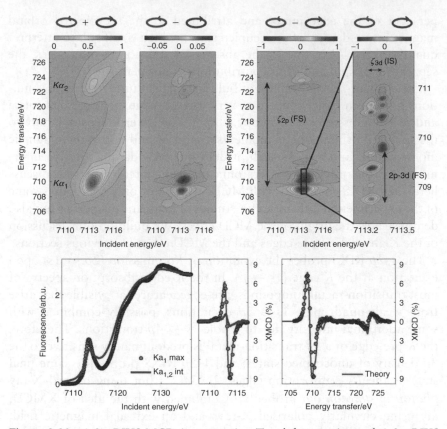

Figure 2.23 $1s2p$ RIXS-MCD in magnetite. Top left: experimental $1s2p$ RIXS plane measured at the Fe K pre-edge in magnetite and averaged over the two circular polarisations, and an experimental RIXS-MCD plane of magnetite, plotted as the difference between the RIXS planes measured for the opposite helicities of circularly polarised light. Top right: theoretical RIXS-MCD plane calculated for tetrahedral Fe^{III} using LFMT. The magnification shows the theoretical plane plotted for the $2p_{3/2}{}^5 3d^6$ final state, with a small broadening. Bottom: line scans extracted from the RIXS-MCD plane: comparison of the Fe K-edge spectra (left) and their magnetic circular dichroism (centre and right), acquired using the maximum of $K\alpha_1$ fluorescence line (dots) and $K\alpha_{1,2}$ yield, which is comparable to TFY detection. The experimental $K\alpha_1$-detected MCD spectrum shows the enhancement of the dichroism compared to $K\alpha_{1,2}$ detection (centre). Reproduced with permission from Sikora *et al.* (2010) [121]. Copyright (2010) American Physical Society.

MCDs are opposite. The weak feature visible at 7112 eV incident energy and 707 eV energy transfer is ascribed to octahedral Fe^{II}.[47]

The experimental data are compared with the theoretical RIXS-MCD calculated in the ligand field multiplet approach, where only the contribution of tetrahedral Fe^{III} is considered. The calculation involves an electric quadrupole excitation from $1s^2 2p^6 3d^5$ ground state to $1s^1 2p^6 3d^6$ intermediate state, followed by an electric dipole emission to the $1s^2 2p^5 3d^6$ final state. The two RIXS-MCD planes are in good agreement in terms of energy splittings and relative transition strengths of the left and right polarised channels. This allows for a profound understanding of the RIXS-MCD effect on analysing the different multiplet interactions in the calculations.

The theoretical RIXS-MCD plane is plotted at the $K\alpha_1$ resonance, where the core hole lifetime broadenings are set to very small (unrealistic) values (Figure 2.23, top right) in order to reveal the underlying interactions. The combination of the exchange field and the $3d$ SOC in the intermediate state implies that a different set of $1s^1 3d^6$ intermediate states is reached from left- and right-polarised X-rays (*cf.* Figure 2.5). The splitting of the MCD features in horizontal and vertical directions is mainly caused by the $3d$ SOC in the intermediate state and the $2p-3d$ Coulomb repulsions in the final state, respectively. The latter effect, coupled with crystal field, tends to split various final states with different L and S kinetic momentums, which are reached with different transition strengths from the intermediate states that are achieved by left and right excitation. The splitting of the MCD features in the intermediate state is much smaller because of the weak $3d$ spin–orbit interaction. The splitting induced by the $2p$ SOC and the Coulomb repulsions involving the $2p$ and $3d$ open shells thus acts as an effective enhancer for the XMCD effect.

We have thus identified the interaction mechanisms that influence the strength of the MCD signal. We have to set the MCD line splittings in the context of the core hole lifetime broadenings. The $2p$ core hole is a factor 2–3 longer lived than the $1s$ hole and thus provides sharper spectral features. This further enhances the MCD signal. Figure 2.23 bottom left compares the Fe K-edge XAS spectra of magnetite, measured using the maximum of $K\alpha_1$ emission and the total $K\alpha_{1,2}$ fluorescence, which is comparable to a conventional absorption spectrum. The $K\alpha_1$-detected one is relatively sharper, due to smaller lifetime broadening of the combined two-photon RIXS process.[96] Comparison of the XMCD spectra reveals that the shape of the pre-edge XMCD is similar for both detection types, in agreement with band structure calculations[117] and previous experiments.[118] However, the intensity of $K\alpha_1$-detected MCD shows a peak-to-peak amplitude as large as 16% of the pre-edge maximum, comparable to the Fe L_3-edge MCD in magnetite.[119,120]

2.4.4.3.2   Applications of Hard X-ray RIXS-MCD

The increase in intensity observed in $1s2p$ RIXS-MCD with respect to K-edge XMCD is a significant advantage as it allows data to be measured with better statistics. However, a strong enhancement is only expected for systems which show well-defined pre-edge structures, such as in iono-covalent compounds: oxides, molecular complexes *etc*. Indeed, when $3d$ states are strongly mixed with p bands and heavily delocalised, such as in metals, it has been observed experimentally that the gain in intensity is lost, *e.g.* in metallic Fe the intensity of RIXS-MCD has been measured as comparable to that of K-edge XMCD.[122]

Ferro- and ferrimagnetic samples to which soft X-ray XMCD would be blind (at least partially), such as buried layers, can readily be investigated using RIXS-MCD. A RIXS-MCD signal can be detected from a 40 nm thick layer of magnetite buried under 60 nm of Pt and Au. Another important observation is a significant reduction of the amplitude of the MCD signal in the thin layer to ~70% of the bulk value, which is in good agreement with the reduction of saturation magnetisation reported in literature.[123] Thus, the RIXS-MCD can be considered a quantitative probe of net magnetisation in thin-layer samples. The technique can also be applied to more complex systems such as bimetallic core-shell nanoparticles, which may additionally show the presence of inter-diffused layers. For such systems, the structural information provided by TEM-EELS may not be representative of the assembly of particles. RIXS-MCD can be used as a complementary magnetospectroscopy, which provides an average quantitative picture of the structure in terms of composition and thickness distribution.

An interesting aspect of RIXS-MCD, as compared to K-edge XMCD, is the potential to select a region of the plane in which the MCD effect is maximised for certain features of interest, *e.g.* in the case of magnetite, the peak at 707 eV energy transfer and 7112 eV incident energy arises from octahedral Fe^{2+}, while the double feature at 7114 eV incident energy and 712 eV energy transfer is dominated by the contribution from tetrahedral Fe^{3+}. The RIXS plane can therefore be used to perform site-selective studies. By monitoring the changes in the MCD as a function of space, pressure, temperature or time, RIXS-MCD can be adopted for element- and site-selective magnetometry and magnetic microscopy with hard X-rays. As an illustration, we show in Figure 2.24 the hysteresis loop measured on a thin buried layer of magnetite using a RIXS-MCD feature selective of tetrahedral Fe^{3+}, which is compared to the vibrating sample magnetometer (VSM) curve.[122]

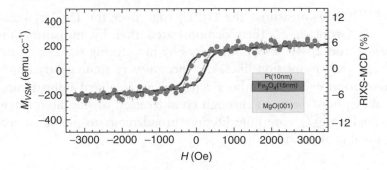

Figure 2.24 Element- and site-selective $Fe^{3+}(T_d)$ hysteresis loop (circles) of a thin buried layer of magnetite, compared with the vibrating sample magnetometry (solid line) results. Reproduced with permission from Sikora *et al.* (2012) [122]. Copyright (2012) American Institute of Physics.

An important question is whether sum rules similar to XAS-MCD can be derived for RIXS-MCD. Borgatti *et al.*[124] have derived sum rules for RIXS-MCD under the fast collision approximation. One assumes that the second-order process is so fast that excitation and decay are simultaneous steps, which leads to neglect of the energy dispersion of the intermediate states. The fast collision approximation provides mathematical simplifications, but the approximation may be too strong. Therefore, the validity of sum rules must still be verified by comparison with the expected values obtained from advanced ligand field multiplet calculations. Future studies of the dependence of RIXS-MCD on the direction of the wavevector, the polarisation vector, the magnetic field or the transfer momentum, possibly combined with a polarisation analysis of the scattered X-rays, will enable the full potential of the technique to be exploited.

2.4.4.4 High-Energy Resolution Fluorescence-Detected XAS

We return to the line plot that results from the diagonal cut through the RIXS plane (Figure 2.18). The spectra can be related to XAS with reduced spectral broadening. The nomenclature for HERFD XAS should contain the intermediate- and final-state core hole, *e.g.* $1s2p_{3/2}$ for $K\alpha_1$-detected HERFD XAS. The technique of reducing the apparent core hole lifetime broadening seen in X-ray absorption spectra by using an emission spectrometer was first suggested by Hämäläinen *et al.* in

1991,[125] who measured the Dy L_3 edge over the Dy $L\alpha_1$ emission line. de Groot *et al.* then demonstrated that by measuring the Pt $2s, 2p_{1/2}$ and $2p_{3/2}$ absorption edges by monitoring the $2s5p$, $2p_{1/2}4d$ and $2p_{3/2}3d_{5/2}$ emission lines, the spectra were significantly improved in resolution.[126] The authors reported a calculated improvement of the absorption spectrum through consideration of the intermediate Γ_i and final-state Γ_f core hole lifetime broadenings to give an apparent measured peak broadening, Γ_{App}:

$$\Gamma_{App} \approx \frac{1}{\sqrt{\left(\frac{1}{\Gamma_n}\right)^2 + \left(\frac{1}{\Gamma_f}\right)^2}} \qquad (2.19)$$

This technique of HERFD XAS has been employed to observe better electronic structure changes during chemical reactions,[127−129] to better quantify the relative proportion of different oxidation states of V in mixed-valence natural samples[130] and to reveal subtle angular-dependent core hole effects in the Co K pre-edge.[51] It is important to note that terminology such as 'lifetime suppressed', which is sometimes found in the literature, does not mean that the lifetime broadenings are reduced or disappear in resonant spectroscopies. The lifetime broadenings are always there, as the RIXS plane shows. What the experimentalist can do is elegantly move through the RIXS plane, trying to find the smallest line broadening. In many cases, this is achieved using a diagonal cut through the RIXS plane (*cf.* Figure 2.18).

The differences between standard and HERFD XAS were first discussed by Carra *et al.*,[94] who pointed out problems in using this technique when attempting to record a true absorption spectrum. Their ideas have since been illustrated by several other authors.[93,95] Kotani *et al.* discussed the correspondence between HERFD and XAS for CeO_2 and $CeFe_2$.[20,61] They found that $2p_{3/2}3d_{5/2}$ HERFD in CeO_2 provides a good approximation to an absorption scan. It appears that two necessary but not always sufficient prerequisites can be identified:

(i) The magnitude of core hole potentials in the intermediate and final state of the RIXS process should be similar. This is the case when both core holes are deep, *e.g.* $2p$ and $3d$ in rare earths or $5d$ transition metals. This has been discussed in references [61, 93, 131], for example.

(ii) The electron–electron interactions (expressed *e.g.* by the Slater integrals) between the core hole and the photoexcited electron should be smaller than the respective core hole lifetime broadenings.

A consequence of (ii) is that a HERFD approximation to XAS may fail only for part of the spectrum (usually the low-energy range, *i.e.* the pre-edge). For example, the $1s-3d$ $(2p-4f)$ excitations in $3d$ transition metals (rare earths) may show strong final-state interactions in $1s2p$ or $1s3p$ $(2p3d)$ RIXS.[16] Thus, pre-edges measured using HERFD in many cases cannot be reproduced just by XAS calculations: they require consideration of the full RIXS process.[94] The experimentalist may still want to apply this technique and analyse the data based on a fingerprint approach where it is important to ensure that the chemical sensitivity of the emission line is considered (*e.g.* energy shifts). The electron–electron interactions between the core hole and the photoexcited electron rapidly decrease in magnitude when moving well above the Fermi level, and the spectral features in HERFD can be understood by considering the absorption only. Deviations may also occur for so-called shake-up spectral features. Their intensity may be reduced in HERFD and comparison between HERFD and XAS can help to identify multi-electron excitations.[61,132,133]

Visual inspection of a RIXS plane provides an already reliable indication of whether HERFD is a good approximation to XAS: if the maxima of all spectral features appear along a diagonal streak in a RIXS plane where incident energy is plotted *vs* the energy transfer, intermediate- and final-state interactions are very similar and HERFD and XAS will show the same spectral features. The intensities, however, may differ.

It has been found that HERFD XAS on $5d$ transition elements can be interpreted to good accuracy as XAS spectra measured below the $2p_{3/2}$ core hole lifetime broadening, due to the delocalised character of the $5d$ electrons.[134] Figure 2.25 shows spectra of Pt systems, which illustrate the dramatic spectral sharpening.

2.5 EXPERIMENTAL CONSIDERATIONS

2.5.1 Modern Sources of X-rays

X-ray spectroscopy today is mainly performed at dedicated (third-generation) synchrotron radiation sources.[135] These make use of the fact that an accelerated charge emits photons and thus loses energy.

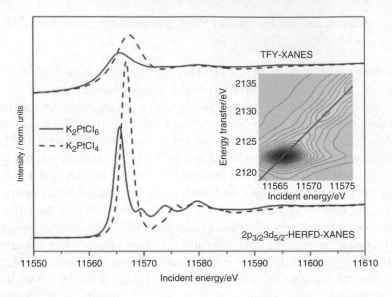

Figure 2.25 Comparison between total fluorescence yield and HERFD XAS recorded simultaneously; $K_2[Pt^{II}Cl_4]$ (dashed) and $K_2[Pt^{IV}Cl_6]$. The inset shows the full $2p_{3/2}3d_{5/2}$ RIXS plane of $K_2[Pt^{IV}Cl_6]$, with a diagonal line indicating the HERFD cut. Reproduced with permission from Glatzel *et al.* (2013) [97]. Copyright (2013) Elsevier Ltd.

A synchrotron is a device in which the magnetic field that keeps the electrons on a circle with fixed radius is adjusted (synchronised) to the kinetic energy of the electrons, which is increased with every turn. A storage ring keeps electrons (or positrons) orbiting at a fixed energy. The lattice of a storage ring consists of devices that create magnetic fields and a so-called resonance frequency cavity, which injects energy into the orbiting electrons. The electrons lose energy with every turn because of the accelerated motion in the magnetic fields. The energy is reinjected when the electrons pass through the cavity. The magnetic devices are used to shape the electron beam, keep the electrons in the orbit, and to create high-brilliance X-rays.

A bending magnet diverts the electron beam to keep it in the orbit, and X-rays are emitted into a wedge that is vertically rather collimated, but horizontally very broad. Insertion devices are installed on straight sections of the storage ring; they create alternating magnetic fields that force the electrons on an undulating trajectory. As a result, the emitted X-rays interfere constructively only at certain energies, which are defined by the period of the magnetic field, the strength of the field and the energy of the electrons. The emitted X-rays are collimated in both directions, resulting in a highly brilliant beam.

Transport of the X-rays from the source (the bending magnet, wiggler or undulator) to the sample is achieved by X-ray optics that create *e.g.* monochromatic and focused X-rays. This is an entire field of research and some excellent textbooks exist.[136,137] A typical XAS beamline uses a pair of Si single crystals cut along the <111>, <220> or <311> crystallographic direction as a monochromator, which is followed by focusing mirrors. If XES is also to be performed, a secondary monochromator is necessary to analyse the scattered X-rays.[138]

The number of photons on the sample depends on whether a bending magnet or an insertion device generates the X-rays. Focusing optics increase the photon density on the samples. Typical numbers are 10^{11}–10^{14} photons s^{-1}, in spot sizes ranging from $100\,nm^2$ to $1\,mm^2$.

2.5.2 Ultrafast X-ray Spectroscopy

Synchrotron radiation has a pulsed structure, due to the fact that the cavity runs at a certain resonance frequency. This results in 'buckets' along the electron beam trajectory that can be filled with electrons. In order to use the pulsed structure for time-resolved experiments, only a few buckets are filled, in order to leave time for the detection electronics to read out the recorded values. As an example, a timing mode at the European Synchrotron Radiation Facility (ESRF, Grenoble, France) has 16 filled bunches around the ring with 176 ns between each, corresponding to 5.68 MHz. The length of one X-ray pulse is about 50–100 ps. The time resolution can be improved significantly in fourth-generation sources that no longer use a storage ring, but instead have a linear accelerator followed by a very long straight section with undulators. These 'free electron lasers' use stimulated emission of X-rays by the electron beam to obtain X-ray pulses that are fully transversally coherent, diffraction-limited collimated and have pulse length down to 10 fs.

Time-resolved X-ray absorption spectroscopy (XAS) and XES have been performed at synchrotron radiation sources, mainly using an optical pump and X-ray probe scheme (Figure 2.26).[139,140] An X-ray pulse follows an optical pulse with a delay Δt. The X-ray pulse probes the state of the system that was created by the laser pulse. The lifetime of the optically excited state determines the required time resolution, and the delay must be on the order of the lifetime. The optical pulse is produced by a laser with frequencies up to several MHz and pulse length between tens of fs and a few ps. The time resolution is given by the length of the X-ray pulse. In an effort to achieve higher time resolutions, efforts

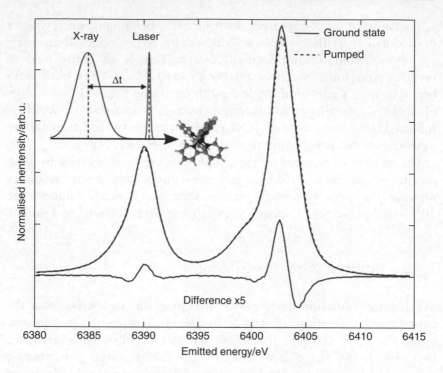

Figure 2.26 Pump and probe XES on $[Fe(bpy)_3]^{2+}$ (bpy $= 2,2'$-bipyridine). The principle is shown in the inset: a laser pulse hits the sample a time Δt before the X-ray probe pulse that results in the emission of the Fe $K\alpha$ fluorescence. $[Fe(bpy)_3]^{2+}$ is low spin ($S = 0$) in its ground state. Laser excitation creates an excited high-spin state ($S = 2$) that lives for 665 ps. The Fe $K\alpha$ lines show the difference in spin state between ground and excited state.[142]

have been made at synchrotron radiation sources to reduce the lengths of the X-ray pulses, *e.g.* using laser slicing. However, such schemes always come with a punishing loss in X-ray intensity. Free electron lasers provide the way out of this dilemma, and XES experiments have been performed at the first hard X-ray free electron laser, the Linac Coherent Light Source (LCLS) at the Stanford Linear Accelerator Center (SLAC).[141]

2.5.3   Measuring XAS/XES

The choice of experimental technique is guided by the information one is looking for (macroscopic magnetic properties, crystal structure, local atomic structure, element-selective electronic structure *etc.*).

Experimental constraints often reduce the number of possible techniques. Experiments that require a certain sample environment are most easily measured using a technique that is not limited by the cell or atmosphere around the sample. Hard X-rays penetrate sufficiently through air and windows and are thus ideal for studies under extreme conditions (*e.g.* high-pressure) and *in situ* experiments. The probing depth can be modified by detection of either outgoing electrons (surface-sensitive) or photons (bulk-sensitive). Also, the incident angle can be modified to adjust the surface or bulk sensitivity.

The choice to use secondary process detection is often motivated by experimental considerations. Samples in which the element of interest (the 'absorber') is very dilute are difficult to measure in transmission, and the detection sensitivity can be increased by a secondary process detection. We have discussed the fact that secondary process detection may not give a signal that is proportional to the linear absorption coefficient.[143,144] This can lead to 'wrong' spectra in the sense that one does not obtain the spectra, *i.e.* the absorption coefficient, that one was aiming for. There is of course nothing fundamentally wrong here, but the theoretical analysis is more complex because the decay channel has to be considered in addition to the absorption coefficient.

The experimentalist must decide on the most appropriate technique and assess its feasibility. In order to do so, the expected spectral quality has to be estimated and the required statistics determined. It is possible that the recorded spectra are distorted in the sense that the relative spectral intensities do not reflect the true variation of cross-section. A practical guide to XAS is presented in a book by Bunker.[12] We address some aspects of particular relevance to XAS combined with XES in the following.

2.5.3.1 Estimate of X-ray Emission Count Rates

The count rate in the emission detection is estimated using Equation 2.20,[12,145,146] with incident energy ω_{in}, fluorescence energy ω_{out} and captured solid angle Ω.

$$y = \frac{\Omega}{4\pi}\kappa\mu^e_{abs}(\omega_{in})\frac{1 - e^{-\frac{d}{\sin\theta}(\mu_{tot}(\omega_{in})+g\mu_{tot}(\omega_{out}))}}{\mu_{tot}(\omega_{in}) + g\mu_{tot}(\omega_{out})} \quad \text{with} \quad g = \frac{\sin\theta}{\sin\phi}$$

(2.20)

The geometry is shown in Figure 2.27. For a full treatment of the solid angle Ω, one should consider variation of the angle ϕ across the detector surface, which will influence the self-absorption (*cf.* Section 2.5.3.3.1).

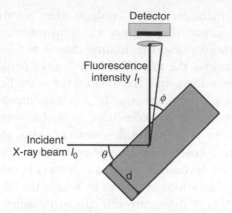

Figure 2.27   Geometry in a photon-in/photon-out experiment.

The absorption coefficient of the sample is the sum of the coefficient of the element of interest ('absorber') and all other elements: $\mu_{tot} = \mu_{abs} + \mu_{else}$. Only the absorber subshell (*i.e.* the edge of interest) with fluorescence yield κ that gives rise to the recorded emission line is considered in μ_{abs}^e (ω_{in}). One can use the jumping ratio J to scale $\mu_{abs}(\omega_{in})$ in order to consider only the photoelectric cross-section of the shell of interest (and not shells with lower binding energy that do not contribute to the fluorescence): $\mu_{abs}^e = \mu_{abs}\tau$, with $\tau = (J-1)/J$.[147]

The total number of incident photons I_0 (flux) is the starting point from which to obtain an estimate of the absolute counts on the emission detector. One obtains a number $I_f = I_0 \cdot y \cdot K$, where the factor K considers some mechanisms that reduce the intensity. For example, the fluorescence yield is given for a subshell and, depending on the energy bandwidth in the emission detection, not all emission lines for the given subshell will necessarily be recorded, *e.g.* when $\Delta\omega_{out} = 1\,eV$, only part of the $K\alpha$ lines is covered. The detector efficiency or analyser crystal reflectivity further reduces the count rate, and possible absorption in air or windows should be included. The captured solid angle is determined using the radius for each detector element, the distance from the sample and the number of elements.

2.5.3.2   *Spectral Differences*

Equation 2.20 provides the starting point from which to assess the feasibility of an experiment. Pump and probe experiments are challenging with respect to the spectral quality and experimental accuracy because usually only a fraction of the atoms is excited by the laser pulse. They measure a difference spectrum, much like in dichroism experiments. Some simple considerations can help the experimentalist in estimating the required spectral quality and thus assessing the feasibility of an experiment. We emphasise that the following considerations hold for all experiments in which spectral differences need to be observed, *e.g.* *in situ* studies of chemical reactions. We assume single-photon counting and the error is determined by Poisson statistics. The counts C are given for (e)xcited, (p)umped and (g)round states. We first require a certain signal-to-noise ratio (*SN*) ratio in the measured difference spectrum:

$$\frac{C_p - C_g}{\sqrt{C_p + C_g}} \cong \frac{C_p - C_g}{\sqrt{2 \times C_g}} > SN \tag{2.21}$$

The relative spectral difference, *DF*, between the ground and excited state is:

$$\frac{C_e - C_g}{C_g} > DF \tag{2.22}$$

The measured spectrum, C_p, may only be a fraction, f, of the excited state spectrum, C_e:

$$C_p = f \times C_e + (1 - f) \times C_g \tag{2.23}$$

All this combines to give an equation for the required counts:

$$C_g = \frac{2}{(f \times DF)^2} \times SN^2 \tag{2.24}$$

If simply observing the spectral difference is enough to answer the scientific question, $SN = 2$ may be sufficient. If the difference spectrum has still to be analysed in detail, values of 10 or 100 may be necessary. In practice, the counts in the experimental spectrum should be at least as high as that estimated in Equation 2.24, in order to allow observation of the spectral difference. For dichroism experiments, the excited-state fraction may be 1. We note that we have neglected any background, statistical error in the incoming flux measurement and systematic errors (*e.g.* drifts of the energy).

2.5.3.3 Spectral Distortion

Measurements of the absorption coefficient in transmission mode may show spectral distortions due to sample inhomogeneity (thickness effect) or higher harmonic contamination in the incident beam.[146] Furthermore, X-ray detection electronics may show a non-linear response to the number of incoming photons. Absorption of the emitted X-rays in the sample (emitted beam self-absorption), absorption in the beam path between sample and detector (air, windows) and detector efficiency may cause a nonuniform detection as a function of emitted energy. This is similar to a partial detection scheme that uses an energy or wavelength dispersive detector which allows the selection of a defined energy window for the emitted X-rays, but is more difficult to evaluate. The most important reason for spectral distortions in photon-in/photon-out experiments is incident-beam self-absorption (IBSA), which we will discuss in Section 2.5.3.3.1.

The common problem for all spectral distortions caused by the experimental conditions is that they are not easily identified in unknown samples unless an experienced experimentalist has a 'hunch' that something is wrong. One is therefore easily led into analysing distorted data that give wrong results and never realising it, although some distortions can be corrected for after data acquisition if exact knowledge of the sample composition and/or detection electronics is available.

2.5.3.3.1 Evaluation of Incident Beam Self-Absorption

Fluorescence-detected absorption spectroscopy is a secondary-process detection technique that aims at recording the linear absorption coefficient. The recorded signal y (cf. Equation 2.20) and $\mu_{abs}(\omega_{in})$, however, do not have a linear relation. The absorption of the element of interest, $\mu_{abs}(\omega_{in})$, also appears in $\mu_{tot}(\omega_{in})$, i.e. in the denominator and in the exponential function of Equation 2.20, because the intensity of the emitted fluorescence also depends on how deep the incoming X-rays penetrate into the sample, i.e. how many atoms the incident beam interacts with. IBSA thus compresses the recorded spectral features and prevents obtaining the real variation of the absorption coefficient.

IBSA affects fluorescence-detected XAS measurements.[148−153] It is important to note that absorption of the **emitted** X-rays in the sample is also often referred to as 'self-absorption' or 're-absorption', but this mechanism does not give rise to any spectral distortion in XAS. Other authors have proposed 'thickness effect' and 'over-absorption' to describe the same mechanism that we refer to as IBSA.[154] Some authors use the term 'saturation' to describe the spectral distortion due to IBSA. In our opinion, 'saturation' more appropriately describes the behaviour of e.g. data acquisition electronics that fail to handle all X-ray events

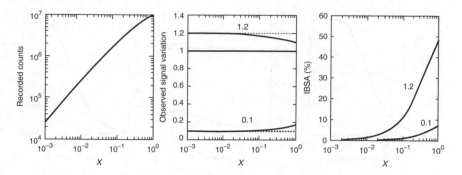

Figure 2.28 IBSA estimation in a thick (1 mm) sample of Mn_xGaN at the Mn K edge, detected using the Mn $K\alpha_1$ line and $\theta = \phi = 45°$ as a function of Mn concentration, as given by x. The Mn absorption coefficient is varied by a factor $m = 0.1$ and $m = 1.2$ (indicated as dashed lines in the centre panel). Reproduced with permission from Bianchini and Glatzel (2012) [155]. Copyright (2012) IUCr.

in cases of high count rates, *i.e.* they 'saturate'. The resulting spectral distortions may look similar in both cases (but they are not).

A simple way to estimate the expected self-absorption effect is to use tabulated values for the absorption coefficients and to plug them into Equation 2.20.[155] A variation in the absorption coefficient of the element of interest by a factor m should result in the same variation of the observed signal y. Figures 2.28 and 2.29 show examples of IBSA effects in Mn doped into GaN and in TiO_2. The recorded counts y become nonlinear as a function of Mn concentration when IBSA is significant, and the variation of the observed signal when the Mn absorption coefficient is varied by a factor m visibly deviates from m. IBSA at the Mn K edge in Mn_xGaN is estimated to be about 51% in the white line ($m = 1.2$) for $x = 1$ and becomes negligible below $x = 0.01$. IBSA is greater than 70% in a thick sample of TiO_2 and decreases below 10% only for samples thinner than 1 μm. The grain size is usually larger than 1 μm, and recording an IBSA free spectrum at the Ti K edge is often very challenging. The recorded counts plateau at sample thicknesses above about twice the attenuation length (~5 μm).

2.6 CONCLUSION

We have discussed some fundamental aspects of X-ray spectroscopic techniques that are related to the linear absorption coefficient, putting

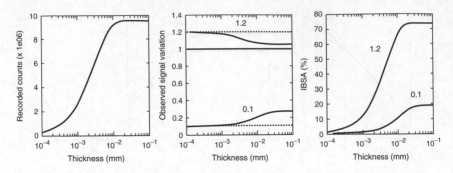

Figure 2.29 IBSA estimation in TiO_2 at the Ti K edge, detected using the Ti $K\alpha_1$ line and $\theta = \phi = 45°$ as a function of sample thickness. The Ti absorption coefficient is varied by $m = 0.1$ and $m = 1.2$. Reproduced with permission from Bianchini and Glatzel (2012) [155]. Copyright (2012) IUCr.

the emphasis on photon-in/photon-out spectroscopy. While the analysis of EXAFS data is well advanced and the information content rather well defined, we are only beginning to understand how to extract information on the electronic structure from spectroscopic data. The bottleneck is the theoretical modelling, due to the complexity of calculating the electronic structure. However, even in cases where the spectrum can be calculated, it is often not obvious what terms to use to describe the electronic structure, *i.e.* how to quantify properties of the electron density. We believe that the full potential of X-ray spectroscopy to study electronic structure is far from being fully explored. The significant interest by experimentalists has triggered important theoretical work that can be readily applied. This mutual encouragement is still gaining momentum, with interesting developments to be expected in the future.

ACKNOWLEDGEMENTS

The authors had many fruitful discussions with colleagues. We thank D. Cabaret, C. Brouder, P. Sainctavit, M. Casula, F. Mauri, M.A. Arrio, F.M.F. Groot, M. Sikora, A. Kotani, E. Gallo and M. Rovezzi.

REFERENCES

[1] M. O. Krause and J. H. Oliver, *J. Chem. Phys. Ref. Data*, **8**, 329 (1979).

[2] J. J. Sakurai, *Advanced Quantum Mechanics*, Addison-Wesley Pub. Co., Redwood City, CA, 1992.

[3] J. Goss, http://www.staff.ncl.ac.uk/j.p.goss/symmetry/index.html (last accessed 22 May 2013).

[4] M. Wilke, F. Farges, P. E. Petit, G. E. Brown and F. Martin, *Am. Mineral.*, **86**, 714 (2001).

[5] E. Balan, J. P. R. De Villiers, S. G. Eeckhout, P. Glatzel, M. J. Toplis, E. Fritsch, T. Allard, L. Galoisy and G. Calas, *Am. Mineral.*, **91**, 953 (2006).

[6] F. Farges, *Phys. Chem. Miner.*, **36**, 463 (2009).

[7] H. A. Kramers and W. Heisenberg, *Z. Phys.*, **31**, 681 (1925).

[8] W. Schülke, *Electron Dynamics by Inelastic X-ray Scattering*, Oxford University Press, Oxford, 2007.

[9] L. J. P. Ament, M. van Veenendaal, T. P. Devereaux, J. P. Hill and J. van den Brink, *Rev. Mod. Phys.*, **83**, 705 (2011).

[10] J. P. Rueff and A. Shukla, *Rev. Mod. Phys.*, **82**, 847 (2010).

[11] S. P. Collins and A. Bombardi, *in Resonant X-Ray Scattering and Absorption, Vol. 133*, E. Beaurepaire, H. Bulou, F. Scheurer and J. P. Kappler (Eds), Springer-Verlag, Berlin, 2010.

[12] G. Bunker, *Introduction to XAFS: A Practical Guide to X-ray Absorption Fine Structure Spectroscopy*, Cambridge University Press, Cambridge, 2009.

[13] A. Kotani and S. Shin, *Rev. Mod. Phys.*, **73**, 203 (2001).

[14] C. Brouder, *J. Phys.-Condens. Mat.*, **2**, 701 (1990).

[15] A. Juhin, C. Brouder, M. A. Arrio, D. Cabaret, P. Sainctavit, E. Balan, A. Bordage, A. P. Seitsonen, G. Calas, S. G. Eeckhout and P. Glatzel, *Phys. Rev. B*, **78**, 195 103 (2008).

[16] K. O. Kvashnina, S. M. Butorin and P. Glatzel, *J. Anal. At. Spectrom.*, **26**, 1265 (2011).

[17] M. H. Krisch, C. C. Kao, F. Sette, W. A. Caliebe, K. Hämäläinen and J. B. Hastings, *Phys. Rev. Lett.*, **74**, 4931 (1995).

[18] U. Bergmann, J. Bendix, P. Glatzel, H. B. Gray and S. P. Cramer, *J. Chem. Phys.*, **116**, 2011 (2002).

[19] J. Herrero-Martin, A. Mirone, J. Fernandez-Rodriguez, P. Glatzel, J. Garcia, J. Blasco and J. Geck, *Phys. Rev. B*, **82**, 075 112 (2010).

[20] A. Kotani, K. O. Kvashnina, S. M. Butorin and P. Glatzel, *Eur. Phys. J. B*, **85**, 257 (2012).

[21] D. Manuel, D. Cabaret, C. Brouder, P. Sainctavit, A. Bordage and N. Trcera, *Phys. Rev. B*, **85**, 224 108 (2012).

[22] C. Brouder, D. Cabaret, A. Juhin and P. Sainctavit, *Phys. Rev. B*, **81**, 115 125 (2010).

[23] A. B. P. Lever, *Inorganic Electronic Spectroscopy*, Elsevier, Amsterdam, 1984.

[24] E. Konig and S. Kremer, *Comput. Phys. Commun.*, **13**, 89 (1977).

[25] C. J. Ballhausen, *Introduction to Ligand Field Theory*, McGraw-Hill, New York, NY, 1962.

[26] J. S. Griffith, *The Theory of Transition-Metal Ions*, Cambridge University Press, Cambridge, 1964.

[27] B. N. Figgis, *Introduction to Ligand Fields*, Interscience, New York, NY, 1967.

[28] E. Stavitski and F. M. F. de Groot, *Micron*, **41**, 687 (2010).

[29] G. van der Laan, *in Hitchhiker's Guide to Multiplet Calculations*, Vol. 697, E. Beaurepaire, H. Bulou, F. Scheurer and J. P. Kappler (Eds), Springer-Verlag, Berlin, 2006.

[30] F. M. F. de Groot and A. Kotani, *Core Level Spectroscopy of Solids*, Taylor and Francis, New York, NY, 2008.

[31] P. H. Butler, *Point Group Symmetry Applications: Methods and Tables*, Plenum Press, New York, NY, 1981.

[32] R. D. Cowan, *The Theory of Atomic Structure and Spectra*, University of California Press, Berkeley, CA, 1981.

[33] C. McGuinness, http://www.tcd.ie/Physics/people/Cormac.McGuinness/Cowan/ (last accessed 22 May 2013).

[34] A. Uldry, F. Vernay and B. Delley, *Phys. Rev. B*, **85**, 125 133 (2012).

[35] A. Mirone, M. Sacchi and S. Gota, *Phys. Rev. B*, **61**, 13 540 (2000).

[36] A. Mirone, *ChemPhysChem*, **13**, 3172 (2012).

[37] P. E. Blochl, *Phys. Rev. B*, **50**, 17 953 (1994).

[38] M. Taillefumier, D. Cabaret, A. M. Flank and F. Mauri, *Phys. Rev. B*, **66**, 195 107 (2002).

[39] V. I. Anisimov, F. Aryasetiawan and A. I. Lichtenstein, *J. Phys.: Condens. Matter*, **9**, 767 (1997).

[40] H. Ikeno, F. M. F. de Groot, E. Stavitski and I. Tanaka, *J. Phys.: Condens. Matter*, **21**, 104 208 (2009).

[41] J. J. Rehr, *Phys. Scr.*, **T115**, 19 (2005).

[42] K. Held, *Adv. Phys.*, **56**, 829 (2007).

[43] E. L. Shirley, *Phys. Rev. Lett.*, **80**, 794 (1998).

[44] J. J. Rehr and R. C. Albers, *Rev. Mod. Phys.*, **72**, 621 (2000).

[45] J. E. Muller and J. W. Wilkins, *Phys. Rev. B*, **29**, 4331 (1984).

[46] S. D. Mo and W. Y. Ching, *Appl. Phys. Lett.*, **78**, 3809 (2001).

[47] T. E. Westre, P. Kennepohl, J. G. DeWitt, B. Hedman, K. O. Hodgson and E. I. Solomon, *J. Am. Chem. Soc.*, **119**, 6297 (1997).

[48] F. de Groot, G. Vanko and P. Glatzel, *J. Phys: Condens. Matter*, **21**, 104 207 (2009).

[49] Y. Joly, *in Interaction of Polarized Light with Matter*, Vol. 133, E. Beaurepaire, H. Bulou, F. Scheurer and J. P. Kappler (Eds), Springer-Verlag, Berlin, 2010.

[50] D. Cabaret, A. Bordage, A. Juhin, M. Arfaoui and E. Gaudry, *Phys. Chem. Chem. Phys.*, **12**, 5619 (2010).

[51] A. Juhin, F. de Groot, G. Vanko, M. Calandra and C. Brouder, *Phys. Rev. B*, **81**, 115 115 (2010).

[52] G. Onida, L. Reining and A. Rubio, *Rev. Mod. Phys.*, **74**, 601 (2002).

[53] L. Hedin, *Phys. Rev.*, **139**, A796 (1965).

[54] O. Bunau, M. A. Arrio, P. Sainctavit, L. Paulatto, M. Calandra, A. Juhin, V. Marvaud and C. C. D. Moulin, *J. Phys. Chem. A*, **116**, 8678 (2012).

[55] S. D. George, T. Petrenko and F. Neese, *J. Phys. Chem. A*, **112**, 12 936 (2008).

[56] E. L. Shirley, *J. Electron Spectrosc. Relat. Phenom.*, **136**, 77 (2004).

[57] H. Ikeno, T. Mizoguchi and I. Tanaka, *Phys. Rev. B*, **83**, 155 107 (2011).

[58] M. W. Haverkort, M. Zwierzycki and O. K. Andersen, *Phys. Rev. B*, **85**, 165 113 (2012).

[59] P. Krüger, *Phys. Rev. B*, **81**, 125 121 (2010).

[60] I. Josefsson, K. Kunnus, S. Schreck, A. Fohlisch, F. de Groot, P. Wernet and M. Odelius, *J. Phys. Chem. Lett.*, **3**, 3565 (2012).

[61] A. Kotani, K. O. Kvashnina, S. M. Butorin and P. Glatzel, *J. Electron Spectrosc. Relat. Phenom.*, **184**, 210 (2011).

[62] M. P. Seah and W. A. Dench, *Surf. Interface Anal.*, **1**, 21 (1979).

[63] C. R. Natoli, *3rd EXAFS Conference*, Stanford, CT, 1984.

[64] A. H. de Vries, L. Hozoi and R. Broer, *Int. J. Quantum Chem.*, **91**, 57 (2003).

[65] P. Glatzel, G. Smolentsev and G. Bunker, *J. Phys. Conf. Ser.*, **190**, 012 046 (2009).

[66] I. Mayer, *Theochem-J. Mol. Struct.*, **34**, 81 (1987).

[67] F. M. F. de Groot, M. Grioni, J. C. Fuggle, J. Ghijsen, G. A. Sawatzky and H. Petersen, *Phys. Rev. B*, **40**, 5715 (1989).

[68] P. Glatzel, A. Mirone, S. G. Eeckhout, M. Sikora and G. Giuli, *Phys. Rev. B*, **77**, 115 133 (2008).

[69] A. Shukla, M. Calandra, M. Taguchi, A. Kotani, G. Vanko and S. W. Cheong, *Phys. Rev. Lett.*, **96**, 077 006 (2006).

[70] J. Danger, P. Le Fevre, H. Magnan, D. Chandesris, S. Bourgeois, J. Jupille, T. Eickhoff and W. Drube, *Phys. Rev. Lett.*, **88**, 243 001 (2002).

[71] A. Bordage, C. Brouder, E. Balan, D. Cabaret, A. Juhin, M. A. Arrio, P. Sainctavit, G. Calas and P. Glatzel, *Am. Mineral.*, **95**, 1161 (2010).

[72] C. Brouder, A. Juhin, A. Bordage and M. A. Arrio, *J. Phys.: Condens. Matter*, **20**, 455 205 (2008).

[73] R. Stumm von Bordwehr, *Ann. Phys. Fr.*, **14**, 377 (1989).

[74] D. C. Koningsberger and R. Prins, *in X-ray Absorption: Principles, Applications, Techniques of EXAFS, SEXAFS, and XANES*, Vol. 92, John Wiley & Sons, Ltd, New York, NY, 1988.

[75] P. Wernet, D. Nordlund, U. Bergmann, M. Cavalleri, M. Odelius, H. Ogasawara, L. A. Naslund, T. K. Hirsch, L. Ojamae, P. Glatzel, L. G. M. Pettersson and A. Nilsson, *Science*, **304**, 995 (2004).

[76] S. Huotari, T. Pylkkanen, R. Verbeni, G. Monaco and K. Hamalainen, *Nat. Mater.*, **10**, 489 (2011).

[77] S. Huotari, T. Pylkkanen, J. A. Soininen, J. J. Kas, K. Hamalainen and G. Monaco, *J. Synchrotron Radiat.*, **19**, 106 (2012).

[78] R. A. Gordon, G. T. Seidler, T. T. Fister, M. W. Haverkort, G. A. Sawatzky, A. Tanaka and T. K. Sham, *EPL*, **81**, 26 004 (2008).

[79] R. A. Mori, E. Paris, G. Giuli, S. G. Eeckhout, M. Kavcic, M. Zitnik, K. Bucar, L. G. M. Pettersson and P. Glatzel, *Anal. Chem.*, **81**, 6516 (2009).

[80] K. Tsutsumi, *J. Phys. Soc. Jpn.*, **14**, 1696 (1959).

[81] S. D. Gamblin and D. S. Urch, *J. Electron Spectrosc. Relat. Phenom.*, **113**, 179 (2001).

[82] J. Messinger, J. H. Robblee, U. Bergmann, C. Fernandez, P. Glatzel, H. Visser, R. M. Cinco, K. L. McFarlane, E. Bellacchio, S. A. Pizarro, S. P. Cramer, K. Sauer, M. P. Klein and V. K. Yachandra, *J. Am. Chem. Soc.*, **123**, 7804 (2001).

[83] G. Vanko, T. Neisius, G. Molnar, F. Renz, S. Karpati, A. Shukla and F. M. F. de Groot, *J. Phys. Chem. B*, **110**, 11 647 (2006).

[84] R. Lengsdorf, J. P. Rueff, G. Vanko, T. Lorenz, L. H. Tjeng and M. M. Abd-Elmeguid, *Phys. Rev. B*, **75**, 180 401 (2007).

[85] P. E. Best, *J. Chem. Phys.*, **44**, 3248 (1966).

[86] G. Smolentsev, A. V. Soldatov, J. Messinger, K. Merz, T. Weyhermuller, U. Bergmann, Y. Pushkar, J. Yano, V. K. Yachandra and P. Glatzel, *J. Am. Chem. Soc.*, **131**, 13 161 (2009).

[87] U. Bergmann, C. R. Horne, T. J. Collins, J. M. Workman and S. P. Cramer, *Chem. Phys. Lett.*, **302**, 119 (1999).

[88] S. G. Eeckhout, O. V. Safonova, G. Smolentsev, M. Biasioli, V. A. Safonov, L. N. Vykhodtseva, M. Sikora and P. Glatzel, *J. Anal. At. Spectrom.*, **24**, 215 (2009).

[89] M. U. Delgado-Jaime, B. R. Dible, K. P. Chiang, W. W. Brennessel, U. Bergmann, P. L. Holland and S. DeBeer, *Inorg. Chem.*, **50**, 10 709 (2011).

[90] V. A. Safonov, L. N. Vykhodtseva, Y. M. Polukarov, O. V. Safonova, G. Smolentsev, M. Sikora, S. G. Eeckhout and P. Glatzel, *J. Phys. Chem. B*, **110**, 23 192 (2006).

[91] K. M. Lancaster, M. Roemelt, P. Ettenhuber, Y. L. Hu, M. W. Ribbe, F. Neese, U. Bergmann and S. DeBeer, *Science*, **334**, 974 (2011).

[92] Y. Pushkar, X. Long, P. Glatzel, G. W. Brudvig, G. C. Dismukes, T. J. Collins, V. K. Yachandra, J. Yano and U. Bergmann, *Angew. Chem., Int. Ed.*, **49**, 800 (2010).

[93] P. Glatzel, M. Sikora and M. Fernandez-Garcia, *Eur. Phys. J.: Spec. Top.*, **169**, 207 (2009).

[94] P. Carra, M. Fabrizio and B. T. Thole, *Phys. Rev. Lett.*, **74**, 3700 (1995).

[95] P. W. Loeffen, R. F. Pettifer, S. Mullender, M. A. van Veenendaal, J. Rohler and D. S. Sivia, *Phys. Rev. B*, **54**, 14 877 (1996).

[96] P. Glatzel and U. Bergmann, *Coord. Chem. Rev.*, **249**, 65 (2005).

[97] P. Glatzel, T. C. Weng, K. Kvashnina, J. C. Swarbrick, M. Sikora, E. Gallo, N. Smolentsev and R. A. Mori, *J. Electron Spectrosc. Relat. Phenom.*, in press (2013).

[98] G. Ghiringhelli, N. B. Brookes, E. Annese, H. Berger, C. Dallera, M. Grioni, L. Perfetti, A. Tagliaferri and L. Braicovich, *Phys. Rev. Lett.*, **92**, 117 406 (2004).

[99] S. G. Chiuzbaian, G. Ghiringhelli, C. Dallera, M. Grioni, P. Amann, X. Wang, L. Braicovich and L. Patthey, *Phys. Rev. Lett.*, **95**, 197 402 (2005).

[100] M. van Veenendaal, X. S. Liu, M. H. Carpenter and S. P. Cramer, *Phys. Rev. B*, **83**, 045 101 (2011).

[101] S. Huotari, T. Pylkkanen, G. Vanko, R. Verbeni, P. Glatzel and G. Monaco, *Phys. Rev. B*, **78**, 041102 (2008).

[102] B. C. Larson, W. Ku, J. Z. Tischler, C. C. Lee, O. D. Restrepo, A. G. Eguiluz, P. Zschack and K. D. Finkelstein, *Phys. Rev. Lett.*, **99**, 026 401 (2007).

[103] P. Glatzel, A. Mirone, S. G. Eeckhout, M. Sikora and G. Giuli, *Phys. Rev. B*, **77**, 115 133 (2008).

[104] J. N. Hancock, G. Chabot-Couture, Y. Li, G. A. Petrakovskii, K. Ishii, I. Jarrige, J. Mizuki, T. P. Devereaux and M. Greven, *Phys. Rev. B*, **80**, 092 509 (2009).

[105] J. Jimenez-Mier, J. van Ek, D. L. Ederer, T. A. Callcott, J. J. Jia, J. Carlisle, L. Terminello, A. Asfaw and R. C. Perera, *Phys. Rev. B*, **59**, 2649 (1999).

[106] P. Glatzel, J. Singh, K. O. Kvashnina and J. A. van Bokhoven, *J. Am. Chem. Soc.*, **132**, 2555 (2010).

[107] N. Smolentsev, M. Sikora, A. V. Soldatov, K. O. Kvashnina and P. Glatzel, *Phys. Rev. B*, **84**, 235 113 (2011).

[108] G. Ghiringhelli, M. Le Tacon, M. Minola, S. Blanco-Canosa, C. Mazzoli, N. B. Brookes, G. M. De Luca, A. Frano, D. G. Hawthorn, F. He, T. Loew, M. M. Sala,

D. C. Peets, M. Salluzzo, E. Schierle, R. Sutarto, G. A. Sawatzky, E. Weschke, B. Keimer and L. Braicovich, *Science*, 337, 821 (2012).

[109] J. Kim, D. Casa, M. H. Upton, T. Gog, Y. J. Kim, J. F. Mitchell, M. van Veenendaal, M. Daghofer, J. van den Brink, G. Khaliullin and B. J. Kim, *Phys. Rev. Lett.*, 108, 177 003 (2012).

[110] J. Schlappa, K. Wohlfeld, K. J. Zhou, M. Mourigal, M. W. Haverkort, V. N. Strocov, L. Hozoi, C. Monney, S. Nishimoto, S. Singh, A. Revcolevschi, J. S. Caux, L. Patthey, H. M. Ronnow, J. van den Brink and T. Schmitt, *Nature*, 485, 82-U108 (2012).

[111] B. T. Thole, G. Vanderlaan and G. A. Sawatzky, *Phys. Rev. Lett.*, 55, 2086 (1985).

[112] G. Schutz, W. Wagner, W. Wilhelm, P. Kienle, R. Zeller, R. Frahm and G. Materlik, *Phys. Rev. Lett.*, 58, 737 (1987).

[113] B. T. Thole, P. Carra, F. Sette and G. van der Laan, *Phys. Rev. Lett.*, 68, 1943 (1992).

[114] J. Badro, G. Fiquet, F. Guyot, J. P. Rueff, V. V. Struzhkin, G. Vanko and G. Monaco, *Science*, 300, 789 (2003).

[115] J. F. Lin, H. Watson, G. Vanko, E. E. Alp, V. B. Prakapenka, P. Dera, V. V. Struzhkin, A. Kubo, J. Y. Zhao, C. McCammon and W. J. Evans, *Nat Geosci*, 1, 688 (2008).

[116] W. A. Caliebe, C. C. Kao, J. B. Hastings, M. Taguchi, A. Kotani, T. Uozumi and F. M. F. de Groot, *Phys. Rev. B*, 58, 13 452 (1998).

[117] V. N. Antonov, B. N. Harmon and A. N. Yaresko, *Phys. Rev. B*, 67, 024 417 (2003).

[118] K. Matsumoto, F. Saito, T. Toyoda, K. Ohkubo, K. Yamawaki, T. Mori, K. Hirano, M. Tanaka and S. Sasaki, *Jpn. J. Appl. Phys., Part 1*, 39, 6089 (2000).

[119] D. J. Huang, C. F. Chang, H. T. Jeng, G. Y. Guo, H. J. Lin, W. B. Wu, H. C. Ku, A. Fujimori, Y. Takahashi and C. T. Chen, *Phys. Rev. Lett.*, 93, 077 204 (2004).

[120] C. Carvallo, P. Sainctavit, M. A. Arrio, N. Menguy, Y. H. Wang, G. Ona-Nguema and S. Brice-Profeta, *Am. Mineral.*, 93, 880 (2008).

[121] M. Sikora, A. Juhin, T. C. Weng, P. Sainctavit, C. Detlefs, F. de Groot and P. Glatzel, *Phys. Rev. Lett.*, 105, 037 202 (2010).

[122] M. Sikora, A. Juhin, G. Simon, M. Zajac, K. Biernacka, C. Kapusta, L. Morellon, M. R. Ibarra and P. Glatzel, *J. Appl. Phys.*, 111, 07e301 (2012).

[123] J. Orna, P. A. Algarabel, L. Morellón, J. A. Pardo, J. M. de Teresa, R. López Antón, F. Bartolomé, L. M. García, J. Bartolomé, J. C. Cezar and A. Wildes, *Phys. Rev. B*, 81, 144 420 (2010).

[124] F. Borgatti, G. Ghiringhelli, P. Ferriani, G. Ferrari, G. van der Laan and C. M. Bertoni, *Phys. Rev. B*, 69, 134 420 (2004).

[125] K. Hämäläinen, D. P. Siddons, J. B. Hastings and L. E. Berman, *Phys. Rev. Lett.*, 67, 2850 (1991).

[126] F. M. F. de Groot, M. H. Krisch and J. Vogel, *Phys. Rev. B*, 66, 195 112 (2002).

[127] M. Makosch, C. Kartusch, J. Sa, R. B. Duarte, J. A. van Bokhoven, K. Kvashnina, P. Glatzel, D. L. A. Fernandes, M. Nachtegaal, E. Kleymenov, J. Szlachetko, B. Neuhold and K. Hungerbuhler, *Phys. Chem. Chem. Phys.*, 14, 2164 (2012).

[128] M. Hubner, D. Koziej, M. Bauer, N. Barsan, K. Kvashnina, M. D. Rossell, U. Weimar and J. D. Grunwaldt, *Angew. Chem., Int. Ed.*, 50, 2841 (2011).

[129] D. Friebel, D. J. Miller, C. P. O'Grady, T. Anniyev, J. Bargar, U. Bergmann, H. Ogasawara, K. T. Wikfeldt, L. G. M. Pettersson and A. Nilsson, *Phys. Chem. Chem. Phys.*, **13**, 262 (2011).

[130] A. Bordage, E. Balan, J. P. R. de Villiers, R. Cromarty, A. Juhin, C. Carvallo, G. Calas, P. V. S. Raju and P. Glatzel, *Phys. Chem. Miner.*, **38**, 449 (2011).

[131] J. J. Kas, J. J. Rehr, J. A. Soininen and P. Glatzel, *Phys. Rev. B*, **83**, 235 114 (2011).

[132] M. Kavcic, M. Zitnik, K. Bucar, A. Mihelic, M. Stuhec, J. Szlachetko, W. Cao, R. A. Mori and P. Glatzel, *Phys. Rev. Lett.*, **102**, 143 001 (2009).

[133] P. Glatzel, F. M. F. de Groot, O. Manoilova, D. Grandjean, B. M. Weckhuysen, U. Bergmann and R. Barrea, *Phys. Rev. B*, **72**, 014 117 (2005).

[134] J. A. van Bokhoven, C. Louis, J. T. Miller, M. Tromp, O. V. Safonova and P. Glatzel, *Angew. Chem., Int. Ed.*, **45**, 4651 (2006).

[135] J. Als-Nielsen and D. McMorrow, *Elements of Modern X-ray Physics*, John Wiley & Sons, Ltd, Chichester, 2001.

[136] D. Attwood, *Soft X-rays and Extreme Ultraviolet Radiation – Principles and Applications*, Cambridge Universiy Press, New York, NY, 2000.

[137] Y. Shvyd'ko, *X-ray Optics*, Springer-Verlag, Berlin, 2004.

[138] P. Glatzel, F. M. F. de Groot and U. Bergmann, *Synchrotron Radiation News*, **22**, 12 (2009).

[139] L. X. Chen, *Angew. Chem., Int. Ed.*, **43**, 2886 (2004).

[140] C. Bressler and M. Chergui, *Chem. Rev.*, **104**, 1781 (2004).

[141] R. Alonso-Mori, J. Kern, R. J. Gildea, D. Sokaras, T. C. Weng, B. Lassalle-Kaiser, R. Tran, J. Hattne, H. Laksmono, J. Hellmich, C. Glockner, N. Echols, R. G. Sierra, D. W. Schafer, J. Sellberg, C. Kenney, R. Herbst, J. Pines, P. Hart, S. Herrmann, R. W. Grosse-Kunstleve, M. J. Latimer, A. R. Fry, M. M. Messerschmidt, A. Miahnahri, M. M. Seibert, P. H. Zwart, W. E. White, P. D. Adams, M. J. Bogan, S. Boutet, G. J. Williams, A. Zouni, J. Messinger, P. Glatzel, N. K. Sauter, V. K. Yachandra, J. Yano and U. Bergmann, *Proc. Natl. Acad. Sci. U.S.A.*, **109**, 19 103 (2012).

[142] G. Vanko, P. Glatzel, V. T. Pham, R. Abela, D. Grolimund, C. N. Borca, S. L. Johnson, C. J. Milne and C. Bressler, *Angew. Chem., Int. Ed.*, **49**, 5910 (2010).

[143] F. M. F. de Groot, M. A. Arrio, P. Sainctavit, C. Cartier and C. T. Chen, *Solid State Commun.*, **92**, 991 (1994).

[144] R. Kurian, K. Kunnus, P. Wernet, S. M. Butorin, P. Glatzel and F. M. F. de Groot, *J. Phys.: Condens. Matter*, **24**, 452 201 (2012).

[145] J. Jaklevic, J. A. Kirby, M. P. Klein, A. S. Robertson, G. S. Brown and P. Eisenberger, *Solid State Commun.*, **23**, 679 (1977).

[146] J. Goulon, C. Goulon-Ginet, R. Cortes and J. M. Dubois, *J. Phys.*, **43**, 539 (1982).

[147] A. Brunetti, M. Sanchez del Rio, B. Golosio, A. Simionovici and A. Somogyi, *Spectrochim Acta, Part B*, **59**, 1725 (2004).

[148] C. H. Booth and F. Bridges, *Phys. Scr.*, **T115**, 202 (2005).

[149] P. Pfalzer, J. P. Urbach, M. Klemm, S. Horn, M. L. denBoer, A. I. Frenkel and J. P. Kirkland, *Phys. Rev. B*, **60**, 9335 (1999).

[150] L. Tröger, D. Arvanitis, K. Baberschke, H. Michaelis, U. Grimm and E. Zschech, *Phys. Rev. B*, **46**, 3283 (1992).

[151] E. Zschech, L. Tröger, D. Arvanitis, H. Michaelis, U. Grimm and K. Baberschke, *Solid State Commun.*, **82**, 1 (1992).

[152] S. Eisebitt, T. Boske, J. E. Rubensson and W. Eberhardt, *Phys. Rev. B*, **47**, 14 103 (1993).

[153] D. Haskel, http://www.aps.anl.gov/~haskel/fluo.html (last accessed 22 May 2013).

[154] M. A. Marcus and A. Manceau http://xafs.org/Experiment/OverAbsorption?action =AttachFile&do=view&target=overabsorption.pdf (last accessed 22 May 2013).

[155] M. Bianchini and P. Glatzel, *J. Synchrotron Radiat.*, **19**, 911 (2012).

3

Neutrons and Neutron Spectroscopy

A. J. Ramirez-Cuesta[a,b] and Philip C. H. Mitchell[c]

[a]*ISIS Facility, Rutherford Appleton Laboratory, STFC, Oxford, UK*
[b]*Neutron Sciences Directorate, Oak Ridge National Laboratory, Oak Ridge, Tennessee, USA*
[c]*Department of Chemistry, University of Reading, Reading, UK*

Neutron spectroscopy is less familiar as a local structural probe than other spectroscopies. A flux of neutrons is required, and, therefore, so too are access to a central facility laboratory, the submission of a research proposal and funding. One must take one's experiment – material and equipment – to the laboratory before undertaking a neutron scattering experiment. An essential preliminary is an understanding of neutron spectroscopy, its possibilities and limitations, and the unique information that can be obtained from it. Accordingly, in this chapter our purpose is to describe the benefits of neutron spectroscopy as a probe of local structure, outlining the principles of the technique and its means of measurement. We shall concentrate on inelastic neutron scattering (INS), which is the most widely used neutron scattering local structural probe.

INS is a powerful tool by which to study hydrogen-containing materials. With the development of neutron spallation sources, and the use of epithermal neutrons, incoherent INS can measure the vibrational spectra of materials on the whole energy range of vibrational motions $(0-4400 \, cm^{-1})$, effectively opening up the field of neutron spectroscopy.

Local Structural Characterisation, First Edition. Edited by Duncan W. Bruce,
Dermot O'Hare and Richard I. Walton.
© 2014 John Wiley & Sons, Ltd. Published 2014 by John Wiley & Sons, Ltd.

The main spallation facilities are ISIS in the UK, SNS in the USA and JPARC in Japan; the European Spallation Source (ESS) in Lund, Sweden is at the design stage. These sources have increased neutron fluxes and are making it possible to increase the number of neutron studies that can be conducted. In this chapter, we will discuss the advantages and limitations of INS at the present time, as well as forthcoming developments, including improvements in instrumentation and data acquisition electronics.

3.1   THE NEUTRON AND HOW IT IS SCATTERED

The neutron is a subatomic particle. Its properties are listed in Table 3.1. In its interaction with matter/other particles, it can be treated as a particle or as a wave. It is convenient first to describe its behaviour as a particle. When a neutron strikes an atom, it interacts with the nucleus and can be either absorbed or scattered. (Absorption, which leads to nuclear transformation, does not concern us here.) Scattering is a general physical process in which some forms of radiation, such as light, sound and moving particles, are forced to deviate from a straight trajectory by one or more localised nonuniformity in the medium through which they pass.

Scattering can be elastic or inelastic. In inelastic scattering there is an exchange of energy and momentum between the neutron and the scattering nucleus (the scatterer). (In elastic scattering there is no energy exchange.) The scattered neutron may gain or lose energy. The energy transferred from the neutron appears as rotational, vibrational or translational energy of the scatterer.

Scattering can be coherent or incoherent. Spectroscopy with neutrons – observing rotations and vibrations of a single atom – relies

Table 3.1   Properties of the neutron.

Property	Value
Rest mass:	
m_n/kg	$1.6749286(10) \times 10^{-27}$
m_n/u	1.008664904(14)
$m_n c^2$/MeV	939.56563(28)
Spin, I	1/2
Charge number, z	0
Mean life/s	889.1 (21)

on incoherent scattering: the positions of an atom, j, at time zero and at a later time, t, are correlated; scattered waves from different nuclei do not interfere. Generally, the incoherent scattering intensity is isotropic, *i.e.* it is the same for any scattering angle. (Coherent scattering describes interference between waves from the scattering of a single neutron from all nuclei: diffraction.)

The scattering intensity (the number of scattered neutrons) depends on three variables: the cross-section of the scattering atom (the scatterer), the number of scattering atoms in the neutron beam and the amplitude of the atomic displacements. The scattering intensity is expressed by the scattering law (or equation).

3.1.1 The Scattering Law

The scattering law describes the spectral intensity. For a given atom j in a vibrational mode v, with vibrational frequency ω_v involving an amplitude of vibration U_{v_j}, the spectral intensity is:

$$S(Q, \omega_v)_j^n \propto \sigma_j \frac{(Q \cdot U_{v_j})^{2n}}{n!} \exp\left(-\left(Q \cdot \sum_v U_{v_j}\right)^2\right) \tag{3.1}$$

where Q is the momentum transfer, σ_j is the cross-section of atom j and n is the order of the final state of the mode excited by the neutron.

The amplitude of vibration of the harmonic oscillator is given by:

$$\left\langle U_{v_j}\right\rangle^2 = \frac{\hbar}{2\mu\omega_v} \coth\left(\frac{\hbar\omega_v}{T}\right) \tag{3.2}$$

where μ is the reduced mass of the mode and T is the sample temperature.[1]

The simplicity of Equation 3.1 is that it directly relates the amplitude of the vibrational displacement of a given mode with the intensity of the spectral line for a given frequency and momentum transfer. It is easy to compare experimental data with any type of dynamic calculation; nowadays the most common methods are density functional theory (DFT)-based calculations. The neutron interaction with a nucleus is a kinematic collision; this is why the interpretation of vibrational scattering from atoms and molecules does not have selection rules and is directly proportional to the amplitude of the atomic displacement.

The drawback of Equation 3.1 is that the exponential term contains the sum of the displacements of the atom j and the amplitude depends explicitly on the temperature. In order to reduce the impact of the exponential term on the measured spectra (Equation 3.2), INS is usually measured below 30 K. The exponential term is called the Debye–Waller factor.

$$\exp\,(-2W) = \exp\,\left(-\left(Q \cdot \sum_v U_{v_j}\right)^2\right) \tag{3.3}$$

We see that the strength of a transition is a function of the atomic displacement during the vibration (the eigenvector) and the momentum transferred by the neutron. The atomic displacements are determined by the molecule's structure and the intramolecular forces, and so we see different intensities for different transitions. The peak intensity is taken as the peak height or, better, as the area under the peak (the integrated intensity). The number of neutrons scattered out of the beam is constant for a fixed mass of a given element, following a law similar to the Beer–Lambert law for optical absorption. Linear approximations to this law can be exploited.[2]

For a neutron to be useful as a spectroscopic probe, its energy and wavelength must be comparable to interatomic and intermolecular distances, *e.g.* C–H, 1.09 Å, and must be comparable to molecular vibrational energies (30–3000 cm^{-1}). The energy and momentum of the neutron are related through the neutron velocity. A neutron of velocity v has a wave vector $k = mv$ and kinetic energy $E = m|v| = \hbar|k|^2/2m$, where m is the mass of the neutron and $\hbar$ is the reduced Planck constant. For example, a neutron of mass 1.675×10^{-27} kg with a speed $|v| = 1000$ ms^{-1} has an energy 5.23 meV or 42.2 cm^{-1} and wavelength 3.96 Å, which are in the required spectroscopic range.[2] Neutrons generated in a nuclear reactor or a spallation source have velocities and energies much greater than these, ~ 2 MeV. Such fast neutrons are brought to useful energies (moderated) by multiple inelastic collisions within tanks of *e.g.* water, liquid methane or liquid hydrogen. The neutrons achieve thermal equilibrium with the moderator temperature, 300 K for water, and mean energy 200 cm^{-1}. Such moderated neutrons are called 'thermalised neutrons'. The wavelengths and energies of thermalised neutrons are comparable to interatomic and intermolecular distances and molecular vibrational energies. This is why neutron scattering experiments can yield simultaneously both structural and dynamic information on a scatterer.

In a neutron scattering experiment, a material is exposed to a beam of neutrons and the scattered neutrons are detected. An INS spectrum is presented as a plot of scattering intensity against wavenumber. The INS spectrum is an energy-loss spectrum: energy is transferred from the incident neutron to the scattering atom, thereby exciting the vibrational and rotational modes of a molecule.

Spectra are accumulated over several hours, or longer for a weak scatterer. For example, at the ISIS Facility at the Rutherford Appleton Laboratory, with a beam current of $180\,\mu A$, the count rate on the TOSCA instrument is in the range $0.1-1.0$ neutrons detector^{-1} s^{-1}. An acceptable signal-to-noise ratio spectrum from ~ 1 g of a typical organic compound can be recorded in about 6 hours. The counting errors are governed by Poisson statistics: the error on a data point of n counts is $\sqrt{n}$.

For a typical INS spectrometer, *e.g.* the ISIS TOSCA spectrometer, the beam area is $4 \times 4\,cm^2$. This is the area sampled by the beam. INS is a macroscopic technique: the number of species detected depends on their concentrations and the sensitivity of the instrument.

As an example, we show the INS spectrum of potassium trichloro-(ethene) platinate(II), $K[PtCl_3(C_2H_4)]$ (Zeise's salt), in Figure 3.1. Infrared and Raman spectra are included for comparison.

The features of an INS spectrum, such as those in Figure 3.1, are as follows:

(i) *Scattering intensity* vs *energy*: An INS spectrum is a plot of the function $S(Q, \omega)$, the scattering intensity, proportional to the number of scattered neutrons, against energy that is transferred from the neutrons to the scatterer. Since the neutrons lose energy, the spectrum is a neutron energy-loss spectrum.

(ii) *Energy range*: Access to the low-energy region ($<300\,cm^{-1}$) is usual in INS spectroscopy. In infrared and Raman spectroscopy, because of instrumental limitations, the cut-off is $200-400\,cm^{-1}$. Beyond $\sim 1600\,cm^{-1}$ the INS spectrum becomes weak and poorly resolved, due to the effect of the Debye–Waller factor (Equation 3.3).

(iii) *Observed transitions*: In the INS spectrum of Zeise's salt we see all transitions involving displacement of hydrogen atoms. In the infrared and Raman, we see only those transitions allowed by the optical selection rules. INS however is not subject to the optical selection rules. So, for example, in INS the torsion of the ethene ligand near $200\,cm^{-1}$ is particularly strong. We see also a Pt–C stretching vibration, because the Pt–$(\eta^2-C_2H_4)$

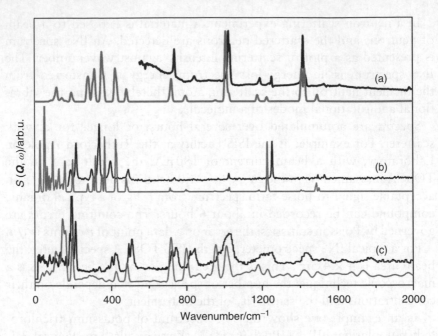

Figure 3.1 Zeise's salt $K[PtCl_3(C_2H_4)]$: (a) infrared spectra, (b) Raman, (c) INS (ISIS, TOSCA spectrometer). Relative intensities are not to scale. Superimposed (grey traces) are the corresponding spectra calculated using CASTEP.

vibration is accompanied by displacement of hydrogen atoms of the bound ethene molecule. In the Raman spectrum, we see a strong peak at $336 \, cm^{-1}$. This is assigned to the Pt–Cl stretching vibration; it is very weak in the INS spectrum because of the weak scattering from chlorine. Also seen in the INS spectrum is the factor group splitting (the splitting of bands in the electronic or vibrational spectra of crystals due to the presence of more than one (interacting) equivalent molecular entity in the unit cell). In the infrared and Raman spectra, some of the factor group components are either forbidden or have zero intensity.

(iv) *Peak positions*: The observed positions of the transitions (the peak maxima, the eigenvalues) are a function of a molecule's structure and the intramolecular forces, as in optical spectroscopy. The atoms are embedded in the molecule; they can gain energy only in the vibrational quanta characteristic of the molecular structure. Corresponding modes observed in the INS, infrared and Raman spectra are at the same energies: the vibrational energies are determined by the molecule, not by the technique.

(v) *Peak intensity*: The peak intensity is the area under the peak (the integrated intensity). The strength of a transition is a function of the atomic displacement during the vibration (the eigenvector) and the momentum lost by the neutron. Atomic displacements are determined by a molecule's structure and the intramolecular forces, and so different transitions have different intensities. Because the intensities are directly proportional to the number of scatterers in INS, we can easily manipulate spectra by adding or subtracting spectra or selected regions. Commonly, the background spectrum of a containing vesel in the neutron beam is subtracted, *e.g.* a steel pressure vessel used in catalyst studies.

(vi) *Peak shape*: Analysis of peak shapes in the INS of solids can provide additional structural and dynamic understanding (see Section 3.2.3).

(vii) The total integrated intensity of the INS spectrum of an instrument like TOSCA is directly proportional to the total amount of hydrogen in the system. This is because an indirect geometry spectrometer measures close to the hydrogen recoil line.

(viii) The spectra may be accurately modeled; see Section 3.2.2.

Note that INS is not to be thought of as a 'fingerprinting' technique (as is infrared). An INS spectrum is rich in dynamic and structural information, which can be extracted by sophisticated computer modelling of the spectrum.

3.2   WHY NEUTRONS?

The neutron is an elementary particle with zero charge, rest mass close to the proton and magnetic moment; its basic properties are given in Table 3.1. It was discovered by Chadwick in 1932. The neutron exhibits wave–particle duality and whether it is treated as a particle or as a wave depends on the observed phenomenon. In incoherent INS it is regarded as a particle.

INS is sensitive to the vibrations of hydrogen atoms; hydrogen, owing to its large neutron cross-section, is 10 times more visible to INS than other elements; see Table 3.2.

Neutron scattering is not a surface-sensitive technique *per se*. The neutron penetrates a material and the resulting spectrum is the sum of the contribution of all atoms in the substance under scrutiny, both surface

Table 3.2 Total neutron scattering cross-section and absorption cross-section of selected elements (1 barn $= 10^{-28}$ m^2).

Element	σ_{total}/barn	σ_{abs}/barn
^{1}H	82.03	0.3326
^{2}H	7.64	0.0005
^{10}B	3.1	3835.0
^{11}B	5.77	0.0055
C	5.551	0.0035
O	4.232	0.0001
N	11.51	1.9
Al	1.503	0.231
Au	7.75	98.65
Si	2.167	0.171
Fe	11.62	2.56
Ni	18.5	4.49
Ru	6.6	2.56
Pd	4.48	6.9
Pt	11.71	10.3

and bulk. If the surface-to-volume ratio is large and the composition of the surface differs from the bulk composition, and in particular if hydrogen concentration is larger at the surface than in the bulk, as when the surface is protonated or hydrogenous compounds are adsorbed upon it, we see surface scattering.

INS can distinguish between molecular and atomic hydrogen. It can also be used to determine adsorption sites and their relative concentrations. For adsorbed molecules, INS can be used to fingerprint species, in a similar way to infrared and Raman spectra, with the difference that it has special selectivity towards hydrogen and that the band intensities are better understood. The subtraction of the background signal is trivial.

Although in INS we do not measure neutron absorption, we need to be aware of the impact of a high absorption cross-section of certain isotopes, for example ^{10}B; see Table 3.2. Consequently, with ^{10}B and other highly absorbing isotopes, neutrons are mostly absorbed rather than scattered, and an INS spectrum is too weak to generate an acceptable signal-to-noise ratio.

3.2.1 The $S(Q, \omega)$ Map

In the interaction of neutrons with matter, some of the neutrons exchange energy and momentum with the sample.

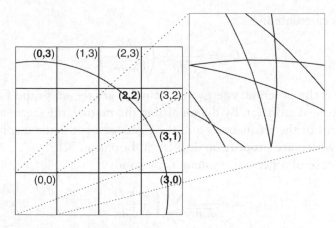

Figure 3.2   Two-dimensional representation of the momentum transfer in a poly-crystalline material. The circle represents a $|Q|$ of 2.1 zone widths (left) and its projection onto the first Brillouin zone (right). Reproduced with permission from Mitchell *et al.* (2005) [2]. Copyright (2005) World Scientific.

Because of the nature of the interaction between neutrons and matter, the neutron has the ability to sample simultaneously energy and momentum transfer, giving rise to the distinction between coherent and incoherent scattering. In the case of hydrogenous materials, the scattering is mostly incoherent, so the net effect for a powder sample is to integrate uniformly the vibrational spectra in the first Brillouin zone, as represented in Figure 3.2.

3.2.2   Modelling of INS Spectra

Computer modelling of experimental observations is fundamental to modern science. The object in modelling INS spectra is to calculate an INS spectrum from a set of initial assumptions: the atom positions.

In a solid or a molecule, atoms will undergo vibrations around their equilibrium configuration. A phonon describes a vibrational motion in a crystal, in which all atoms in a lattice oscillate at the same frequency. Classically, this is the normal mode. From a starting structural configuration, the atom positions are relaxed, *i.e.* are allowed to move towards their minimum energy arrangement. This step is usually called 'geometry optimisation'. Once the equilibrium structure has been obtained, the Hellman–Feynman forces are determined and the dynamic matrix is calculated:[2]

$$D_{ij} = \frac{1}{\sqrt{m_i m_j}} \left(\frac{\partial^2 V}{\partial x_i \partial x_j} \right) \qquad (3.4)$$

where V is the potential energy and m and x are the mass and Cartesian coordinates of atom i, j. By diagonalising the matrix, the eigenvalues are the squares of the frequencies and the eigenvectors are the displacement vectors, which are proportional to U_{v_i} in Equation 3.1.

In the case of a periodic system, $i.e.$ a solid:

$$D_{ij}(q) = \frac{1}{\sqrt{m_i m_j}} \sum_r \left(\frac{\partial^2 V(r)}{\partial x_i \partial x_j} \right) e^{(-i \cdot q \cdot r)} \qquad (3.5)$$

where q is an arbitrary point in the first Brillouin zone and r is the relative positions of atoms i and j in real space. The first Brillouin zone is a uniquely defined primitive cell in reciprocal space. The importance of the Brillouin zone stems from the Bloch wave description of waves in a periodic medium, in which it is found that the solutions can be completely characterised by their behaviour in a single Brillouin zone.

In order to calculate the vibrational properties inside the Brillouin zone, useful interpolation algorithms based on the work of Gonze are available.[7-9] Using this methodology, it is possible to calculate the dynamic matrix on a selected group of points in reciprocal space and to sample the phonons on a grid of points in the Brillouin zone, giving a good representation of the phonon density of the states of the system. The INS spectra can be calculated using the aClimax software[1] from the calculated frequencies and eigenvectors.

Alternatively, it is possible to use molecular dynamics methods to determine atomic displacements by calculating the density of states $G(\omega)$ as the power spectrum of the Fourier transform of the autocorrelation function $\langle v_i(0) \cdot v_i(t) \rangle$ of velocities $v_i(t)$ for atom i:[2]

$$G_i(\omega) = \frac{1}{2\pi} \int_{-\infty}^{\infty} \exp(-i\omega t) \langle v_i(0) \cdot v_i(t) \rangle dt \qquad (3.6)$$

this correlates with the scattering function for atom i via:

$$\lim_{|Q| \to 0} S(Q, \omega)_i = \frac{|Q|^2}{\omega^2} G_i(\omega) \cdot \sigma_i \qquad (3.7)$$

The problem with this approach is that it only gives information to compare at the Γ point ($\lim |Q| \to 0$). In order to model properly the INS spectra, it is necessary to include the contributions of the higher overtones and combinations and to include momentum transfer. This can be done by determining the amplitude of motion of each atom within a range of vibrational frequencies:

$$U_{v_i} = \int_{\omega_v - \Delta\omega}^{\omega_v + \Delta\omega} G_i(\omega) d\omega \qquad (3.8)$$

where $G_i(\omega)$ is now a vector determined as the Fourier transform of the velocity autocorrelation function for each of the Cartesian components of the velocity of atom i. The numbers from Equation 3.8 can be inserted back into Equation 3.2 and the same procedure applied to obtain the total $S(Q, \omega_v)$.

There are few considerations to be made when using these methods; using the dynamic matrix methodology (Equation 3.4), the harmonic approximation is assumed and anharmonic effects are not taken into account; it is necessary to introduce them separately if required. On the other hand, the sampling of the first Brillouin zone is correct and the calculation of the multiple quantum transitions effects on the spectra is also exact within the harmonic approximation.

When using the molecular dynamics approach and calculating velocity autocorrelation functions, the anharmonic effects are better taken into consideration. There are some practical limitations: in order to sample the low-energy modes, it is necessary to sample the molecular dynamic (MD) trajectories for long periods of time while simultaneously sampling the high-energy modes, which requires short time steps to be introduced into the calculation. In order to properly sample the Brillouin zone, it may be necessary to do the calculations on relatively large supercells. The use of the result of Equation 3.8 is not completely rigorous and can give rise to unrealistic spectral amplitudes for some anisotropic systems.

3.2.3 Example of the Effects of Sampling of the Brillouin Zone

As a worked example, we can look at the case of Raney nickel, a hydrogenation catalyst.[10−13] Calculation of the dynamic properties of hydrogen adsorbed on Ni(111) was performed using DFT[14] as implemented in the CASTEP code. In order to sample the phonon dispersion curves, we calculated the phonons in a super cell of $8 \times 8 \times 1$ (the surface calculations lack periodicity on the z-axis). The effect on the shape of the INS spectrum due to the sampling of the dispersion of the

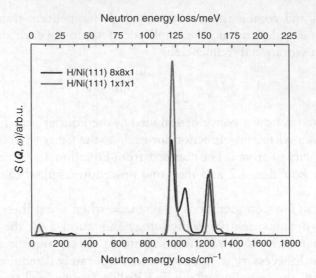

Figure 3.3 Calculated INS spectra of H/Ni(111) for two different samplings of the first Brillouin zone: $1 \times 1 \times 1$ and $8 \times 8 \times 1$.

surface modes can therefore be studied. The INS spectra were generated using the aClimax software.[1]

Figure 3.3 shows the effect on the calculated INS spectrum of the sampling of the Brillouin zone. The required separation for a grid in the reciprocal space is of the order of $0.041/\text{Å}^{-1}$. As a consequence, in the case of large unit cells, the effect is less noticeable; it is thus possible to rely on the vibrational modes determined at the centre of the zone, *i.e.* the Γ point. This fact, linked with the lack of selection rules, means that the shape of the INS peak has a physical significance, which can be tested against the calculations and which underpins the quantitative nature of the integrated spectral intensity.

3.2.4 INS Spectrometers

There are two methods for generating neutron beams: from a nuclear reactor and by spallation (spallation is the ejection or vaporisation of particles from a target during impact by a projectile). In neutron spallation sources, neutrons are produced by firing a high-energy beam of protons at a metal target, generating nuclear debris which includes neutrons; approximately 15–20 neutrons are produced by each proton impact.

The process is usually repeated at frequencies of 10–60 Hz in a very short pulse. The neutrons produced have a very high energy and are not useful for materials science; their energies may be reduced by moderation.

In order to perform neutron spectroscopy, it is necessary to know the energies of the incident and scattered neutrons. For pulsed neutron sources, we use time-of-flight (ToF) techniques, taking advantage of the fact that all neutrons produced in one frame are produced at practically the same time. Neutron detectors measure the time of arrival of the neutron, but cannot measure its energy.

The incident beam of a neutron may be monochromatic, as when its initial energy is determined by a chopper device; such instruments are called 'direct geometry' instruments. At ISIS, the UK neutron spallation source, there are presently four direct geometry instruments: MARI,[15,16] MAPS,[17] MERLIN[18] and LET[19] at SNS there are three: ARCS,[20] SEQUOIA[21] and CNCS;[22] and at JPARC there are three: 4SEASONS,[23,24] HRC and AMATERAS.[25]

Alternatively, the neutron beam may comprise neutrons with a range of energies, a so-called 'white beam'. If a filter device is used to fix the energy of the scattered neutrons, such an instrument is called an 'indirect geometry' instrument. TOSCA is the high-resolution INS spectrometer at ISIS;[26] at SNS, VISION is under construction and is expected to be operational during 2013.[27]

3.2.4.1 Direct Geometry Instruments

Direct geometry instruments at spallation sources have choppers to fix the incident energy. A consequence of the monochromation process is a low incident flux, which requires a large detector area. The resolution of the instruments is improved by increasing the distance between the sample and the detector, demanding a larger vacuum tank and a higher price tag. These instruments usually have a large number of detectors at a range of scattering angles so that they can measure Q and ω independently.

The resolution of an instrument changes depending on the choice of chopper and incident energy settings. In order to span the whole range of energy, a series of spectra of different resolutions must be combined. At a first approximation, the resolution, Δ-ω, can be considered constant for a given incident energy, and for the high-resolution settings it is of the order of 1% of $E_{incident}$. Note that the resolution of any spectral feature can be improved by choosing an optimal incident energy, usually 10–20% higher than the energy of the transition.

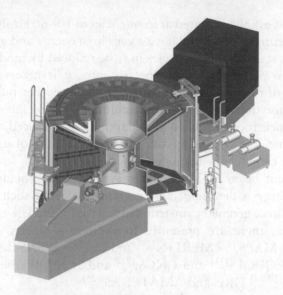

Figure 3.4 Engineering drawing of MERLIN, with cutaways showing the inside of the tank. The scale is relative to the human figure shown to the right. Reproduced with permission from Bewley *et al.* (2006) [18] Copyright (2006) Elsevier Ltd.

The trajectory that describes the path followed by each neutron at an angle θ in $\mathbf{Q}$, ω space is:

$$\frac{\hbar^2 |\mathbf{Q}|^2}{2m_n} = 2E_i - \hbar\omega - 2\sqrt{(E_i(E_i - \hbar\omega))} \cos\theta \qquad (3.9)$$

where ω is the neutron energy transfer, E_i is the incident energy of the neutron, m_n is the mass of the neutron, θ is the angle of the detector and $\mathbf{Q}$ and ω are the momentum and energy transfers of the scattered neutron respectively. Figure 3.4 shows the MERLIN spectrometer at ISIS.

3.2.4.2 Indirect Geometry Instruments

In an indirect geometry spectrometer, the sample is illuminated with a white beam of neutrons, the fast neutrons arriving earlier and the slower ones arriving later. After scattering, the neutrons travel through a secondary spectrometer, at angles of 45 and 135°. The secondary spectrometer consists of highly orientated pyrolytic graphite, which selects the final energy of the neutrons as $30\,cm^{-1}$. The neutrons are diffracted and only those that satisfy the Bragg condition are reflected

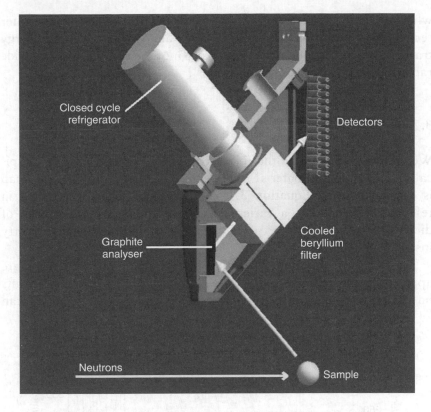

Figure 3.5 Analyser module of the TOSCA spectrometer.

back, to later be filtered by a beryllium block cooled to 30 K. A diagram of the analyser module including the filter assembly is shown in Figure 3.5. The whole assembly operates as a monochromator of the scattered beam. The initial energy of the neutron is calculated from the ToF and the spectra are then determined.

The resolution of a direct geometry instrument is determined by the characteristics of the secondary spectrometer and the distance between the sample and the moderator. The larger the initial flight path, the higher the resolution. Indirect geometry spectrometers are compact compared with direct geometry ones. In general, their sampling of Q and ω is along a defined trajectory.[26]

The trajectory in Q, ω space for this type of instrument is given by:

$$\frac{\hbar^2 |Q|^2}{2m_\mathrm{n}} = \hbar\omega + 2E_\mathrm{f} - 2\sqrt{((E_\mathrm{f} + \hbar\omega)E_\mathrm{f})} \cos\theta \qquad (3.10)$$

where E_f is the final energy of the neutron, determined by the filter set-up. The resolution of TOSCA varies almost linearly with energy transfer, and it can be approximated to $\Delta\omega/\omega = 1.5\%$ over a wide range of energies from 50 to 4000 cm^{-1}.

3.2.4.3 INS Spectra

We take as our example the INS spectrum of polyethylene,[28] a typical C–H-containing material (see Figure 3.6). The total intensity map is calculated using Equation 3.1, following the scheme presented in reference[38], including a variety of cuts following the trajectories of direct (energy resolution $\Delta\omega/E_{incident} = 3\%$) and indirect geometry instruments ($\Delta\omega/\omega = 1.5\%$).

The ranges of optimum operation for the different instruments are apparent just from inspecting Figure 3.6. Indirect geometry instruments have a good resolution below 1600 cm^{-1}, while direct geometry can

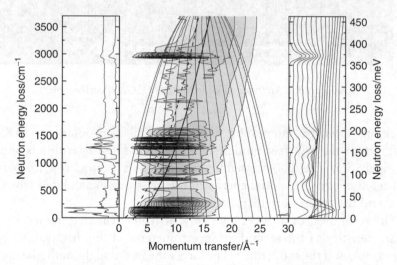

Figure 3.6 Illustration of the different trajectories in Q, ω space, calculated for polyethylene $(–CH_2–CH_2–)_n$ at 15 K. Central panel: contour map representing the intensity of $S(Q, \omega)$; thick lines are the trajectories for TOSCA in forward and backscattering, thin lines are the trajectories for a direct geometry instrument with detectors at 2, 10, 20, 30, 40, 50, 60, 70, 80, 90, 100, 110, 120, 130 and 135° and an incident energy of 4500 cm^{-1}. Left panel: the calculated INS spectra as measured in TOSCA (forward and backscattering). Right panel: the calculated INS spectra along each of the trajectories represented in the central panel for a direct geometry INS spectrometer.

adjust the resolution, but at any given incident energy ($E_{incident}$) their resolution degrades below $E_{incident}/2$, the sharp features of the spectra becoming lost, as shown in the right panel of Figure 3.6.

It can be seen that the INS spectra in a TOSCA-like spectrometer are above the so-called hydrogen recoil line. This has a very important consequence, as it allows the quantitative determination of the relative amount of hydrogen. The correlation of the integrated intensity with the hydrogen content has been demonstrated to be linear in activated carbons.[29] It is imperative that the effective background of the instrument is very low. In the case of a direct geometry instrument, it is not necessarily the case that the integration along any particular cut is proportional to the total amount of hydrogen, but sensitivity in a well-defined region of the spectrum is good for direct geometry instruments, as has been recently shown.[30]

3.2.5 Measurement Temperature

Because neutrons are highly penetrating, engineering materials like aluminium and steel can be used for sample containment and for other equipment which may be in the neutron beam, simplifying the design and manufacture of a complex sample environment if desired. In general, neutron scattering measurements can be undertaken at temperatures in the range 0.05–2000 K, pressures from 10^{-9} to 10^{5} bar and magnetic fields up to 7 T. However, the Debye–Waller factor, which enters into the equation *via* the temperature dependence of the amplitude of vibrations (Equation 3.2), is strongly dependent on the sample temperature and severely suppresses the INS signal, particularly when dealing with hydrogen. This in turn means that most of the INS studies in indirect geometry instruments are performed at temperatures of 30 K or lower. In the case of direct geometry instruments, the effect of temperature can be reduced by measuring at the lower momentum transfers[31] (see Section 3.3).

3.2.6 Amount of Sample Required

With the present instrumentation (for example, TOSCA), the amount of hydrogen required to obtain a signal above background levels is of the order of 5 mmol of hydrogen atoms in the beam. So for organic materials, an amount of 1 g can be measured easily within a few hours. Equally, if desired, INS spectra can be obtained on bulk (tens of grams) quantities of a substance.

3.3 MOLECULAR HYDROGEN (DIHYDROGEN) IN POROUS MATERIALS

3.3.1 The Rotational Spectrum of Dihydrogen

Pure rotational transitions of symmetrical diatomic molecules are forbidden in infrared spectroscopy but active in Raman spectroscopy. They are in principle observable in INS, but with the exception of dihydrogen they are difficult to observe because of the weak scattering from elements other than hydrogen.

Rotational transitions of dihydrogen can act as a molecular probe of the local environment that the hydrogen molecule experiences in interacting with a surface or active site.

Dihydrogen has two nuclear spin isomers, *para*-hydrogen (p-H_2), with spins paired or antiparallel ($\uparrow\downarrow$), and *ortho*-hydrogen (o-H_2), with spins unpaired or parallel ($\uparrow\uparrow$). For a dihydrogen molecule *in vacuo*, the rotational wavefunctions are the spherical harmonics: $Y_{JM}(\theta, \phi)$. The integers J and M are the angular momentum quantum numbers, with allowed values:

$$J = 0, 1, \ldots ; \quad M = 0, \pm 1, \pm 2, \ldots, \pm J \qquad (3.11)$$

and the eigenvalues are:

$$E_{JM} = E_J = J(J+1)\frac{\hbar^2}{2I} = J(J+1)B_{rot} \qquad (3.12)$$

where I is the moment of inertia, $B_{rot} = \hbar^2/2I = \hbar^2/m\, r_{H-H}^2$ is the rotational constant (7.35 meV), m is the mass of the hydrogen atom and r is the hydrogen molecule bond length.

Quantum mechanical restrictions on the symmetry of the rotational wavefunction produce two molecular species for dihydrogen: p-H_2 and o-H_2. The spin statistics are different for each of the isotopic configurations (H_2, HD and D_2). The hydrogen nucleus, a particle with half-integer spin, is a fermion. Its quantum mechanical description must obey Fermi statistics, as expressed through the Pauli principle. This imposes a symmetry requirement on molecular wavefunctions and forbids the occupation of certain states. Identical nuclei are indistinguishable; their exchange can change the sign of the total wavefunction. The relevant wavefunctions are those for molecular rotation, $Y_{JM}(\theta, \phi)$, and nuclear spin, ψ_{spin}. The symmetry of the rotational wave functions under exchange is:

$$Y_{JM}(\pi - \theta, \pi - \phi) = (-1)^J Y_{JM}(\theta, \phi) \tag{3.13}$$

since the total wave function has to be antisymmetric under the exchange of indistinguishable particles. Also, we may represent the spin states by ↑, spin 'up', and ↓, spin 'down', if we represent the wavefunction in the Dirac notation $|J, M\rangle$. The spin wavefunctions are thus:

$$\Psi_{spin}^{symm} = \tfrac{\uparrow_1\downarrow_2 - \downarrow_1\uparrow_2}{2} \quad |0, 0\rangle \tag{3.14}$$

$$\Psi_{spin}^{symm} = \begin{cases} \uparrow_1\uparrow_2 & |+1, +1\rangle \\ \tfrac{\uparrow_1\downarrow_2 + \downarrow_1\uparrow_2}{2} & |+1, 0\rangle \\ \downarrow_1\downarrow_2 & |+1, -1\rangle \end{cases} \tag{3.15}$$

The even values of J correspond to antisymmetric wavefunctions and must be combined with symmetric spin wavefunctions. The only allowed rotational states of spin-paired, or antiparallel, (↑↓) dihydrogen are thus those with $J = 0, 2, 4, \ldots$ This defines *para*-hydrogen. There are $2J + 1$ possible spin states for each acceptable J value and it follows from Equation 3.12 that there is only one spin state for the *para*-hydrogen ground state, $J = 0$.

The detailed theory of the dihydrogen rotational INS spectrum is described in references [32–34]. A consequence of the selection rules for nuclear spin transitions in INS is that the $J(1 \leftarrow 0)$ transition has high intensity (at 14.7 meV in the case of solid hydrogen; see Figure 3.7). This is because, through spin exchange with the neutron, this transition has access to the incoherent scattering cross-section of hydrogen (80 barn).

3.3.2   The Polarising Power of Cations and H_2 Binding

The rotational line can be used to probe the interaction of dihydrogen with a surface site, *e.g.* as in studies of the INS spectra of dihydrogen in zeolites and metal oxides.

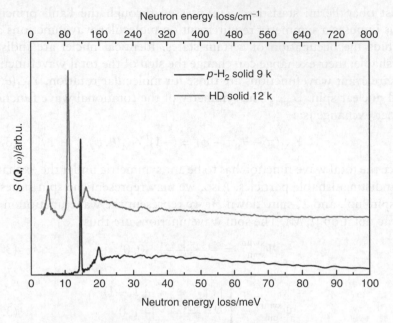

Figure 3.7 Rotational spectrum of solid HD (top trace) and p-H$_2$ (bottom trace). Note the change in the position of the rotational line for HD due to its higher moment of inertia. Also, the vibrational feature below the rotational line is apparent for HD because there are no symmetry considerations, since the molecule is formed from two distinguishable nuclei.[35,36]

When interacting with a surface site, the rotational line can be affected by the nature of the interaction.[32,33,37−44] In the case of dihydrogen in cation-substituted X-type zeolites INS has been used to determine the polarising effects of the cation subtitution.

The INS spectra of dihydrogen sorbed by zeolite X containing sodium, calcium and zinc cations after subtraction of the background (zeolite plus can) are shown in Figure 3.8. There is no evidence for free dihydrogen – no sharp peak at 14.7 meV (118 cm^{-1}) (by 'free dihydrogen', we mean H$_2$ molecules interacting only with each other, as in solid dihydrogen). The INS spectra of dihydrogen in the zeolites are different from the spectrum of dihydrogen in solid dihydrogen. The H$_2$-zeolite spectra are broadened and shifted, clearly a consequence of the interaction of the H$_2$ molecules with the micropore surface. The dihydrogen-in-zeolite spectra show a sequence of bands towards higher energies, with decreasing intensities, most obviously for dihydrogen in CaX. In Figure 3.8 the INS spectra of dihydrogen are decomposed to Gaussian peaks. In CaX, the peaks are in two groups: 1–4, and 5 and

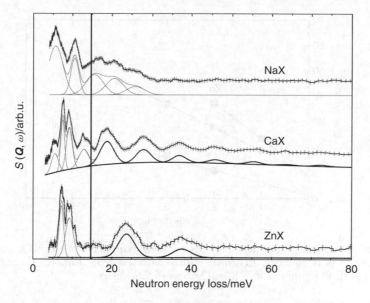

Figure 3.8   From top to bottom: the INS spectra of *para*-hydrogen in ZnX, CaX and NaX, and the decomposition into Gaussians. The ZnX and NaX spectra have had backgrounds removed with a smooth function for clarity. Redrawn with permission from Ramirez-Cuesta *et al.* (2007) [43]. Copyright (2007) Royal Society of Chemistry.

higher. The second group forms a sequence with a constant energy separation of ∼9 meV; their frequencies (centres of gravity, from the Gaussian analysis) are plotted against peak number in Figure 3.9: the plot is linear (correlation coefficient, $r = 0.99985$), having gradient 9.05 meV and intercept 9.55 meV. The intercept makes a positive contribution to each peak, *e.g.* peak 5 is a combination: $9.55 + 9.05 = 18.6$ meV (*cf.* the observed value 18.7 meV) and peak 6: $9.55 + 2 \times 9.05 = 27.7$ meV (observed 27.8 meV).

In the assignment of the H_2–CaX spectrum (see Figure 3.10), peaks (1–4) at 5.49, 7.45, 9.18 and 12.8 meV are associated with the $J(1\ r\ 0)$ transition. The energies are all lower than the $J(1 \leftarrow 0)$ transition of solid dihydrogen (14.7 meV); the downward shift and the splitting are due to the interaction of the dihydrogen molecule with one or more binding sites.

Cations are substituted into zeolites in order to enhance the acidity, the binding of adsorbed molecules or the catalytic activity. There is a considerable literature on the binding of small adsorbate molecules, including H_2 in zeolite cages. The property of the cation most often

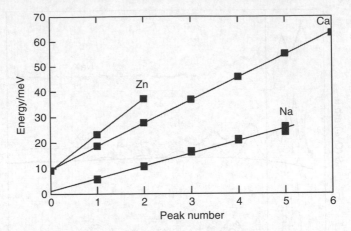

Figure 3.9 Linear plot of peak energies for *para*-hydrogen in CaX, NaX and ZnX against the peak number. Redrawn with permission from Ramirez-Cuesta *et al.* (2007) [43]. Copyright (2007) Royal Society of Chemistry.

invoked to account for enhanced acidity or adsorbate binding is its polarising power, z/r, where z is the formal charge on the ion and r is its radius. The quantity z/r (also called the electrostatic or ionic potential) measures the charge density at the surface of the ion.[45] Enhanced binding due to the substituted cation has generally been related to its ability to polarise an adsorbed molecule, creating a dipole and hence an attractive cation-induced dipole interaction.[46−50]

The analysis of the INS spectra of the substituted zeolites X enables us to comment on the binding of the H_2 molecule in the zeolites. That the spectra are different for each cation tells us immediately that dihydrogen is interacting with the substituted cation (or cation plus oxide anion combination). The trend in H_2–(zeolite) vibrational frequencies with cation is Na^+, Ca^{2+} and Zn^{2+}. In Figure 3.10 (bottom-right panel), the H_2–(zeolite) vibrational frequencies are plotted against the polarising potential of the cations as expressed by the function z/r. We see that the relationship is indeed linear; therefore, the interaction of H_2 molecules with the zeolite X increases with the polarising potential of the cation. This result is consistent with theoretical studies of the interaction of H_2 molecules with gaseous cations, increasing with the increase of polarising potential, *e.g.* from Rb^+ to Li^+ for the bare alkali metal cations.[46]

In the INS study of the rotational−vibrational spectrum of dihydrogen sorbed by zeolite X with substituted sodium, calcium and zinc cations, in the low-energy region ($<200\,cm^{-1}$) of the spectra of adsorbed H_2 (Figure 3.8), we can observe the rotational−vibrational spectrum of H_2,

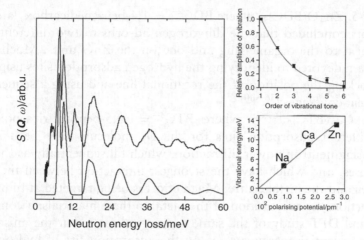

Figure 3.10 Spectrum of *para*-hydrogen in CaX zeolites and a fit of the vibrational contribution of *para*-hydrogen motion. Top right: relative amplitude of the vibrational line following the TOSCA backscattering trajectory. The best fit is obtained for a vibrational amplitude of 0.195 Å using aClimax.[1] Bottom right: H_2–(zeolite) vibrational frequencies plotted against the polarising potential of the cations, z/r, where z is the charge on the cation and r is the ionic radius. Redrawn with permission from Ramirez-Cuesta *et al.* (2007) [43]. Copyright (2007) Royal Society of Chemistry.

where the vibration is that of the H_2 molecule against the binding site (*i.e.* H_2X, not H–H). This vibration frequency is directly proportional to the polarising power of the cation ($Na^+ < Ca^{2+} < Zn^{2+}$). This experiment directly demonstrates that polarisation of the H_2 molecule dominates the interaction of H_2 with the binding site.

3.3.3   Hydrogen in Metal Organic Frameworks

Metal organic frameworks (MOFs) are crystalline structures formed by linking metal ions with organic ligands.[51−53] Their structures contain extended three-dimensional (3D) networks of channels and pores. These materials are very versatile in their structures, pore shapes and pore sizes. They also have very large surface areas, once the solvents used for their synthesis have been removed from their structures. The large surface areas and pore volumes make MOFs interesting candidate materials for hydrogen storage, and consequently they are currently one of the most active areas of materials research in energy storage.[54] In an early paper, Rosi *et al.* [55] used INS to determine the adsorption sites of hydrogen in

MOF-5, $Zn_4O(BDC)_3$, where $BDC^{2-} = 1,4$-benzenedicarboxylate. The authors concluded that the dihydrogen adsorbs to two different sites, one next to the organic ring and one on the ZnO units. Much work has been devoted to identifying the hydrogen adsorption sites using INS by tracking the position of the rotational line and using a sequence of different coverages.[56−60]

For $Cu_3(BTC)_2$,[61,62] where $BTC^{3-} = 1,3,5$-benzenetricarboxylate, six different adsorption sites for the D_2 molecule have been identified using neutron power diffraction, which fill sequentially as the load increases, and which have the strongest interaction between the dihydrogen and the metal centre. Much effort is being targeted at tuning the interactions by substitution of the metal in these materials; a combined INS and DFT study of the same system indicates that the anisotropy of the potential energy surface on the adsorption site of hydrogen on copper is responsible for the very strong shift of the rotational line in this system.[63] The packing of dihydrogen in some of these materials achieves densities that surpass the density of liquid hydrogen.[64] A comprehensive review of the field of MOFs and hydrogen storage can be found in reference [65].

Compounds of the type M_2(dobdc) (M = Mg, Mn, Fe, Co, Ni, Zn; $dobdc^{4-} = 2,5$-dioxido-1,4-benzenedicarboxylate), also commonly referred to as either M-MOF-74 or CPO-27-M, constitute an isostructural family of metal–organic frameworks with one-dimensional (1D) hexagonal channels.[67−73] The framework walls consist of helical chains of oxo- and carboxylato-bridged M^{2+} cations interlinked through $dobdc^{4-}$ ligands. Upon thermal activation of the framework, a bound solvent molecule is liberated from each pseudo-octahedral M^{2+} cation, leaving an open coordination site that points directly into one of the channels. These highly reactive, electron-deficient sites allow for strong polarisation of incoming H_2 molecules, which bind to the framework surface, leading to zero-coverage isosteric heats ranging from -8.8 to $-13.5\,kJ\,mol^{-1}$.[64,71,74]

These values are among the largest reported and show significant improvement over the isosteric heats of H_2 adsorption observed for frameworks lacking exposed metal cation sites.[53,75−78] While a number of metal–organic frameworks incorporating iron(II)[79−81] or iron(III)[82−85] have been reported, the gas adsorption properties of only a few such compounds featuring open metal coordination sites have been studied.[79,81,84,85] The hydrogen storage characteristics of the redox-active metal–organic framework Fe_2(dobdc) (i.e. Fe-MOF-74) and its oxidised counterpart $Fe_2(O_2)$(dobdc) have been described.[66]

The combination of H_2 adsorption studies with neutron scattering measurements affords important insight into the H_2 loading characteristics of the materials, allowing us to draw correlations between binding energy and $M–D_2$ distances for the homologous M_2(dobdc) series.

Figure 3.11 (top) shows the INS data obtained from TOSCA over an energy range of 2–20 meV, highlighting the low-energy rotational transitions observed for the three adsorption sites. At loadings above

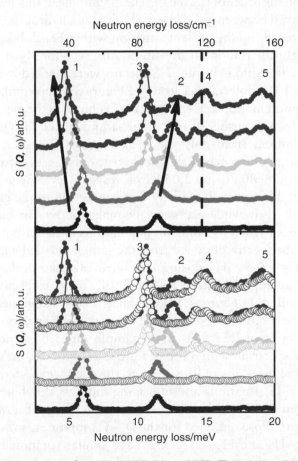

Figure 3.11   INS spectra from Fe-MOF-74 at 20 K. Top: increasing loadings of p-H_2 *per* Fe^{2+}–0.5, 1.0, 2.25, 3.0 and 3.75 (from bottom to top). The data shown were taken after subtraction of the spectrum of the evacuated framework. All rotational lines, especially those associated with site I (denoted by arrows), shift as a function of loading. The dashed line, at 14.7 meV, represents the *para–ortho* transition of bulk H_2. Bottom: same increasing loadings as in the top frame, the open circles represent the following loadings: $1 D_2$, $0.25 H_2 + 1 D_2$, $1.25 H_2 + 1 D_2$, $2 H_2 + 1 D_2$ and $2.75 H_2 + 1 D_2$. Reproduced with permission from Queen *et al.* (2012) [66]. Copyright (2012) Royal Society of Chemistry.

1.0 H_2 molecules per iron, as sites II and III are populated the additional molecules are close enough to each other to affect the rotational potential at site I, which consequently adjusts the rotational energy level, as seen in the shift of peak 1 to lower frequencies and peak 2 to higher frequencies. The lowest frequency line shifts by almost 1.4 meV over the entire loading range, much more than is observed for the other members of the M_2(dobdc) series, which show only small effects on the lowest rotational transition (on the order of 0.2 meV). The magnitude of this shift is similar to that observed between gas phase and solid bulk hydrogen,[34,35] which is very different from the present situation within Fe_2(dobdc).

To highlight the rotational transitions observed for H_2 at sites II and III, the site I transitions in the INS spectra were masked by dosing the sample with 1 D_2 molecule per iron (see Figure 3.11 bottom). Due to the lower zero-point motion and entropy, the D_2 binds to the metal preferentially over H_2. Any further dosing of the sample with p-H_2 yields spectra with contributions from only the weaker adsorption sites, II and III, due to the large difference in the incoherent cross-section for hydrogen versus deuterium: 80.26 and 2.05 barns, respectively. The use of p-H_2 is important in this case as it is likely that conversion to the ground state cannot occur easily while access to the magnetic Fe^{2+} is blocked, and the objective is to directly compare populations in each rotational state. Comparing the spectra obtained from the samples loaded with 1.5 p-H_2 to the 1.0 D_2 + 0.5 p-H_2 dosing clearly reveals two peaks associated with site II at approximately 10.6 and 14.2 meV, respectively (peaks labeled '3' and '4' in Figure 3.11), enabling a clear assignment of peaks 1 and 2 to the rotational levels of p-H_2 bound at the open metal site. An additional dosing of 1.25 p-H_2 to the sample with 1.0 D_2 molecules per iron yields a spectrum with a broad feature around 15.0 meV that must originate from p-H_2 at site III, similar to the energy expected for almost unhindered p-H_2 rotations and indicating more weakly bound p-H_2. It is evident that there is little difference between the spectra obtained for pure p-H_2 and those obtained for the D_2–H_2 mixtures, so we conclude that the H_2–H_2 and H_2–D_2 '*pairs*' have similar rotational levels and hence that the interactions between sites I and II must be weak.[61]

3.3.4 Hydrogen Trapped in Clathrates

A clathrate consists of a lattice of one type of molecule, *e.g.* water, trapping and containing a second type of molecule, *e.g.* methane. Water, in the presence of a small amount of other molecular substances, can form several crystalline compounds, different from common ice, known as clathrate hydrates. The water molecules held in place by hydrogen bonding can host different guest molecules in cages of various sizes and geometries. The anharmonic motion of the guest molecules differs from the other normal modes of the crystal and is responsible for the anomalous thermal conductivity of these materials.[86] Besides being present in nature, these compounds can be prepared in the laboratory and have recently been proposed as effective, safe and economical materials for energy storage.[87−89] Hydrogen clathrate hydrates, *i.e.* made of H_2O and H_2, require 2 kbar of pressure to be formed at $T \sim 273$ K. However, it has recently been shown that the ternary compound with tetrahydrofuran (THF) can store significant amounts of hydrogen with a much lower formation pressure.[88] How dihydrogen interacts with the host framework has been studied by INS spectroscopy.[90−92]

The INS spectra from the hydrogen-charged clathrates following background subtraction showed the presence of several narrow and intense bands caused by the motions of the dihydrogen molecule. The bands were assigned to combinations of rotation and translations of the p-H_2 molecule, mostly $E \geq 14.7$ meV, to the fundamental transition of quantum rattling $E \approx 10$ meV for o-H_2 and to their combinations. Dihydrogen trapped in the clathrate cages is a nonequilibrium mixture of o-H_2 and p-H_2, and the use of different ratios of both species allows the decomposition of the o-H_2 and p-H_2 signals.

Considering the neutron scattering cross-section for both species and its dependence on the rotational transitions,[34] we see that the spectrum consists of centre-of-mass excitations for the o-H_2 species. In the case of p-H_2, these vibrational excitations of the centre-of-mass are replicated and shifted by the energy of any possible rotational excitation of the molecule. The scattering law in this case will be the convolution between the vibrations of the centre-of-mass of the molecule and the rotational energy levels; this expression holds if the hypothesis of decoupling of the internal rotational motion of the molecule from the centre-of-mass motion is satisfied.

Since INS is not subject to selection rules, all rotational transitions and centre-of-mass vibrations are allowed. At base temperature, as was the case in this work, only the lowest rotational states are populated: $J = 0, 1$. Only a few transitions contribute to the spectrum in the observed frequency region, namely the rotationally elastic $J = 1 \leftarrow 1$ and the inelastic $J = 2 \leftarrow 1$ transition of o-H_2 and the inelastic $J = 1 \leftarrow 0$ of p-H_2. The $J = 0 \leftarrow 0$ transition of the p-H_2 molecule, being weighted by the coherent scattering length only, does not contribute appreciably to the spectrum, and consequently the observed fundamental rattling transition $E \approx 10\,\text{meV}$ is mainly caused by the *ortho* molecules.

Using two samples with different concentrations of o-H_2 and p-H_2, it was found that the intensity ratio of the two bands at $E \approx 10\,\text{meV}$ and $E \approx 14.7\,\text{meV}$ is lower in the p-H_2-rich sample. By combining linearly the p-rich and o-rich spectra, the spectra of pure p-H_2 and pure o-H_2 were determined (see Figure 3.12). The spectrum of o-H_2 (solid line) contains contributions from the fundamental transition of rattling vibration and the overtones and combinations, while the p-H_2 spectrum (dashed line) contains the pure rotational $J = 1 \leftarrow 0$ transition at about $14.7\,\text{meV}$, plus combinations of this with the centre-of-mass transitions. By shifting the p-H_2 spectrum in the energy of the rotational transition (central graph, red, in the right panel of Figure 3.12), obtaining a very convincing match of the translational bands.

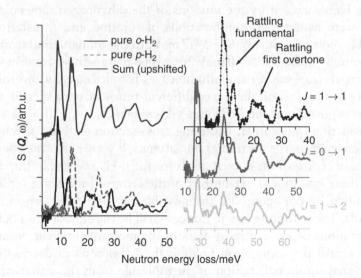

Figure 3.12 Left panel: the clathrate spectrum (top, vertically displaced) decomposed into the sum of the o-H_2 (black, solid line) and p-H_2 (grey, dashed) contributions. Evident in the p-H_2 spectrum is the strong $J = 1 \leftarrow 0$ rotational band (a triplet at $E = 14\,\text{meV}$), and in the o-H_2 spectrum, the (split) band around 10 meV, arising from the quantum rattling motion of the molecule in the cage. Right panel: from top, o-H_2 spectrum (black), p-H_2 spectrum (dark gray) shifted by $\Delta E_{01} = 14.7\,\text{meV}$, and again o-H_2 spectrum (light gray) shifted by $-\Delta E_{12} = 29.0\,\text{meV}$. The coincidence, in frequency and shape, of the rotational-rattling combination bands with the rattling fundamental and first overtone is evident. Reproduced with permission from Ulivi *et al.* (2007) [90]. Copyright (2007) American Physical Society.

The spectrum of o-H_2 also contains the band arising from the $J = 2 \leftarrow 1$ transition at $\sim$29 meV and the combination of this with the centre-of-mass spectrum. The anharmonicity of the potential energy surface was determined for the molecular rattling transitions by solving numerically the Schrödinger equation for the centre-of-mass motion. The potential energy for one H_2 molecule, as a function of the centre-of-mass displacement $\vec{R}$ from the center of the dodecahedral cage, has been calculated summing the pair H_2/D_2O potential (assumed equal to the H_2/H_2O potential) over 514 molecules of D_2O around the centre of the cage on a rigid lattice, and averaging over H_2 molecular orientations. The disorder of the water deuterons was been taken into account by performing several averages over random configurations. A further average over the direction of $\vec{R}$ gives the isotropic part of the potential. The potential energy of the hydrogen molecule as a function of $\vec{R}$, obtained using the two potential models available from the literature,[90,93,94] which coincide for the purpose of this analysis, is represented in Figure 3.13 (left panel). The potential is strongly anharmonic and rather flat in the centre. As a consequence, the calculated energy levels (labeled by the principal quantum number $N = 1, 2, 3, \cdots$) are split; the energies depend on the orbital quantum number $L = 0, 1, 2, \cdots$ and are not equally spaced. The residual $(2L + 1)$-fold degeneracy of each level is not removed in this model due to the assumed isotropy of the potential. Knowledge of the wavefunction for the translational degrees of freedom, where all coefficients J, J', Q are known, permits the calculation of the intensity of transitions, determining the whole neutron energy loss spectrum. For each rotational transition $J \to J'$ of the molecule, we obtain a spectrum given by the rattling transition energies plus the rotational energy transition $E_{JJ'}$. The calculated spectra are shown in Figure 3.13, where all transitions are represented by Gaussian lines with the same arbitrary width. This tells us where the hydrogen molecules are and what energy landscape they experience.

3.4 INS AND CATALYSIS

INS is not, in itself, a surface technique. Neutrons penetrate and so an INS spectrum is the sum of the spectra from neutrons scattered in the bulk of a substance and at its surface.[2] We see surface scattering when a substance has a large signal-to-noise ratio (surface area $\sim$20 m^2 g^{-1}) and when the surface composition differs from the bulk composition,

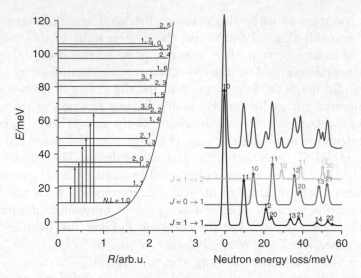

Figure 3.13 Left panel: isotropic part of the potential energy for one H_2 molecule as a function of the distance R (atomic units) from the centre of the dodecahedral cage and calculated energy levels. Right panel: calculated incoherent INS spectra for one H_2 molecule (top line). This is the sum of the three contributions shown below, where the same transition lines of the centre-of-mass motion (labeled by the final values of N,L) are shifted by the amount $\Delta_{jj''}$ of each rotational transition (from bottom to top): $\Delta_{11} = 0$, $\Delta_{01} = 14.7$ meV, $\Delta_{12} = 29$ meV. Reproduced with permission from Ulivi *et al.* (2007) [90]. Copyright (2007) American Physical Society.

especially when the hydrogen concentration at the surface is greater than in the bulk, *e.g.* if the surface is protonated or a hydrogenous adsorbate is present.

In order to measure with the current sensitivity, it is necessary to use at least $500\,m^2$ of active surface on *e.g.* a catalyst. As a consequence, in the case of a supported catalyst, high loadings of catalyst on the support are common. Generally up to $30\,g$ of catalyst are routinely used. The collection time is usually of the order of 12 hours *per* spectrum for most experiments.

In order to measure the signal of adsorbed species on the surface of a catalyst, it is necessary to determine the background spectrum. The background is the spectrum of the sample before any treatment or dosing. The resulting spectrum, after subtracting the background, is called the difference spectrum. This is the spectrum reported and analysed. It is common practice, when possible, to measure the spectrum of the substance to be adsorbed.

It is also good practice, when possible, to remove extra sources of scattering, such as water adsorbed during catalyst treatment. This is generally done with the catalyst in the cell by flowing inert gases and evacuating while monitoring the desorbed products with a mass spectrometer. This procedure can be very time-consuming.

In order to understand the catalytic chemistry, we need to know, *inter alia*, the local structure of the active site, how it is changed (if at all) by interaction with a reactant molecule and how the structures of the bound reactant molecules change (if at all) through their interaction with the active site. With regard to the use of the neutron as a spectroscopic probe in catalytic studies, neutrons, strictly, do not see single sites. What is seen in the INS spectrum is the cumulative effect on the energies and numbers of the scattered neutrons of all collisions with the catalyst and adsorbed species. The challenge is to decompose the spectrum into the different scattering species by intelligent modelling of the spectrum. The aim of the modelling is to generate a calculated spectrum by *e.g.* DFT calculations of the vibrations of supposed components added together. Such modelling is a vital part of the interpretation of INS spectra; although, before the advent of powerful computers and computational methods a qualitative interpretation of an INS spectrum using knowledge of the characteristic infrared and Raman spectra may have been performed. It should also be noted that the catalytically significant structure of a species may not be seen because its concentration is usually low. But this is a problem in all research on catalysts, overcome to some degree by pooling knowledge from a number of techniques and from chemical and theoretical understanding. Here we show, with well-chosen examples, how the neutron can be beneficial as a probe in studies of catalysts. But first we make a short digression on experimental technique.[95]

INS is a macroscale technique: $10-50\,g$ of catalyst are held in a containing vessel ('can'), which may serve as a reactor for pretreatment, *e.g.* by activation in hydrogen, and for dosing with reactant. The containing vessel is commonly made of aluminium, steel or zirconium alloy, all of which are transparent to neutrons. The catalyst samples to be studied can be prepared offsite and brought to the neutron facility in sealed cans ready for positioning in the neutron beam, or else are prepared in a reactor system onsite. It is possible, should the investigation require it, for a can to be loaded directly from a commercial reactor. The use of bulk quantities of catalyst is possible because neutrons are highly penetrating. Scattering is caused almost entirely by the hydrogen component of the catalyst or reactants.

Catalysis takes place on the surface; it is possible to observe hydrogenous surface species by subtracting the spectrum of the catalyst before reaction. It is common practice to record the spectrum of the catalyst (and the can) before pretreatment or reaction (giving the so-called 'background spectrum') and to subtract this spectrum from the spectrum of the pretreated or reacted catalyst thereby observing species of interest. The practice of containing the catalyst in sturdy metal cans facilitates the use of elevated pressures, temperatures and *in situ* experiments.[96,97] But there are two drawbacks: the time needed to accumulate the scattering for a well-resolved spectrum and the low temperature (below 30 K) needed to minimise Debye–Waller broadening. For samples with more than $\sim 10^{22}$ protons, the acquisition of well-resolved spectra requires that the sample be exposed to the neutron beam for up to 24 hours.[98] For example, a Pt/C fuel cell catalyst requires 15 g of hydrogen-dosed catalyst in the beam. The demands of the technique and the cost of experimental time at a central neutron scattering facility make imperative careful planning of an experiment, including computational modelling!

How INS is used in studies of catalysts is illustrated by in a series of classic experiments on the reaction between hydrogen and carbon monoxide catalysed by Raney nickel:[12,99,100]

$$3H_2 + CO \rightarrow CH_4 + H_2O \qquad (3.16)$$

The Raney nickel catalyst (85 g, $14\,m^2\,g^{-1}$) was loaded into the INS cell, which was used for gas dosing and as the reactor. The cell could be attached to a gas line for dosing and to monitor pressure changes, and to a gas chromatograph in order to analyse reaction products. The sealed cell was removed from the gas line and transferred to the INS spectrometer. By dosing with hydrogen and carbon monoxide, the reaction was carried out at various temperatures and the INS spectra were recorded at 80 K: effectively a stopped flow procedure. Because of the time taken to accumulate an INS spectrum, the INS technique cannot be used directly to follow a chemical reaction. The gas dosing and the reaction was allowed to proceed for a certain time under the required conditions and then frozen prior to the INS measurement. The catalyst in the cell was first heated in hydrogen to remove surface oxide. The cell was then evacuated and the INS spectrum was recorded. This is the background or blank spectrum, which is subtracted from all subsequent spectra. Typical spectra are shown and described in Figure 3.14. The main observations from these are:

- From surface hydrogen atoms, strong scattering at $900-1200\,cm^{-1}$ is assigned to displacements (vibrations) of hydrogen atoms at

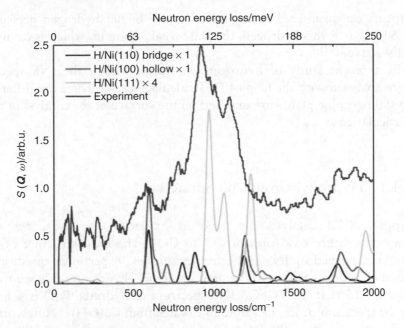

Figure 3.14 INS spectrum of Raney nickel with hydrogen adsorbed.[14] The *ab initio* calculations of hydrogen adsorbed on Ni (111), Ni(110) and Ni(100) are also shown. Reproduced under the terms of the STM agreement from Parker *et al.* (2010) [14]. Copyright (2010) Royal Society of Chemistry.

nickel adsorption sites and in threefold coordination: $940\,cm^{-1}$, E mode, displacement parallel to surface; $1130\,cm^{-1}$, A mode, perpendicular.

- At 80 K dosing, dihydrogen-saturated Raney nickel with carbon monoxide hardly changes the spectrum; carbon monoxide and dihydrogen adsorbed independently and did not interact.
- At 273 K, after dosing with dihydrogen carbon monoxide, hydrogen evolved and new hydrogenous species appear on the surface.
- After reaction in a flowing H_2/CO mixture at 300 and 450 K, new peaks near 450 and $600\,cm^{-1}$ can be assigned to surface C–H species (by comparison with the spectra of adsorbed ethyne and ethane). Gas-phase products (desorbed CH_4 and C_2, C_3, C_4 and water) are observed only after reaction at 450 K.

In a later study of hydrogen on Raney nickel (hydrogenation of acetonitrile), a peak at $1800\,cm^{-1}$ was observed and assigned to the Ni–H stretching vibration of on-top hydrogen (terminal hydrogen). This peak appears on unsintered Raney nickel under higher hydrogen pressures (1.5 kPa). It is this hydrogen, the least weakly bound, which is catalytically active.[101]

In a recent study of hydrogen on Raney nickel, the INS spectra were analysed with the help of DFT calculations.[14] There are different crystallographic planes are exposed on the surface of the catalyst in this particular case.

3.4.1   Hydroxyl Groups on Surfaces

Supported Pd catalysts are used in oxidation reactions, *e.g.* the low-temperature oxidation of CO to CO_2. The local structure of Pd has been probed by INS and other techniques. A pertinent question is whether hydrous PdO is a hydrated oxide or a hydroxide. We see from Figure 3.15 that the TOSCA INS spectrum of hydrated PdO is similar to the spectrum of ice, *i.e.* more like water than $Ca(OH)_2$. Noteworthy is the absence from the Pd spectrum of a strong peak near $300 \, cm^{-1}$ in the spectrum of $Ca(OH)_2$ assigned to an OH lattice translational mode.[102] This conclusion is supported by DFT calculation of the INS spectra of Pd hydroxy and Pd *aqua* species (Figure 3.16).

The INS spectra were also recorded on the MARI instrument, which enables us to record spectra at ambient temperatures and has a better resolution than the TOSCA instrument at higher frequencies (Figure 3.17). We see that the spectra are comparable at ambient temperature and 5 K, giving us confidence in the modelling and reassuring us that the system does not change within that temperature range.

The results show that the formulation $PdO \cdot H_2O$ better describes the experimental results than the previously proposed $Pd(OH)_2$ model. This model provides a quantitative explanation for why hydrous palladium oxide deactivates three times faster at 25 than at $100 \,°C$ in the low-temperature oxidation of CO to CO_2. From Figure 3.16, the conclusion is that within the simplicity of the model, the surface is best represented as a periodic slab of PdO terminated with hydroxyls and later a water molecule is added on the surface. The periodicity of the water molecules in the model gives a more structured set of bands between 450 and $800 \, cm^{-1}$, whereas the observed spectrum contains broad features and a higher onset of the librational modes from the observed $350 \, cm^{-1}$ to the calculated $470 \, cm^{-1}$. The calculation correctly predicts the upshift of the hydroxyl mode from $\sim865 \, cm^{-1}$ for the hydroxyl-terminated

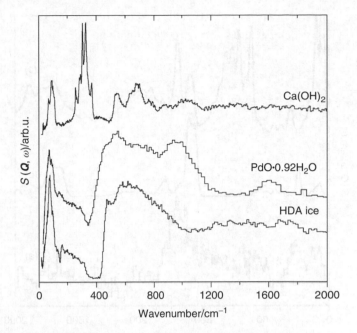

Figure 3.15 INS spectra at 20 K of Ca(OH) (top), PdO·0.92H$_2$O (middle) and high-density amorphous (HDA) ice (bottom), recorded on TOSCA. Reproduced under the terms of the STM agreement from Parker *et al.* (2010) [31]. Copyright (2010) American Chemical Society.

surface to 968 cm^{-1} for the water-covered one. The evidence of the simultaneous presence of water molecules and hydroxyls comes from of bending modes of the hydroxyls at around 960 cm^{-1}, of water features at around 1620 cm^{-1} and of the bonded OH signal at 3480 cm^{-1} (see Figure 3.17).

The water molecules associated with the Pd provide a supply of hydroxyl groups to replenish hydroxyl groups used up in the low-temperature oxidation of CO.

3.5   CO$_2$ AND SO$_2$ CAPTURE

The efficient removal of gases such as CO$_2$ and SO$_2$ is a major challenge in the development of the 'low-carbon economy'.[103] At present, the removal of CO$_2$ generated by power plants uses solutions of organic amides. The cost and energy penalty associated with the regeneration of the amine solutions, as well as their corrosive properties, limit such

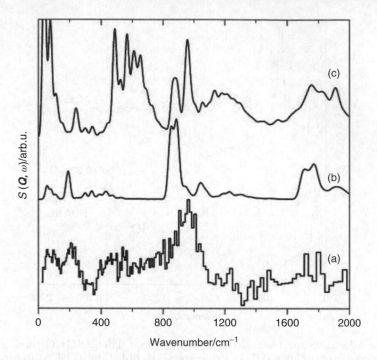

Figure 3.16 (a) Experimental TOSCA INS spectrum of PdO · 0.31(H₂O). (b) Calculated INS spectrum of [PdO]₃(OH)₂. (c) Calculated spectrum of [PdO]₃(OH)₂(H₂O)₂. Note the split peak near 950 cm⁻¹ in (a) and (c), and a feature near 500 cm⁻¹ in (a) and (c) Reproduced under the terms of the STM agreement from Parker *et al.* (2010) [31]. Copyright (2010) American Chemical Society.

applications in the long term.[104,105] There are major drives to develop efficient strategies for removing CO_2 using alternative materials with high selectivity and easy regeneration at a viable cost. MOF complexes are candidate materials for gas storage and separation, and are being studied for the purpose of CO_2 capture.[106]

INS has particular sensitivity in the detection of the motion of hydrogen atoms, so its use in the study of CO_2 and SO_2 is not immediately obvious. Figure 3.18 (top left) shows the spectra of the porous MOF NOTT-300 before and after CO_2 adsorption. The small differences between spectra are due to the presence of CO_2 in the porous space. Exploiting the fact that INS can be readily and rigorously determined from *ab initio* dynamic calculations, the corresponding INS spectra for both the empty and the CO_2-dosed NOTT-300 were calculated (Figure 3.18, bottom left) using a structural model from synchrotron

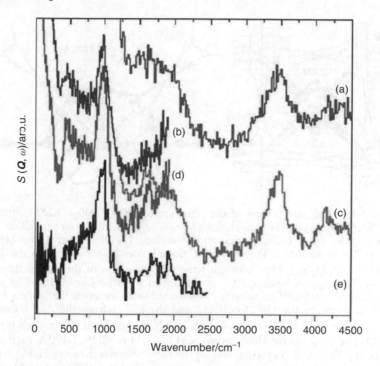

Figure 3.17 INS spectra of PdO · 0.31(H$_2$O) recorded on MARI at ambient temperature with (a) 5243 cm^{-1} and (b) 2017 cm^{-1} incident energy, at 5 K with (c) 5243 cm^{-1} and (d) 2017 cm^{-1} incident energy and (e) on TOSCA at 20 K. Reproduced under the terms of the STM agreement from Parker *et al.* (2010) [31]. Copyright (2010) American Chemical Society.

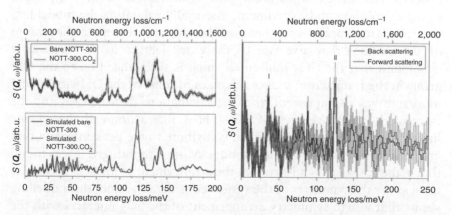

Figure 3.18 *In situ* inelastic neutron scattering INS spectra for bare and CO$_2$-loaded NOTT-300. Left panel: comparison of the experimental (top) and DFT simulated (bottom) spectra. Right panel: difference plot for experimental spectra. Two distinct energy transfer peaks are labelled as I and II. Reproduced with permission from Yang *et al.* (2012) [106]. Copyright (2012) Nature Publishing Group.

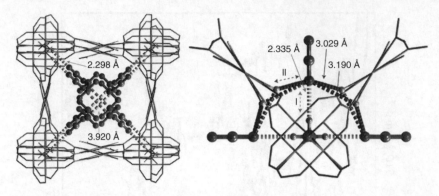

Figure 3.19 Left panel: view of the structure of NOTT-300 − 3.2CO$_2$ obtained from PXRD analysis. The adsorbed CO$_2$ molecules in the pore channel are highlighted by the use of ball-and-stick representation. The dipole interaction between CO$_2$ (I, II) molecules is shown by the dotted lines at the centre of the image [O=C=O···CO$_2$ = 3.920 Å]. (Right panel: detailed view of the role of the −OH and −CH groups in binding CO$_2$ molecules in a pocket-like cavity. The model was obtained from DFT simulation. The dotted lines represent the modest hydrogen bond between the O(δ^-) of CO$_2$ and the H(δ^+) from the Al−OH moiety [O···H = 2.335 Å] and the weak cooperative hydrogen bond interactions between the O(δ^-) of CO$_2$ and the H(δ^+) from −CH [O···H = 3.029, 3.190 Å, each occurring twice]. Each O(δ^-) centre therefore interacts with five different H($\delta+$) centres. Reproduced with permission from Yang *et al.* (2012) [106]. Copyright (2012) Nature Publishing Group.

X-ray diffraction data (Figure 3.19) as a starting point. It may be observed that the extra feature appearing around 1000 cm^{-1} is present in the calculation and experiment, essentially validating the modelling; the calculated spectra can therefore be used to determine the nature of the motions that give rise to the extra feature in the spectra. It was found that the CO$_2$ molecule, which is sitting head-on to the OH group in the framework, induces changes in some of the OH vibrational modes; it was also apparent from the calculations that the CO$_2$ and the framework interact via an O−C−O···H−C interaction with the organic linker; there are new modes which contribute to the peak intensity that are clearly associated with these modes. More interestingly, it was found that in the case of SO$_2$ adsorption, the arrangement of the SO$_2$ molecules must lower the symmetry of the system considerably, since the modelling shows that a high-symmetry arrangement of the SO$_2$ interacts with the chiral structure of the framework and makes it collapse. The calculations also show that the CO$_2$ molecule in the adsorption site is sitting in a strongly anharmonic potential well, with a potential energy surface that is calculated as a negative (imaginary) frequency. The system is nonetheless quantum mechanically and thermodynamically stable. In short, *in situ* powder X-ray diffraction and INS studies, combined with modelling, reveal that hydroxyl groups bind CO_2 and SO_2 through the formation of $O=C(S)=O(\delta^-)\cdots H(\delta^+)-O$ hydrogen bonds, which are reinforced by weak supramolecular interactions with C–H atoms on the aromatic rings of the framework. This offers the potential for the application of new 'easy-on/easy-off' capture systems for CO_2 and SO_2, which carry fewer economic and environmental penalties.

In this study, we can imagine the neutrons and the HO groups of the MOF acting together as a probe of the local structure and dynamics of the trapped CO_2 and SO_2 molecules.

3.6   WHAT COULD BE NEXT?

At ISIS, the pulsed spallation process produces neutrons by firing 0.8 GeV protons with an effective current of the order of 200 μA and a power dissipated in the tungsten target of 160 kW. This produces around 2×10^{13} neutrons cm^{-2} s^{-1} average flux and a peak flux of 8×10^{15} neutrons cm^{-2} s^{-1} peak flux. The source operates at 50 Hz.[107] SNS produces an average proton beam current on target of 1400 μA, with a 1.0 GeV proton energy, and operates at 60 Hz, producing an average flux of 1.5×10^{14} neutrons cm^{-2} s^{-1}.[22] JPARC produces a similar number of neutrons as SNS, but the energy of the proton beam is higher at 3 GeV, with a repetition rate of 25 Hz.[108] These numbers show that in the 25 years since ISIS was first operated, the available neutron flux has increased by a factor of the order of 10 or 20. Unless there is a breakthrough in the methods used to produce neutrons, it is difficult to foresee a radical increase in the number of neutrons produced by spallation methods.

3.6.1   How Could We Improve INS?

Table 3.3 presents a comparison between the direct geometry INS spectrometers at ISIS. All these instruments are based at the first target

Table 3.3 Comparison of various spectrometers at ISIS, for a 3% energy resolution and an intensity at 50 meV. From Bewley *et al.* (2010) [19].

	MERLIN	HET	MARI	MAPS
Intensity/neutrons meV^{-1} s^{-1}	400×10^4	50×10^4	58×10^4	113×10^4
Detector pixels	69 000	2000	1000	42 000
Solid angle (Sr)	3.1	0.4	0.4	0.45
Angular range	3–135°	3–30°	3–132°	3–60°
		110–135°		
Best % energy res.	3	3	2	1.5
MERLIN gain	–	62	53	24

station, so the comparison is fair; the improvements are mainly due to better neutron guides and larger detector areas (please note that HET has been decommissioned and is no longer part of the instrument suite at ISIS).

Further operational gains can and have been obtained with the improvement of target, neutron guide and moderator design, as it has been demonstrated at the ISIS second target station (TS2).[19,109,110]

3.6.1.1 INS of Non-hydrogenous Materials

We emphasise the value of INS spectroscopy largely as a local probe for hydrogen; however, with the increasing intensity of neutron sources, it is now becoming feasible to record the INS spectra of nonhydrogenous materials.

The major difficulty in studying non-hydrogenous compounds is the lack of sensitivity. This largely accounts for the lack of studies of non-hydrogenous systems. It is only in the last decade or so that neutron sources have become sufficiently intense that the recording of incoherent INS spectra of non-hydrogenous systems has become feasible. The only realistic way to overcome the sensitivity problem is to use large samples; sample quantities of 0.05–0.10 mol are needed to record a useful spectrum in 24 hours or more. An example of the sensitivity required can be seen in Figure 3.20, in which the spectrum corresponds to 20 g of solid CO_2 measured for 24 hours. The calculated spectrum is also shown. There is a need for a larger neutron flux at sample positions; the resulting flux increase as a function of the neutron incident energy if the TOSCA spectrometer were to have a neutron guide is shown in Table 3.4.

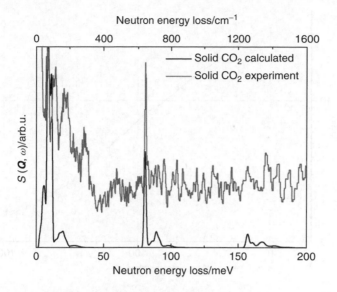

Figure 3.20   The INS spectrum of solid CO_2.

Table 3.4   Gain achieved with TOSCA's guide.

Energy at sample/meV	Gain
10	39.5
20	24.3
50	12.2
100	7.4
150	5.1
200	3.9
250	3.3
300	2.7
350	2.5
400	2.4
566	1.9

3.6.1.2   Multiple Incident Energies Methods

Innovative approaches, such as designing choppers that allow the measurement of different incident energies in a single frame[24,111,112] for direct geometry instruments, have been implemented in the new LET spectrometer at ISIS, using disk choppers,[19] (see Figure 3.21) and in the Fermi chopper spectrometer 4SEASONS at JPARC[23,24] (see Figure 3.22).

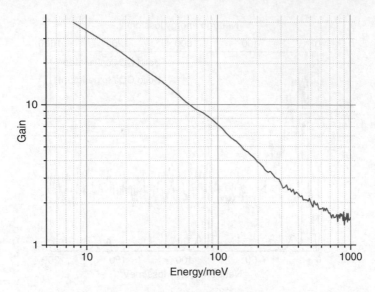

Figure 3.21 Gain achieved with a TOSCA simulated guide between 0 and 1000 meV. Top: linear scales; bottom in: logarithmic plot.

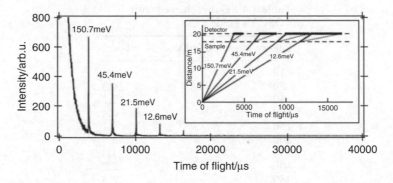

Figure 3.22 ToF spectrum of $CuGeO_3$ single crystal. Data were collected using detector pixels positioned in the horizontal plane. The chopper frequency was 200 Hz and phase delay was tuned to the incident energy of 45.4 meV. Other incident neutrons with different energies were also concomitantly delivered to the sample. The inset shows the ToF diagram of four inelastic processes with multiple E_i. Reproduced with permission from Nakamura *et al.* (2009) [24]. Copyright (2009) Physical Society of Japan.

This allows a more efficient use of the neutrons by taking advantage of the larger time frame, *e.g.* at TS2 in ISIS the frame is 100 ms, as TS2 operates at 10 Hz, and at JPARC the frame is 40 ms, since it operates at 25 Hz. In such situations, the INS spectra of polyethylene can be measured simultaneously and should appear as in Figure 3.23.

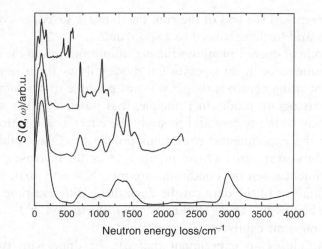

Figure 3.23 Calculated spectra of polyethylene in a direct geometry instrument such as MERLIN with multiple incident energies $E_{incident} = 600$, 1130, 2300 and $4500\,cm^{-1}$, T = 300 K, an energy resolution of $\Delta\omega/E_{incident} = 3\%$ and at an angle of $3°$.

3.6.1.3 Event-Mode Data Collection, Time-Resolved INS?

At a pulsed source, the use of ToF techniques means that the smallest time frame is of the order of 20 ms. This establishes a lower limit for the real-time resolution that can be used to study catalysis with these techniques. To date, all studies of catalysts performed with INS have required collection times of a few hours in order to achieve good statistics on the collected datasets.

So far, all INS experiments on catalysis performed with neutrons have been *post mortem*, *i.e.* the catalyst sample backgrounds have been measured, some chemical reaction has taken place and then another measurement has been performed. Usually the aim is the determination of some intermediate or reacted species. The spectrum of the species of interest is obtained by subtracting the corresponding spectra.

In the older instruments, data are stored in histograms; essentially the time frame is subdivided and the data are stored at the time of arrival at the detector in the corresponding time interval with respect to the origin of the frame. Data are accumulated during a given period of time and further saved and reduced to produce the spectrum.

Data collection in the newer instruments is done in the so-called 'event mode':[113] at the arrival of each neutron, the time of detection is stored for each event. At the end of the run, the data are saved, converted to a histogram and further reduced to a spectrum.

This mode of operation allows further offline analysis of the data after the experiment. Such an operational mode allows for a new type of experiment using neutrons that it was not possible to conduct with the previous histogram mode. In principle, it is possible to use a periodic perturbation of the system and to study the effect as a function of time and steer the experimental conditions on the fly. One could monitor the signal-to-noise ratio as the number of cycles increases, and stop the experiment when the conditions are met. Not only that, but a *post mortem* analysis of the data can be done using different time slicing. It may even be possible to do studies using *in operando*[114,115] conditions with INS on some catalysts.

An example of an experiment that can be done with this operational mode is to periodically pulse gases throughout a cell and collect the spectrum; after a number of periods, the dataset can be folded so that it consists of a series of spectra which give a picture of the reaction on time. Realistically, the time limit will be of the order of minutes in the best-case scenario, since the limitation will be in the macroscopic diffusion of species throughout the material. Figure 3.24 represents schematically the hypothetical study of a catalysed reaction. In this particular case, the model molecules are perhydro-N-ethylcarbazole ($C_{14}H_{25}N$) dehydrogenating through a series of elementary steps into N-ethylcarbazole ($C_{14}H_{13}N$), an organic molecule that has been proposed as an organic liquid hydrogen carrier.[116–119] This is just an numerical example of a molecule containing different types of bond, to show the hypothetical INS spectra.

The fact that gases are flowing and chemical reactions happening means that this type of experiment can only be performed in a direct geometry instrument. The spectra have been calculated with $E_{incident} = 4000\ meV^{-1}$ and a resolution of 1% of $E_{incident}$ at a temperature of 300 K.

3.6.2   A Hypothetical INS Instrument for Catalysis

Neutrons are a very scarce resource. There are few places in the world where they can be used as a probe for INS. Both direct and indirect geometry instruments have a role to play in the understanding of catalysis. The model in Figure 3.23 could have relative good resolution

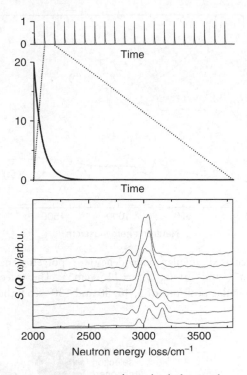

Figure 3.24 Schematic representation of a pulsed chemical reaction. Top panel: a flow of gas is fed periodically through the system. Middle panel: the information is folded into a single frame (note the x- and y-axis scales). Bottom panel: a sequence of hypothetical spectra of an organic molecule as it would look if measured by a single detector at 3° at a temperature of 300 K as it dehydrogenates by losing two hydrogen atoms in each step as a function of time; the absence of particular peaks *e.g.* will give information about the dehydrogenation mechanism.

at all energies, using the multiple incident energies in a direct geometry instrument. However, this is the spectrum calculated at an angle of 3°, and as can be seen from the right-hand panel in Figure 3.6, the spectra degrade with increasing angle.

If we calculate the spectra at a series of angles, as in Figure 3.25, it is clear that the low angular positions, below 40°, will be of more relevance to the study of hydrogen-containing materials. The new direct geometry spectrometers have a large coverage in $|Q|$; because this is fundamental to the study of coherent and magnetic studies, these instruments can sample multiple Brillouin zones,[18,19] but for hydrogenous substances, large $|Q|$ sampling is not necessary. If we also take into consideration that indirect geometry instruments like TOSCA and VISION are very compact, it is possible to envisage an INS spectrometer that has direct

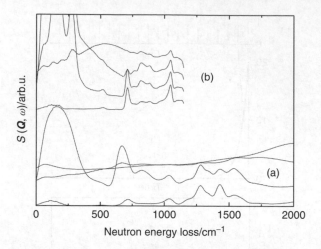

Figure 3.25 Calculated angular dependency for polyethylene at $300\,\mathrm{K}$ at (a) $E_{\mathrm{incident}} = 2300\,\mathrm{cm}^{-1}$ and (b) $E_{\mathrm{incident}} = 1150\,\mathrm{cm}^{-1}$. The traces for each group are calculated at angular positions of 3, 40, 80 and $130°$ from the bottom up.

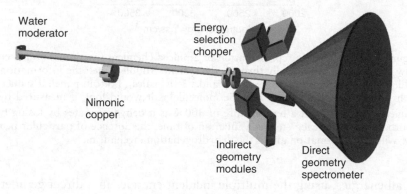

Figure 3.26 A simplified view of a hybrid INS spectrometer.

geometry configuration in forward scattering and an indirect geometry in backscattering, like the one shown in Figure 3.26. This will require that the chopper is moved in and out in order to change operational mode, but that could be done automatically. A higher-temperature measurement could be made with the direct spectrometer, while at low temperatures the indirect geometry spectrometer could be used. When illuminating the sample with a white beam (indirect geometry operation), the direct geometry spectrometer acts as a diffraction bank at lower momentum transfer. The reduced size of the direct spectrometer will also result in a lower price tag. Such an instrument being considered for a possible future upgrade of the first target station at ISIS. A similar concept has already been used in the NOVA[120] instrument at JPARC, where an instrument designed as a diffractometer has a series of choppers that can be used to transform the diffractometer into a low-resolution INS spectrometer. This instrument has been built and is being commissioned at the time of writing.[121] Such an instrument could be built at any of the major spallation neutron sources, SNS, JPARC or ISIS, and make good use of neutrons for catalysis.

3.7 CONCLUSION

INS is an excellent vibrational technique by which to study the local structures of inorganic materials. Because INS is not subject to optical selection rules, all vibrations are active and, in principle, measurable. The whole range of the molecular vibrational and rotational spectra, $0-550$ meV ($0-4400$ cm^{-1}) is covered by INS spectrometers. INS observations are not restricted to the centre of the Brillouin zone (Γ point), as are the optical techniques, so the shape of the bands has a physical meaning. INS spectra are readily and accurately modelled; the intensities are proportional to the concentration of elements in the sample and their cross-sections, the measured INS intensities relate straightforwardly to the associated displacements of the scattering atom and subtraction of backgrounds is straightforward. Therefore, computer modelling plays a fundamental part in the understanding of the INS experiments in catalysis, gas adsorption, encapsulated molecules *etc*.

Because neutrons penetrate deeply into materials and pass easily through the walls of metal sample vessels, such as aluminium and stainless steel containers, neutrons are ideal for measuring the bulk properties of samples within controlled environments; there is no need for optical windows or gas dosing, and the application of magnetic and electric fields is experimentally straightforward. New developments in neutronics (neutron guides and moderator design), together with multiple-incident-energies choppers and data acquisition electronics, will increase the efficiency of INS instruments. This should lead in the near future to the implementation of *in operando* experiments on catalysts, taking full advantage of the unique properties of INS. Thus, a hybrid indirect–direct geometry instrument especially orientated towards catalysis research is envisaged.[114,115]

REFERENCES

[1] A. J. Ramirez-Cuesta, *Comp. Phys. Commun*, **157**, 226 (2004).

[2] P. C. H. Mitchell, S. F. Parker, A. J. Ramirez-Cuesta and J. Tomkinson, *Vibrational Spectroscopy With Neutrons: With Applications in Chemistry, Biology, Materials Science and Catalysis*, World Scientific, London, 2005.

[3] G. E. Bacon, *Neutron Physics*, Wykeham Publications, London and Winchester, 1969.

[4] J. Chadwick, *Nature*, **120** (1932).

[5] P. C. H. Mitchell, *Acta Phys. Hung.*, **75**, 131 (1994).

[6] E. B. Wilson, Jr., J. C. Decius and P. C. Cross, *Molecular Vibrations. The Theory of Infrared and Raman Vibrational Spectra*, Dover, New York, 1980.

[7] X. Gonze and C. Lee, *Phys. Rev. B*, **55**, 10355 (1997).

[8] K. Refson, P.R. Tulip and S.J. Clark, *Phys. Rev. B*, **73**, 125120 (2006).

[9] S. J. Clark, M. D. Segall, C. J. Pickard, P. J. Hasnip, M. J. Probert, K. Refson and M. C. Payne, Z. *Krist.* **220**, 567 (2005).

[10] M. Raney, US Patent 1628190 (1927).

[11] R. D. Kelley and T. Udovic, *Abs. Papers Am. Chem. Soc.*, **188**, 41 (1984).

[12] R. D. Kelley, R. R. Cavanagh and J. J. Rush, *J. Catalysis*, **83**, 464 (1983).

[13] H. Jobic, G. Clugnet and A. J. Renouprez, *J. Electron Spectrosc. Relat. Phenom.*, **45**, 281 (1987).

[14] S. F. Parker, D. T. Bowron, S. Imberti, A. K. Soper, K. Refson, E. S. Lox, M. Lopez and P. W. Albers, *Chem. Commun.*, **46**, 2959 (2010).

[15] M. Arai, *Adv. Colloid Interface Sci.*, **71–72**, 209 (1997).

[16] K. Andersen, *Nucl. Instrum. Methods Phys. Res., Sect. A*, **371**, 472 (1996).

[17] T. G. Perring, A. D. Taylor, D. Paul, A. T. Boothroyd and G. Aeppli, In *Proceedings of the Twelfth Meeting of the International Collaboration on Advanced Neutron Sources, Icans-xii*, RAL Report 94-025, 1994.

[18] R.I. Bewley, R.S. Eccleston, K.A. McEwen, S.M. Hayden, M.T. Dove, S.M. Bennington, J.R. Treadgold and R.L.S. Coleman, *Physica B – Condens. Matter*, **385–386**, 1029 (2006).

[19] R. I. Bewley, J. W. Taylor and S. M. Bennington, *Nucl. Instrum. Methods Phys. Res., Sect. A*, **637**, 128 (2010).

[20] E. S. Božin, P. Juhás, W. Zhou, M. B. Stone, D. L. Abernathy, A. Huq and S. J. L. Billinge, *J. Phys.: Conf. Series*, **251**, 012080, (2010).

[21] G. E. Granroth, D. H. Vandergriff and S. E. Nagler, *Physica B: Condens. Matt.*, **385–386**, 1104 (2006).

[22] SNS Web site http://neutrons.ornl.gov/ (last accessed 22 May 2013).

[23] R. Kajimoto, K. Nakajima, M. Nakamura, K. Soyama, T. Yokoo, K. Oikawa and M. Arai, *Nucl. Instrum. Methods Phys. Res., Sect. A* **600**, 185 (2009).

[24] M. Nakamura, R. Kajimoto, Y. Inamura, F. Mizuno, M. Fujita, T. Yokoo and M. Arai, *J. Phys. Soc.*, **78** (2009).

[25] JPARC Web site, http://j-parc.jp/index-e.html (last accessed 22 May 2013).

[26] D. Colognesi, M. Celli, F. Cilloco, R. J. Newport, S. R. Parker, V. Rossi-Albertini, F. Sacchetti, J. Tomkinson and M. Zoppi, *Appl. Phys. A: Mater. Sci. Process.* **74**, S64 (2002).

[27] P. A. Seeger, L. L. Daemen and J. Z. Larese, *Nucl. Instrum. Methods Phys. Res., Sect. A*, **604**, 719 (2009).

[28] G. D. Barrera, S. F. Parker, A. J. Ramirez-Cuesta and P. C. H. Mitchell, *Macromolecules*, **39**, 2683 (2006).

[29] P. W. Albers, J. Pietsch, J. Krauter and S. F. Parker, *Phys. Chem. Chem. Phys.*, **5**, 1941 (2003).

[30] I. P. Silverwood, N. G. Hamilton, C. J. Laycock, J. Z. Staniforth, R. M. Ormerod, C. D. Frost, S. F. Parker and D. Lennon, *Phys. Chem. Chem. Phys.*, **12**, 3102 (2010).

[31] S. F. Parker, K. Refson, A. C. Hannon, E. R. Barney, S. J. Robertson and P. W. Albers, *J. Phys. Chem. C*, **114**, 14164 (2010).

[32] I. F. Silvera, *Rev. Mod. Phys.*, **52**, 393 (1980).

[33] I. F. Silvera and M. Nielsen, *Phys. Rev. Lett.*, **37**, 1275 (1976).

[34] J. A. Young and J. U. Koppel, *Phys. Rev. A: Gen. Phys.*, **135**, A603 (1964).

[35] M. Celli, D. Colognesi and M. Zoppi, *Eur. Phys. J. B: Cond. Matt. Comp. Syst.*, **14**, 239 (2000).

[36] D. Colognesi, F. Formisano, A. J. Ramirez-Cuesta and L. Ulivi, *Phys. Rev. B*, **79**, (2009).

[37] D. White and E. N. Lassettre, *J. Chem. Phys.*, **32**, 72 (1960).

[38] J. Eckert, J. M. Nicol, J. Howard and F. R. Trouw, *J. Phys. Chem.*, **100**, 10646 (1996).

[39] B. L. Mojet, J. Eckert, R. A. van Santen, A. Albinati and R. Lechner, *J. Am. Chem. Soc.*, **123**, 8147 (2001).

[40] J. Z. Larese, T. Arnold, L. Frazier, R. J. Hinde and A. J. Ramirez-Cuesta, *Phys. Rev. Lett.*, **101** (2008).

[41] A. Ramirez-Cuesta and P. C. H. Mitchell, *Catalysis Today*, **120**, 368 (2007).

[42] A. J. Ramirez-Cuesta, P. C. H. Mitchell, D. K. Ross, P. A. Georgiev, P. A. Anderson, H. W. Langmi and D. Book, *J. Alloys Compd.*, **446**, 393 (2007).

[43] A. J. Ramirez-Cuesta, P. C. H. Mitchell, D. K. Ross, P. A. Georgiev, P. A. Anderson H. W. Langmi and D. Book, *J. Mat. Chem.*, **17**, 2533 (2007).

[44] A. J. Ramirez-Cuesta, P. C. H. Mitchell, S. F. Parker and P. A. Barrett, *Chem. Commun.*, 1257 (2000).

[45] C. A. Coulson, *Electricity*, Oliver & Boyd, Edinburg, 1953.

[46] J. G. Vitillo, A. Damin, A. Zecchina and G. Ricchiardi, *J. Chem. Phys.*, **122** (2005).

[47] J. Ward, *Trans. Faraday Soc.*, **67**, 1489 (1971).

[48] A.Y. Khodakov, L.M. Kustov, V.B. Kazansky and C. Williams, *J. Chem. Soc., Faraday Trans.*, **89**, 1393 (1993).

[49] E. Garrone, B. Bonelli and C. O. Areán, *Chem. Phys. Lett.*, **456**, 68 (2008).

[50] E. Garrone, B. Bonelli, C. Lamberti, B. Civalleri, M. Rocchia,. P. Roy and C. O. Arean, *J. Chem. Phys.*, **117**, 10274 (2002).

[51] M. Eddaoudi, D. B. Moler, H. Li, B. Chen, T. M. Reineke, M. O'Keeffe and O. M. Yaghi *Acc. Chem. Res.*, **34**, 319 (2001).

[52] A. J. Ramirez-Cuesta, M. O. Jones and W. I. F. David, *Mat. Today*, **12**, 54 (2009).

[53] J. L. C. Rowsell and O. M. Yaghi, *Ang. Chem., Int. Ed.*, **44**, 4670 (2005).

[54] K. S. Park, Z. Ni, A. P. Côté, J. Y. Choi, R. Huang, F. J. Uribe-Romo, H. K. Chae, M. O'Keeffe and O. M. Yaghi,, *Proc. Nat. Acad. Sci.*, **103**, 10186 (2006).

[55] N. L. Rosi, J. Eckert, M. Eddaoudi, D. T. Vodak, J. Kim, M. O'Keeffe, and O. M. Yaghi, *Science*, **300**, 1127 (2003).

[56] S. S. Y. Chui, S. M. F. Lo, J. P. H. Charmant, A. G. Orpen and I. D. Williams, *Science*, **283**, 1148 (1999).

[57] F. M. Mulder, T. J. Dingemans, H. G. Schimmel, A. J. Ramirez-Cuesta and G. J. Kearley, *Chem. Phys.*, **351**, 72 (2008).

[58] F. M. Mulder, B. Assfour, J. Huot, T. J. Dingemans, M. Wagemaker and A. J. Ramirez-Cuesta, *J. Phys. Chem. C*, **114**, 10648 (2010).

[59] Y. Liu, C. M. Brown, D. A. Neumann, V. K. Peterson and C. J. Kepert, *J. Alloys Compd.*, **446**, 385 (2007).

[60] S. Yang, S. K. Callear, A. J. Ramirez-Cuesta, W. I. F. David, J. Sun, A. J. Blake, N. R. Champness and M. Schroder, *Faraday Disc.*, **151**, 19 (2011).

[61] V. K. Peterson, Y. Liu, C. M. Brown and C. J. Kepert, *J. Am. Chem. Soc.*, **128**, 15578 (2006).

[62] V. K. Peterson, G. J. Kearley, Y. Wu, A. J. Ramirez-Cuesta, E. Kemner and C. J. Kepert, *Ang. Chem. Int. Ed.*, **49**, 585 (2010).

[63] C. M. Brown, Y. Liu, T. Yildirim, V. K. Peterson and C. J. Kepert, *Nanotech.*, **20**, 204025 (2009).

[64] Y. Liu, H. Kabbour, C. M. Brown, D. A. Neumann and C. C. Ahn, *Langmuir*, **24**, 4772 (2008).

[65] M. Dinca and J. R. Long, *Ang. Chem., Int. Ed.*, **47**, 6766 (2008).

[66] W. L. Queen, E. D. Bloch, C. M. Brown, M. R. Hudson, J. A. Mason, L. J. Murray, A. J. Ramirez-Cuesta, V.K. Peterson and J. R. Long, *Dalton Trans.*, **41**, 4180 (2012).

[67] N. L. Rosi, J. Kim, M. Eddaoudi, B. Chen, M. O'Keeffe, O. M. Yaghi, *J. Am. Chem. Soc.*, **127**, 1504 (2005).

[68] P. D. C. Dietzel, Y. Morita, R. Blom, and H. Fjellvag, *Ang. Chem., Int. Ed.*, **44**, 6354 (2005).

[69] P. D. C. Dietzel, B. Panella, M. Hirscher, R. Blom and H. Fjellvåg, *Chem. Commun.*, 959 (2006).

[70] S. R. Caskey, A. G. Wong-Foy and A. J. Matzger, *J. Am. Chem. Soc.*, **130**, 10870 (2008).

[71] W. Zhou, H. Wu and T. Yildirim, *J. Am. Chem. Soc.*, **130**, 15268 (2008).

[72] P. D. C. Dietzel, R. E. Johnsen, R. Blom and H. Fjellvag, *Chem. Eur. J.*, **14**, 2389 (2008).

[73] S. Bhattacharjee, J.-S. Choi, S.-T. Yang, S. B. Choi, J. Kim and W.-S. Ahn, *J. Nanosci. Nanotechnol.*, **10**, 135 (2010).

[74] J. G. Vitillo, L. Regli, S. Chavan, G. Ricchiardi, G. Spoto, P. D. C. Dietzel, S. Bordiga and A. Zecchina, *J. Am. Chem. Soc.*, **130**, 8386 (2008).

[75] X. Lin, J. Jia, X. Zhao, K. M. Thomas, A. J. Blake, G. S. Walker, N. R. Champness, P. Hubberstey and M. Schroder, *Ang. Chem., Int. Ed.*, **5**, 7358 (2006).

[76] D. J. Collins and H.-C. Zhou, *J. Mat. Chem.*, **7**, 3154 (2007).

[77] L. J. Murray, M. Dinca and J. R. Long, *Chem. Soc. Rev.*, **8**, 1294 (2009).

[78] J. Yang, A. Sudik, C. Wolverton and D. J. Siegel, *Chem. Soc. Rev.*, **39**, 656 (2010).

[79] P. Horcajada, S. Surble, C. Serre, D.-Y. Hong, Y.-K. Seo, J.-S. Chang, J.-M. Greneche, I. Margiolaki and G. Ferey, *Chem. Commun.*, 2820 (2007).

[80] S. Bauer, C. Serre, T. Devic, P. Horcajada, J. Marrot, G. Férey and N. Stock, *Inorg. Chem.*, **47**, 7568 (2008).

[81] A. Fateeva, S. Devautour-Vinot, N. Heymans, T. Devic, J.-M. Greneche, S. Wuttke, S. Miller, A. Lago, C. Serre, G. De Weireld, G. Maurin, A. Vimont and G. Férey, *Chem. Mat.*, **23**, 4641 (2011).

[82] G. J. Halder, K. W. Chapman, S. M. Neville, B. Moubaraki, K. S. Murray, J.-F. Letard and C. J. Kepert, *J. Am. Chem. Soc.*, **130**, 17552 (2008).

[83] S. Ma, D. Yuan, J.-S. Chang and H.-C. Zhou, *Inorg. Chem.*, **48**, 5398 (2009).

[84] K. Sumida, S. Horiki, S. S. Kaye, Z. R. Herm, W. L. Queen, C. M. Brown, F. Grandjean, G. J. Long, A. Dailly and J. R. Long, *Chem. Sci.*, **1**, 184 (2010).

[85] D. Lupu, O. Ardelean, G. Blanita, G. Borodi, M. D. Lazar, A. R. Biris, C. Ioan, M. Mihet, I. Misan and G. Popeneciu, *Int. J. Hydrogen Ener.*, **36**, 3586 (2011).

[86] J. S. Tse, D. D. Klug, J. Y. Zhao, W. Sturhahn, E. E. Alp, J. Baumert, C. Gutt, M. R. Johnson and W. Press, *Nat. Mat.*, **4**, 917 (2005).

[87] W. L. Mao, H. K. Mao, A. F. Goncharov, V. V. Struzhkin, Q. Guo, J. Hu, J. Shu, R. J. Hemley, M. Somayazulu and Y. Zhao, *Science*, **297**, 2247 (2002).

[88] L. J. Florusse, C. J. Peters, J. Schoonman, K. C. Hester, C. A. Koh, S. F. Dec, K. N. Marsh and E. D. Sloan, *Science*, **306**, 469 (2004).

[89] H. Lee, J-W Lee, D. Y. Kim, J. Park, Y-T Seo, H. I. Zeng, L. Moudrakovski, C. I. Ratcliffe and J. A. Ripmeester, *Nature*, **434**, 743 (2005).

[90] L. Ulivi, M. Celli, A. Giannasi, A. J. Ramirez-Cuesta, D. J. Bull and M. Zoppi, *Phys. Rev. B*, **76**, 161401 (2007).

[91] L. Ulivi, M. Celli, A. Giannasi, A. J. Ramirez-Cuesta and M. Zoppi, *J. Phys.: Cond. Matt.*, **20** (2008).

[92] K. T. Tait, F. Trouw, Y. Zhao, C. M. Brown and R. T. Downs, *J. Chem. Phys.*, **127**, 134505 (2007).

[93] M. Celli, D. Colognesi, A. Giannasi, L. Ulivi, M. Zoppi, V.G. Sakai and A. J. Ramirez-Cuesta, *Adv. Sci. Technol.*, **72**, 196 (2011).

[94] L. Ulivi, M. Celli, A. Giannasi, A. J. Ramirez-Cuesta and M. Zoppi, *J. Phys.: Conf. Ser.*, **121**, 042018 (2008).

[95] A. R. McFarlane, L. McMillan, I. P. Silverwood, N. G. Hamilton, D. Siegel, S. F. Parker, D. T. Lundie and D. Lennon, *Catalysis Today*, **155**, 206 (2010).

[96] P. W. Albers, G. Prescher, K. Seibold, S. F. Parker, *Chem. Eng. & Technol.*, **22**, 135 (1999).

[97] S. F. Parker, J. W. Taylor, P. W. Albers, M. Lopez, G. Sextl, D. Lennon, A. R. McInroy and I. W.. Sutherland, *Vibration. Spectrosc.*, **35**, 179 (2004).

[98] P. W. Albers and S. F. Parker, *Adv. Catalysis*, **51**, 99 (2007).

[99] R. D. Kelley, J. J. Rush and T. E. Madey, *Chem. Phys. Lett.*, **66**, 159 (1979).

[100] R. D. Kelley, R. R. Cavanagh, J. J. Rush and T. E. Madey, *Surf. Sci.*, **97**, L335 (1980).

[101] F. Hochard, H. Jobic, J. Massardier and A. J. Renouprez,, *J. Mol. Catal. A: Chem.*, **95**, 165 (1995).

[102] R. Baddourhadjean, F. Fillaux, N. Floquet, S. Belushkin, I. Natkaniec, L. Desgranges and D. Grebille, *Chem. Phys.*, **197**, 81 (1995).

[103] D. W. Keith, *Science*, **325**, 1654 (2009).

[104] G. T. Rochelle, *Science*, **325**, 1652 (2009).

[105] C. Villiers, J.-P. Dognon, R. Pollet, P. Thuery and M. Ephritikhine, *Ang. Chem., Int. Ed.*, **49**, 3465 (2010).

[106] S. Yang, J. Sun, A. J. Ramirez-Cuesta, S. K. Callear, W. I. F. David, D. P. Anderson, R. Newby, A. J. Blake, J. E. Parker, C. C. Tang and M. Schroder, *Nat. Chem.*, **4**, 887 (2012).

[107] ISIS Web site http://www.isis.stfc.ac.uk/ (last accessed 22 May 2013).

[108] F. Maekawa, M. Harada, K. Oikawa, M. Teshigawara, T. Kai, S. Ichiro Meigo, M. Ooi, S. Sakamoto, H. Takada, M. Futakawa, T. Kato, Y. Ikeda, N. Watanabe, T. Kamiyama, S. Torii, R. Kajimoto and M. Nakamura, *Nucl. Instrum. Methods Phys. Res., Sect. A*, **620**, 159 (2010).

[109] D. T. Bowron, A. K. Soper, K. Jones, S. Ansell, S. Birch, J. Norris, L. Perrott, D. Riedel, N. J. Rhodes, S. R. Wakefield, A. Botti, M. A. Ricci, F. Grazzi and M. Zoppi, *Rev. Sci. Inst.*, **81**, 033905 (2010).

[110] S. M. Bennington, *Nucl. Instrum. Methods Phys. Res., Sect. A* **600**, 32 (2009).

[111] F. Mezei and M. Russina, In *Advances in Neutron Scattering Instrumentation*, I. S. Anderson and B. Guerard (Eds), SPIE Optical Engineering Press, Bellingham, WA, 2002.

[112] M. Russina and F. Mezei, *Nucl. Instrum. Methods Phys. Res., Sect. A* **604**, 624 (2009).

[113] S. Satoh, S. Muto, N. Kaneko, T. Uchida, M. Tanaka, Y. Yasu, K. Nakayoshi, E. Inoue, H. Sendai, T. Nakatani and T. Otomo, *Nucl. Instrum. Methods Phys. Res., Sect. A* **600**, 103 (2009).

[114] B. M. Weckhuysen, *Phys. Chem. Chem. Phys.*, **5**, vi (2003).

[115] B. M. Weckhuysen, *Phys. Chem. Chem. Phys.*, **5**, 4351 (2003).

[116] G. P. Pez, A. R. Scott, A. C. Cooper, H. Cheng, F. C. Wilhelm and. A. H. Abdourazak, US Patent 7,351,395 (2008).

[117] G. P. Pez, A. R. Scott, A. C. Cooper and H. Cheng, US Patent 7,101,530 (2006).

[118] K. M. Eblagon, K. Tam, K. M. K. Yu, S. L. Zhao, X. Q. Gong, H. Y. He, L. Ye, L. C. Wang, A. J. Ramirez-Cuesta and S. C. Tang, *J. Phys. Chem. C*, **114**, 9720 (2010).

[119] K. M. Eblagon, D. Rentsch, O. Friedrichs, A. Remhof, A. Zuettel, A. J. Ramirez-Cuesta and S. C. Tsang, *Int. J. Hydrogen Ener.*, **35**, 11609 (2010).

[120] M. Arai, *Pramana: J. Phys.*, **71**, 629 (2008).

[121] H. Ohshit, T. Otomo, K. Ikeda, N. Kaneko, M. Tsubota, T. Seya and K. Suzuya, *Commissioning for the High Intensity Total Diffractometer (Nova) at JPARC* Poster, Imss Symposium '10, Dec 2010

4

Electron Paramagnetic Resonance Spectroscopy of Inorganic Materials

Piotr Pietrzyk, Tomasz Mazur and Zbigniew Sojka

Faculty of Chemistry, Jagiellonian University, Krakow, Poland

4.1 INTRODUCTION

Electron magnetic resonance (EMR) is a group of closely related spectroscopic techniques which includes, apart from the most common continuous-wave electron paramagnetic resonance (CW-EPR),[1−6] high-field electron paramagnetic resonance (HF-EPR),[7] electron nuclear double resonance (ENDOR), electron spin resonance imaging (ESRI)[8] and a number of Fourier transform-based pulsed techniques (FT-EPR), such as electron spin echo envelope modulation (ESEEM) and hyperfine sublevel correlation spectroscopy (HYSCORE).[9,10]

By probing stationary field-dependent Zeeman splitting, $H_0(B)$, field-independent interactions (fine, hyperfine and quadrupole couplings), H_0, and time-dependent magnetic interactions, $H(t)$, all conventionally expressed together in terms of the spin Hamiltonian, $H = H_0(B) + H_0 + H_0(t)$, electron paramagnetic resonance (EPR) provides an in-depth insight into the formation, local structure, dynamics and reactivity of paramagnetic materials. The chemical information deduced from EPR

Local Structural Characterisation, First Edition. Edited by Duncan W. Bruce, Dermot O'Hare and Richard I. Walton.
© 2014 John Wiley & Sons, Ltd. Published 2014 by John Wiley & Sons, Ltd.

varies from simple confirmation of the presence of a given paramagnet to a more detailed description of its energy levels, ground-state wavefunction (singly occupied molecular orbital, SOMO) and spin density distribution, both at the paramagnetic centre and on the neighbouring atoms.

The first applications of EPR in materials chemistry and related fields were reported at the beginning of the 1960s. Since then, review articles have periodically appeared,[11−17] with current literature surveys found in the *Royal Society of Chemistry Specialist Periodical Reports – Electron Paramagnetic Resonance*, which regularly include chapters dedicated to the characterisation of materials. With the advent of modern techniques such as FT-EPR and HF-EPR, in combination with computer simulation of the spectra, EPR has evolved into a highly sophisticated and powerful research tool for the exploration of materials with unpaired electrons.

In this chapter, all equations are consistently given in SI units.[18] The reduced Planck constant appears in these equations because the angular-momentum S and L operators produce the $\hbar$ factor when operating on the corresponding spin and wavefunctions.

4.2   ELECTRON SPIN IN A MAGNETIC FIELD

The spin angular momentum, S, a quantum vector property for which only the lengths $|S| = [S(S + 1/2)]^{1/2}\hbar$ and its z-component, $S_z = m_S\hbar$, are measurable simultaneously in an external magnetic field, B, is a fundamental property of an electron. As a result, S can assume only $(2S + 1)$ allowed orientations with respect to B (space quantisation), dictated by the magnetic spin number $m_S = -S, -S + 1, \ldots, S - 1, S$. For any electron, $S = 1/2$, and then $S_z = 1/2\hbar$ (which corresponds to α or $\uparrow$ state) or $S_z = -1/2\hbar$ (β or $\downarrow$ state). Since determination of the remaining S_x and S_y components is restricted by the Heisenberg principle, possible orientations of the S vector are constrained to positions situated on a cone surface, as indicated in Figure 4.1a.

By virtue of its spin, the electron has an associated magnetic moment, $\mu_e = -g_e(q_e/2m_e)S = -g_e(\mu_B/\hbar)S$, that is collinear but antiparallel to the spin itself, where $g_e = 2(1 + \alpha/2\pi - \cdots) = 2.0023(19304)$ is the so-called free-electron g factor, m_e and q_e stand for the mass and the charge of the electron, respectively, and $\mu_B = 9.274 \times 10^{-24}\,J\,T^{-1}$ is the Bohr magneton. For paramagnets with several unpaired electrons, the total spin is $S = \Sigma(S_i)$. Paramagnets with non-integer values of S are called

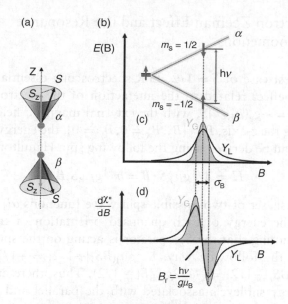

Figure 4.1 (a) Possible orientations of the electron spin ($S = 1/2$) in an external magnetic field. (b) Energy level splitting due to the Zeeman effect and generation of an EPR signal by a sweeping magnetic field at fixed microwave frequency. (c) The corresponding absorption line (Gaussian, Y_G; Lorentzian, Y_L). (d) Its first derivative. The resonance occurs at the field B_r when the energy difference matches the $h\nu$ value.

'Kramers systems', whereas those with integer S are referred to as 'non-Kramers systems'.

In analogy to electrons, magnetic nuclei possessing spin $I \neq 0$ also exhibit a magnetic moment $\boldsymbol{\mu}_n = +g_n(\mu_n/\hbar)I$, where g_n is the nuclear g factor, which, depending on the nucleus nature, has positive and negative values, tabulated elsewhere,[3] and $\mu_n = 5.0508 \times 10^{-27}\,\mathrm{J\,T^{-1}}$ is the nuclear magneton. Since the nuclear spin vector can assume only $2I + 1$ discrete orientations in the magnetic field (labelled by the nuclear magnetic quantum number $m_I = -I, -I + 1, \ldots, I - 1, I$), this property is also conferred on the associated nuclear magnetic moments.

In addition to nuclear spin magnetic moment, $\boldsymbol{\mu}_n$, nuclei with $I > 1/2$ exhibit electric quadrupole moment, Q, associated with a nonspherical charge distribution, and are thus sensitive to the local electric field gradients. Electron spin energy levels become distinct only when the paramagnet is immersed in an external magnetic field. This phenomenon is called the Zeeman effect and provides physical fundamentals for EPR spectroscopy.

4.2.1   Electron Zeeman Effect and the Resonance Phenomenon

In the simplest case of $S = 1/2$, EPR spectroscopy essentially rests on the Zeeman effect related to the interaction of the electron magnetic moment, $\boldsymbol{\mu}_e = -g_e(\mu_B/\hbar)\boldsymbol{S}$, with the external magnetic field, $\boldsymbol{B}$. For $\boldsymbol{B}$ aligned along the z-axis, $\boldsymbol{B} = [B_z, B_x = 0, B_y = 0]$, the energy of such an interaction can be derived using the following spin Hamiltonian:

$$H = \hbar^{-1}g\mu_B \boldsymbol{S} \cdot \boldsymbol{B} = \hbar^{-1}g\mu_B S_z B_z \qquad (4.1)$$

Using the basis set of two possible spin wave functions $\alpha(|+1/2\rangle)$ and $\beta(|-1/2\rangle)$, the energy of each quantised orientation of the spin can be derived, noting that the S_z operator is acting on the spin functions α and β in the following way: $\hbar^{-1}g\mu_B B S_z |+1/2\rangle = +1/2 g\mu_B B|1/2\rangle$ and $\hbar^{-1}g\mu_B B S_z |-1/2\rangle = -1/2 g\mu_B B| - 1/2\rangle$. Thus, there are only two discrete energy sublevels associated with the parallel and antiparallel orientations of the electronic spin ($m_S = \pm 1/2$):

$$E(m_s) = \pm 1/2 g\mu_B B \qquad (4.2)$$

According to Equation 4.2, the resulting Zeeman energy levels linearly diverge with the increasing magnetic field B (Figure 4.1b), and the energy difference between those states equals $\Delta E = g\mu_B B$. Owing to the limited lifetime, t, of the flipping spins in a given state, both sub-levels are diffused because of the Heisenberg uncertainty principle ($t \cdot \Delta E \geq \hbar$), giving rise to line broadening.

In the presence of a steady microwave radiation of a constant frequency, ν, resonance absorption occurs when the magnetic field is linearly changed (field sweeping), to satisfy the energetic condition:

$$\Delta E = E(m_S = +1/2) - E(m_S = -1/2) = g\mu_B B = h\nu \qquad (4.3)$$

In order to conserve angular momentum, the EPR transitions obey the selection rule $\Delta m_s = \pm 1$, *i.e.* spins flip from the lower ($m_S = -1/2$) to the higher ($m_S = +1/2$) energy states, and *vice versa*. Thus, an electron spin resonance experiment consists in measuring the induction of an external magnetic field, B, at which the resonant absorption of the electromagnetic radiation, required to reverse the directions of the magnetic moments (associated with the electron spins), takes place (Figure 4.1b).

In conventional CW-EPR spectrometers, in which magnetic fields of up to 1.5 T can be reached, resonances are observed in the microwave region of the electromagnetic spectrum, usually at 9.4 GHz (3.2 cm) – the so-called X-band – and less frequently at 35 GHz (8 mm) – the Q-band. In such fields, the energy difference between the lower and the upper Zeeman levels is of the order of $0.3-1.3$ cm^{-1}. Other regions of higher microwave frequencies used in commercial EPR spectrometers are the K-band ($\sim$24 GHz), W-band (95 GHz), D-band (140 GHz) and far infrared (HF-EPR). Less important to investigations of materials are the low-frequency bands such as the S- and L-band, corresponding to 3 and 1 GHz, respectively.

As in most spectroscopic techniques, the signal of the absorbed energy (the shadowed curve in Figure 4.1c) is detected, amplified and plotted as a function of the magnetic field. The EPR spectra are actually registered in a form of first derivative (Figure 4.1d), both for instrumental reasons and for resolution enhancement (see Section 4.2.4). The shape of this signal is typically described by the Gaussian (Y_G) or Lorentzian (Y_L) function, whose normalised analytical expressions are as follows:

$$Y_G = \frac{\sqrt{2/\pi}}{\sigma_B} \exp\left[-2\left(\frac{B - B_r}{\sigma_B}\right)^2\right]$$

$$Y_L = \frac{2\sqrt{3}}{\pi}\left[\frac{\sigma_B}{3\sigma_B^2 + 4(B - B_r)^2}\right]$$

(4.4)

where σ_B stands for peak-to-peak linewidth. However, for magnetically interacting centres, dynamic systems or conducting electrons, other lineshapes such as mixed Gauss–Lorentz, Voigt[19] or Dysonian are more appropriate.[20]

The g value, which is a key parameter describing the shift of the resonance line with respect to $g_e = 2.0023$ due to the different chemical environment seen by the unpaired electron, can be determined from the energetic resonance condition $g = h\nu/\mu_B B_r$ (Equation 4.3). The resonant magnetic field, B_r, is defined by the crossing of the EPR signal with the baseline (Figure 4.1d). For ν given in GHz and magnetic field expressed in mT ($1\,T = 10^4$ G), a succinct operational formula can be used for this purpose:

$$g = 71.447\,\nu(\text{GHz})/B_r(\text{mT})$$

(4.5)

The g value is thus derived from an experimental EPR spectrum, by the simultaneous measurement of the resonant magnetic field and the current frequency of the microwaves. In practice, the g value may also be determined by comparing the magnetic field values at resonance for the investigated (B_x) and a reference (B_{ref}) sample of the known g_{ref} value using the following relationship: $g = g_{ref}B_{ref}/B_x$. Both specimens must be inserted into the resonant cavity during spectrum recording in order to experience the same microwave frequency.

4.2.2   Spin Relaxation

Observation of an EPR signal depends on a dynamic balance between the absorption of the microwave energy and the efficiency of the energy dissipation processes through which the excited spins return to the ground state. This latter phenomenon is called 'relaxation'.

In actual EPR experiments, a population of spins, $N = (n_\beta + n_\alpha)$, has to be taken into account in order to describe the resonance in a comprehensive way. When an external magnetic field is not applied, $n_\beta = n_\alpha$, because both spin levels are degenerate (Figure 4.2a). In the presence of the magnetic field, occupation of the lower (n_β) sub-level is higher than that of the upper one (n_α), as implied by the Boltzmann distribution function: $n_\alpha/n_\beta = \exp(-g\mu_B B/kT)$. In such a case, $n_0 = (n_\beta - n_\alpha)_{eq} > 0$, and the sample acquires a net magnetisation M_0. The resultant vector, $M_0 = (0, 0, M_z)$, is orientated along the z-axis of the applied magnetic field $(M_z = -n_0 g\mu_B m_S)$, and the M_{xy} components decay due to random precession of the spins around B with the Larmor frequency $\nu_L = (g\mu_B/h)B$, as shown in Figure 4.2b. In resonance, when the Larmor frequency of the spin precession matches the microwave

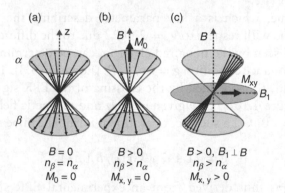

Figure 4.2 Dynamics of the electron spin population (a) without an external magnetic field, (b) in a static magnetic field, B, and (c) in the resonance conditions upon application of a perpendicular continuous microwave field, B_1.

frequency ($\nu_L = \nu$), spin flip takes place. This is accompanied by an increasing phase coherence of the spins under precession, leading to the appearance of net magnetisation in the xy-plane (Figure 4.2c). Upon denoting the actual spin difference between both states as $n = (n_\beta - n_\alpha)$, the spin dynamics can be described by the following equation:

$$\frac{dn}{dt} = -\frac{n - n_0}{T_1} - 2Pn \qquad (4.6)$$

where T_1 is the spin lattice (longitudinal) relaxation time, which controls the rate of dissipation of the absorbed energy into the surroundings (lattice) in order to restore the thermal equilibrium perturbed by the resonant absorption, while $P = P_{\beta\alpha} = P_{\alpha\beta} = 2\pi/\hbar^2 |\langle\alpha|g\mu_B B_1 S_x|\beta\rangle|^2 \delta(E_\alpha - E_\beta - h\nu)$ is the transition probability (proportional to the square root of the microwave power) and B_1 is a microwave magnetic field. At steady state, $dn/dt = 0$, and hence $n = n_0/(1 + 2PT_1)$. As a result, under the steady-state conditions, the rate of energy absorption, dE/dt, to which the intensity of the EPR signal is proportional, is dictated by the actual population difference, n:

$$dE/dt = nP\Delta E = n_0 P\Delta E/(1 + 2PT_1) \qquad (4.7)$$

As long as $PT_1 \ll 1$, the sample does not saturate ($n > 0$) and an undistorted EPR signal can be easily observed. When the microwave field B_1 has sufficiently large amplitude, two energy levels can attain equal populations ($n = 0$) and the net absorption of the energy decays to zero. This condition is known as 'saturation' and forms the basis of pulsed EPR techniques.[16] Since the microwave power absorption of the sample is proportional to the spin population difference, n, the signal intensity is essentially affected by the saturation effect.

When the relaxation time, T_1, is too short, the EPR signal is broadened and its amplitude decreases, sometimes beyond the detection limits. As with most rate processes, temperature has a strong effect here, and cooling the sample will increase the relaxation time, leading to better-resolved spectra. Typically, the temperature dependence of T_1 can be factored into three principal terms:

$$1/T_1 = aT + bT^n + c\Delta^3 \exp(-\Delta/kT) \qquad (4.8)$$

associated with direct spin–phonon, Raman and Orbach processes, respectively. Typical values of n are in the range of five to nine. The parameter Δ defines the energy distance to a close-lying excited electronic state.[5] In the direct process, the coupling with lattice occurs *via* modulation of the crystal field due to the lattice vibrations; therefore, the phonon energy must match exactly the photon energy of the microwaves. The efficiency of this channel strongly depends on the phonon density of the states. In the Raman process, the relaxation occurs *via* virtual phonon states, and the difference between the frequencies of the incident and the scattered (by the paramagnetic centre) phonons must be equal to the microwave frequency. This is more efficient than the direct process, as all lattice phonons are involved. The Orbach mechanism is a thermally activated, field-independent phenomenon and involves low lying-electronic states. Generally, for high-spin S-state ions such as Mn^{2+} or Fe^{3+},[21,22] the couplings to the lattice are weak and T_1 times are long enough to observe the resonance easily even at ambient temperatures (Figure 4.3a).[23] On the other hand, transition-metal ions with the F state are strongly coupled to the lattice (Co^{2+} doped in MgO[24] or ZnO[25]), and the T_1 relaxation is often so short that liquid helium temperatures are required to observe the EPR spectrum (Figure 4.3b).

The absorbed microwave energy can not only be transferred to the phonon modes of the lattice (T_1) but also be passed to other spins *via* dipolar or exchange couplings with a characteristic spin–spin relaxation time T_2 (transverse relaxation). This process is associated with loss of the phase coherence when the magnetisation is flipped into the *xy*-plane. Whereas T_1 strongly depends on temperature, the T_2 relaxation is more influenced by spin concentration and increases with decreasing spin–spin distances in the system. The resulting linewidth, σ_B, resulting from the

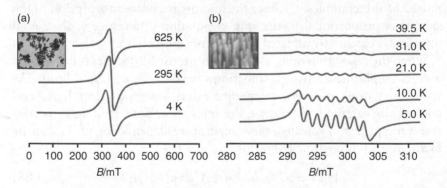

Figure 4.3 Temperature dependence of the X-band EPR spectra of (a) Mn^{2+} in ZnO nanorods (3 mol%) (reproduced with permission from Djerdj *et al.* (2008) [23]. Copyright (2008) Royal Society of Chemistry) and (b) Co^{2+} in ZnO nanowires (5 at.%). Reproduced with permission from Ankiewicz *et al.* (2007) [25]. Copyright (2007) American Institute of Physics.

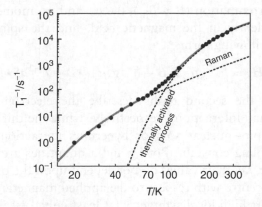

Figure 4.4 Temperature dependence of the longitudinal, T_1, relaxation time of hydrogen atoms trapped in octasilsesquioxane cages. Reproduced with permission from Mitrikas (2012) [26]. Copyright (2012) Royal Society of Chemistry.

simultaneous T_1 and T_2 processes, is often described by introducing time T_2', defined as $1/T_2' = 1/T_2 + 1/(2T_1)$,[5] which leads to $\sigma_B = \hbar/(g\mu_B T_2')$. If the relaxation is dominated by T_1, the resulting EPR signal assumes a Lorentzian lineshape, whereas for leading spin–spin broadening (T_2 relaxation) the signal tends to be Gaussian-like.[5] An example of the temperature dependence of the relaxation characteristics of an atomic hydrogen encapsulated in the octasilsesquioxane cage is given in Figure 4.4.[26]

4.2.3 Electron–Nucleus Hyperfine Interaction

In the case of paramagnetic materials containing magnetic nuclei, their magnetic moments, $\boldsymbol{\mu}_n = g_n(\mu_n/\hbar)\mathbf{I}$, produce an intrinsic quantised magnetic hyperfine field, B_{hf}, which adds to the external field, B, and modifies the resonance condition in the following way: $h\nu = \Delta E = g\mu_B(B + B_{hf})$. Since the nuclear spin can assume only $(2I + 1)$ definite orientations, its spatial quantisation leads to the appearance of discrete sublevels (gauged by the m_I number), giving rise to the so-called 'hyperfine structure' of an EPR spectrum (*i.e.* splitting of a single EPR line into a multiplet). Thus, the updated spin Hamiltonian will include new terms describing the nuclear Zeeman interaction (between the external magnetic field and the nuclear magnetic moment) and the hyperfine interaction (between the electron and the nuclear magnetic moments $\boldsymbol{\mu}_e$ and $\boldsymbol{\mu}_n$). Assuming a high-field approximation ($B > B_{hf}$), the μ_e and μ_n moments are quantised independently in the magnetic field, and the spin Hamiltonian assumes the following form:

$$H_s = \hbar^{-1}g\mu_B \mathbf{S} \cdot \mathbf{B} - \hbar^{-1}g_n\mu_n \mathbf{I} \cdot \mathbf{B} + \hbar^{-2}a\mathbf{S} \cdot \mathbf{I} \qquad (4.9)$$

The first and the second terms describe the electron–Zeeman and nuclear–Zeeman interactions, respectively, while the third accounts for the electron spin–nuclear spin or hyperfine interaction and a is the hyperfine coupling constant. Both g and a quantities are assumed here to be isotropic, *i.e.* their values do not depend upon the orientation of a paramagnetic centre with respect to the applied magnetic field, which is diagnostic of its high local symmetry (at least cubic) (see Section 4.3.6) or rapid tumbling of paramagnetic species (see Section 4.4.2). The term 'hyperfine' is employed when the nucleus with $I \neq 0$ belongs to the atom hosting the unpaired electron, while the term 'superhyperfine' refers to $I \neq 0$ nuclear spins belonging to neighbouring atoms. For non-magnetic nuclei such as common ^{16}O, ^{12}C and ^{59}Ni species, isotopic labelling (^{17}O $I = 5/2$, ^{13}C $I = 1/2$ and ^{61}Ni $I = 3/2$) can be used to enhance the information content of the EPR spectra.[27,28] The magnitude of the isotropic hyperfine coupling – or the Fermi contact interaction – is related to the finite spin density $\rho^{\alpha-\beta}$ at a nucleus A:

$$^{A}a_{iso} = 2/3\mu_0 g\mu_B g_n\mu_n \rho^{\alpha-\beta}(R_A) \qquad (4.10)$$

Both the hyperfine and the superhyperfine interactions are indirect probes of the nature (covalency) of the chemical bonds in inorganic materials (see Section 4.3.2).

In most solid samples, particularly powders, the nuclear Zeeman term, $-\hbar^{-1}g_n\mu_n \mathbf{I} \cdot \mathbf{B}$, can usually be neglected as being too small to influence appreciably the CW-EPR spectra, so the spin Hamiltonian can be truncated into a simplified form, $H_s = \hbar^{-1}g\mu_B B S_z + \hbar^{-2}a S_z I_z$. For nuclei with $I = 1/2$, such as ^{31}P, ^{1}H, ^{115}Sn and ^{195}Pt, there are four spin states, $\alpha_e\alpha_n, \alpha_e\beta_n, \beta_e\alpha_n$ and $\beta_e\beta_n$, defined by all possible combinations of $m_I = \pm 1/2$ and $m_S = \pm 1/2$ quantum numbers. Since the S_z and I_z operators act only on their own electron and nuclear spin functions, giving the corresponding $\pm 1/2\hbar$ eigenvalues (*e.g.* for $S_z I_z |\alpha_e, \alpha_n\rangle = 1/2\hbar \times 1/2\hbar|\alpha_e, \alpha_n\rangle = 1/4\hbar^2|\alpha_e, \alpha_n\rangle$), the following four energy levels can be derived:

$$E_1 = \langle\phi_1|H|\phi_1\rangle = \langle\alpha_e\alpha_n|\hbar^{-1}g\mu_B B S_z + \hbar^{-2}a S_z I_z|\alpha_e, \alpha_n\rangle = 1/2g\mu_B B + 1/4a$$

$$E_2 = \langle\phi_2|H|\phi_2\rangle = \langle\alpha_e\beta_n|\hbar^{-1}g\mu_B B S_z + \hbar^{-2}a S_z I_z|\alpha_e\beta_n\rangle = 1/2g\mu_B B - 1/4a$$

$$\qquad (4.11)$$

$$E_3 = \langle\phi_3|H|\phi_3\rangle = \langle\beta_e\alpha_n|\hbar^{-1}g\mu_B BS_z + \hbar^{-2}aS_zI_z|\beta_e\alpha_n\rangle = -1/2g\mu_B B - 1/4a$$

$$E_4 = \langle\phi_4|H|\phi_4\rangle = \langle\beta_e\beta_n|\hbar^{-1}g\mu_B BS_z + \hbar^{-2}aS_zI_z|\beta_e\beta_n\rangle = -1/2g\mu_B B - 1/4a$$

These are shown as a function of the magnetic field in Figure 4.5a, while Figure 4.5b illustrates the splitting at a constant magnetic field, including the shifts caused by the nuclear Zeeman term of relevance to pulsed EPR measurements (see Section 4.5.2). EPR transitions obey the selection rules $\Delta m_S = \pm 1$ and $\Delta m_I = \pm 0$, and the positions of the two resultant resonances are calculated as follows:

$$h\nu = \left(\frac{1}{2}g\mu_B B_r + \frac{1}{4}a\right) - \left(-\frac{1}{2}g\mu_B B_r - \frac{1}{4}a\right) = g\mu_B B_r + \frac{1}{2}a \quad (4.12a)$$

$$h\nu = \left(\frac{1}{2}g\mu_B B_r + \frac{1}{4}a\right) - \left(-\frac{1}{2}g\mu_B B_r + \frac{1}{4}a\right) = g\mu_B B_r - \frac{1}{2}a \quad (4.12b)$$

In Figure 4.5a, the allowed transitions are indicated by the vertical solid lines, while the dotted line indicates a virtual transition associated with the electronic Zeeman term only. Two resonance lines of equal intensity separating the value of the hyperfine constant, a, constitute the hyperfine structure of the EPR spectrum (Figure 4.5c). In a more succinct form, the resonance condition may be written as:

$$h\nu = g\mu_B B_r + m_I a \quad (4.13)$$

and then rearranged in the following way:

$$B_r = h\nu/g\mu_B - m_I(a/g\mu_B) = B_0 - m_I(a/g\mu_B) \quad (4.14)$$

Thus, these two transitions take place at two different B_r values, separated by $a' = a/(g\mu_B)$ (in magnetic field units) and centred at $B_0 = h\nu/g\mu_B$ (Figure 4.5c).

The hyperfine constant, measured directly from an experimental spectrum (Figure 4.5d) as a separation of the spectral lines in the magnetic field units (*i.e.* Tesla or Gauss), should be transformed into units proportional to energy, such as frequency ($a/\text{MHz} = a/\text{h}$) or wavenumber ($a/\text{cm}^{-1} = a/\text{hc}$), using the following conversion formulae:

$$a/\text{MHz} = 28.025(g/g_e)a/\text{mT} \quad (4.15)$$

$$a/(10^{-4}\,\text{cm}^{-1}) = 9.3480(g/g_e)a/\text{mT}$$

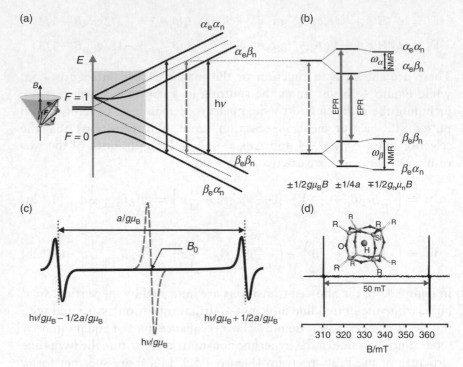

Figure 4.5 (a) Energy-level splitting for an $S = 1/2$ and $I = 1/2$ system as a function of the magnetic field (on the left is shown a coupling of the electron and nuclear spins in the weak fields, $F = J + I$). (b) Energy-level diagram at a constant field (ω_α and ω_β indicate the frequencies of two NMR transitions). (c) The resulting isotropic EPR spectrum (the dotted line shows the virtual Zeeman signal). (d) Experimental EPR spectrum with doublet hyperfine structure, caused by a hydrogen atom ($I = 1/2$) encaged in octasilsesquioxane (spectrum reproduced with permission from Mitrikas (2012) [26]. Copyright (2012) Royal Society of Chemistry).

For nuclei with $I > 1/2$, the hyperfine splitting of the energy levels can be written as:

$$E = m_S(g\mu_B B + m_I a) \qquad (4.16)$$

which gives rise to $(2I + 1)$ lines (resonances) of equal intensity, nicely illustrated by a three-line room-temperature EPR spectrum of an atomic nitrogen ($I = 1$) clathrated in C_{60}[29] (Figure 4.6a) or a spectrum of impurities of vanadium(II) with $I = 7/2$ in MgO[30] (Figure 4.6b). For isotopes with different spin values and abundances, the EPR spectrum is simply a weighted mixture of the component isotopomers.[31] When several $I \neq 0$ nuclei interact with the unpaired electron, the EPR spectrum is more complex. The total number, N, of the hyperfine lines is

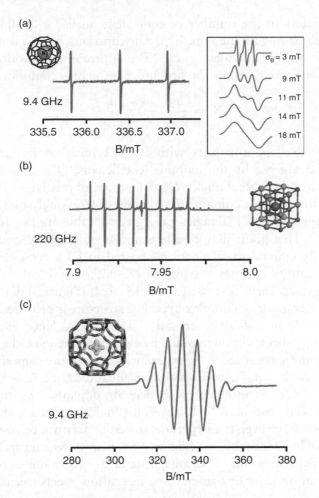

Figure 4.6 Isotropic powder EPR spectra illustrating hyperfine splitting. (a) Triplet spectrum of nitrogen atom in C_{60}. Reproduced with permission from Weidinger *et al.* (1998) [29]. Copyright (1998) Springer Science + Business Media. (b) Hyperfine octet due to V^{2+} ions in MgO matrix. Reproduced from [30]. (c) EPR spectrum of six magnetically equivalent silver atoms hosted in the LTA zeolite. Reproduced with permission from Baldansuren *et al.* (2009) [33]. Copyright (2009) American Chemical Society. In the insert is shown the effect of increasing linewidth on the hyperfine structure resolution.

determined by the product of the possible spin states $(2I_i + 1)$ for each of the involved magnetic nuclei:

$$N = \prod_i (2n_i I_i + 1) \qquad (4.17)$$

where n_i refers to the number of equivalent nuclei, I_i, coupled to the unpaired electron with the same hyperfine constants, a_i, and i enumerates the groups of nonequivalent nuclei. The approximate positions of the hyperfine lines are given by the first-order resonance condition:

$$B_r = \frac{h\nu}{g\mu_B} - \sum_i \frac{a_i}{g\mu_B} \qquad (4.18)$$

For polynuclear paramagnets with $I = 1/2$ nuclei only, the relative intensities are given by the binomial coefficients, $(1 + x)^n$, where n is the number of equivalent nuclei. Determination of relative intensities for $I > 1/2$, although not complicated, becomes increasingly more tedious, and appropriate Pascal triangles for a given I value are helpful for this purpose.[32] This point may be illustrated by an EPR spectrum of six magnetically equivalent silver atoms hosted in LTA zeolite,[33] where the total number of lines is equal to $N = (2 \times 6 \times 1/2 + 1) = 7$, and their intensities vary as $1 : 6 : 15 : 20 : 15 : 6 : 1$ (Figure 4.6c). Another paramount example of complex hyperfine structure is provided by ^{55}Mn ions ($I = 5/2$) in a fluorite crystal (^{19}F $I = 1/2$). Since each of the dopant Mn_{Ca} species is surrounded by eight fluorine ions, the resultant spectrum can be regarded as a (hyperfine) sextet of the (superhyperfine) nonets. Thus, the total number of lines is equal to $(2 \times 5/2 + 1)(2 \times 8 \times 1/2 + 1) = 54$, confirming definitely that Mn dopants substitute for the framework Ca^{2+} cations.[34] However, it should be noted that often the superhyperfine, or even the hyperfine, structure may not be resolved due to excessive linewidth, $\sigma_B > a$ (Figure 4.6a, insert), which can be caused by a high concentration of paramagnetic centres, strain effects induced by structural disorder or dynamic and relaxation effects (see below).

4.2.4   EPR Spectrometers

Design of the most commonly used continuous-wave-mode EPR spectrometers is dictated by the fact that in order to register a spectrum, the microwave frequency is kept constant but the magnetic field is linearly varied (swept) with time. A modern reflection CW-EPR spectrometer (Figure 4.7a) consists of four main functional building blocks: (i) a microwave generation unit composed of a microwave source (Gunn diode) and devices that control and measure the intensity (attenuator) and the frequency of the microwave radiation; (ii) a magnetic field unit that produces a highly homogeneous and stable ($\pm\mu T$), linearly

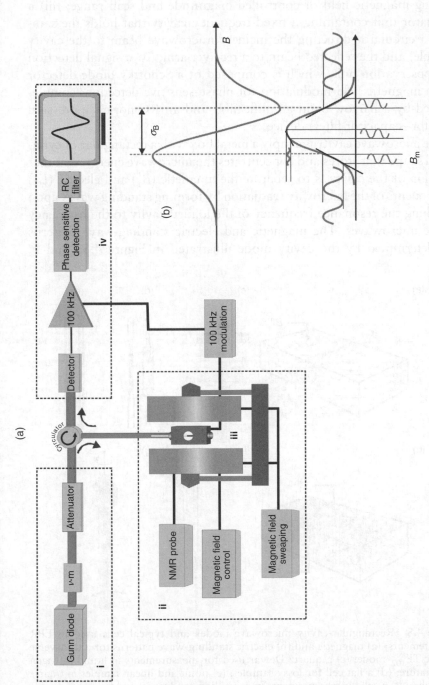

Figure 4.7 (a) Block diagram of a reflection CW-EPR spectrometer with 100 kHz phase-sensitive detection. (b) The effect of small-amplitude field modulation (B_m) on the detector output current, along with the resulting first derivative EPR spectrum.

varying magnetic field of controlled magnitude and scan range; (iii) a resonator unit containing a fixed-frequency cavity that holds the samples, a circulator directing the incident microwave beam to the cavity (sample) and the reflected beam to a receiver arm; (iv) a signal detection and observation unit, which is composed of a Schottky diode detector and a magnetic field modulation for phase-sensitive detection, complemented by a narrow-band amplification and an RC noise suppression (resistor–capacitor filter) device.

The microwave cavity is simply a metal box with a rectangular or cylindrical shape that stores and concentrates the microwave energy. Another function of the cavity is to separate the magnetic (B_1) and electric (E_1) components of the microwave radiation by forming standing waves upon matching the resonance frequency of the loaded cavity to the frequency of the microwaves. The magnetic and electric standing-wave patterns are determined by the cavity mode illustrated in Figure 4.8a and b,

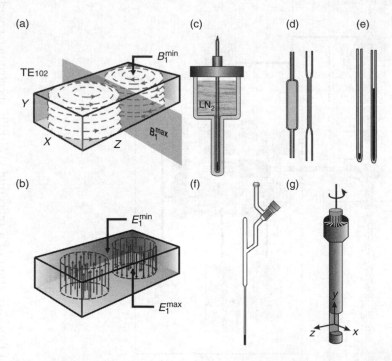

Figure 4.8 Rectangular-cavity microwave modes and typical cells used in EPR measurements: (a) magnetic and (b) electric standing-wave patterns for a transverse electric TE_{102} mode; (c) a quartz Dewar used for measurements in liquid nitrogen temperature; (d) a flat cell for lossy samples; (e) point and linear samples in quartz EPR tubes; (f) a cell for measurements in controlled conditions; and (g) a goniometer for single-crystal studies.

respectively, taking a transverse electric TE_{102} mode of the rectangular cavity as an example. An appropriate cavity mode should permit a high-energy density storage, should allow placement of the sample at a maximum of the magnetic (B_1) and a minimum of the electric (E_1) field and should (usually) provide B_1 perpendicular to the static field B (to fulfil the geometric condition of the resonance).

The key parameters that describe the cavity and essentially control the signal voltage $(V_S = \chi'' \eta Q \sqrt{PZ_0})$ are the Q factor (a figure of merit defined as a quotient of the maximum energy stored in the cavity to the energy dissipated *per* cycle) and the filling factor, η. The latter can be approximated as the ratio of the sample to the cavity volume, providing that the B_1 field is not too perturbed by the sample. The resonance absorption, as well as dielectric and resistive properties of the sample, will decrease the Q value of the cavity in the following way:[35]

$$\frac{1}{Q} = \frac{1}{Q_u} + \frac{1}{Q_\varepsilon} + \frac{1}{Q_\mu} + \frac{1}{Q_\chi} \qquad (4.19)$$

where Q_u describes resistive losses in the walls of the unloaded (empty) cavity, Q_ε and Q_μ correspond to undesired dielectric and ohmic losses, respectively, and $Q_\chi = 1/(\chi'' \eta)$ accounts for target losses due to the resonant absorption of the microwaves. The actual Q_χ value controls the reflected microwave radiation that is directed by the circulator to the receiver arm, carrying the information about the resonant absorption of the investigated sample (χ'').

The phase-sensitive detection technique commonly employed in EPR spectroscopy utilises small sinusoidal amplitude modulation of the magnetic field to suppress the noise by rejecting all noise components except those in a very narrow range (± 1 Hz) about the frequency of the modulation. Since a diode detector exhibits an inherent noise, inversely proportional to frequency, a high-frequency 100 kHz field modulation is widely used for this purpose. A basic concept of the modulation is shown in Figure 4.7b. Its effect is that the detector output is proportional to the slope of the absorption line at the midpoint of the modulating field; therefore, rather than the absorption signal itself, its first derivative is recorded. Although the signal-to-noise ratio of the recorded first derivative spectrum is proportional to the modulation amplitude, B_m, oversised modulation ($B_m > 0.3\,\sigma_B$) leads to appreciable lineshape distortions.

In pulsed EPR techniques, the electron/nuclear energy levels are probed by employing microwave pulses of a defined sequence of durations and time intervals at constant B. A rectangular pulse of duration τ_p (*e.g.* 20 ns) in the time domain is equivalent upon Fourier transformation to a $\sin x/x$ band (of 50 ± 25 MHz width) in the frequency domain.[7] Thus, application of a short pulse generates a band of frequencies, subjecting the sample to multifrequency excitation. As a result, in contrast to CW-EPR, the entire spectrum is generated in one pulse (as a time-domain free induction decay (FID) response), rather than by a slow magnetic field sweeping. The major advantages of such measurements include dramatic enhancement of the signal:noise ratio, direct measurements of the relaxation times and, most importantly, simplification of the spectra by judicious design of various multiple-pulse sequences.[7]

The key component of a pulse spectrometer is the pulse programmer unit, which converts the microwave radiation into controlled sequences of pulses of appropriate amplitude and duration, with a time resolution of a few nanoseconds. The pulses are amplified (typically to 10^3 W) and sent by the circulator to the cavity. The important characteristic of the cavity is the cavity ring-down time, t_r:

$$t_r = \log\left(\frac{P_0}{P_{noise}}\right)\frac{Q}{2\pi\nu} \qquad (4.20)$$

where P_0 is the pulse power and P_{noise} is the average power of noise. The cavity ring-down time controls the shortest time interval (dead time, t_d) between the last pulse and the possibility of undisturbed FID measurements. The FID signal is then recorded and converted by Fourier transformation into the frequency domain (EPR spectrum). The whole pulse sequence can be repeated many times, at intervals dictated by how fast the sample can recover its original equilibrium state (M_z magnetisation), and the response signals are averaged to enhance the signal:noise ratio. As in the case of CW-EPR, the most common are the X-band and the Q-band pulsed spectrometers; however, because the sensitivity and resolution increase with frequency, pulse instruments working at up to 300 GHz have been developed.[7] More comprehensive descriptions of pulsed spectrometers can be found elsewhere.[9,10]

4.2.5   Samples, Sample Holders and Registration of EPR Spectra

Given the vast diversity of inorganic materials that can be studied by EPR, it is nearly impossible to provide a comprehensive set of recommendations concerning sample handling. Generally, inorganic materials are studied in the form of single crystals or powders.

Two main practical factors have to be taken into account with regards to sample handling for EPR measurements: optimisation of the signal: noise ratio (sensitivity) and maximisation of the signal resolution. For solid samples, a major source of line broadening, limiting the resolution, is the dipole–dipole interaction between the neighbouring spins (T_2 relaxation) and the measurement temperature (T_1 relaxation). In order to minimise the resolution problems, dilution of the paramagnetic centres in an isomorphous diamagnetic host is recommended. Since limiting linewidths in solids are usually >0.1 mT, spin concentrations of 10^{-3} to 10^{-4} are acceptable.[2]

For maximum sensitivity and, in most cases, resolution, EPR spectra, especially of the samples containing transition-metal or rare-earth ions, should preferably be recorded at the lowest temperatures possible (77 K, or even 4 K). Liquid nitrogen (LN_2) spectra are usually recorded using a special quartz Dewar inserted in the cavity (Figure 4.8c). Variable-temperature EPR spectra can be measured by using a flow of cooled nitrogen in LN_2-cryostat from about 90 up to 450 K. A liquid helium (LHe) cryostat allows spectra to be recorded down to $\sim$ 4–5 K. For the investigation of ferromagnetic materials, solid-state reactions or diffusion processes, high-temperature measurements can be carried out using appropriate accessories described in the literature.[35]

The factors controlling intensity of the measured EPR signals (Q, η) impose several limitations on the properties of the samples. In general, the signal increases with increasing size of the sample and Q value of the cavity. Yet if a sample is lossy (exhibits high dielectic constant ε''), an increase of its size will decrease Q, and both factors will have opposite effects. Two limiting cases of a sample may be distinguished: (i) insulators and semiconductors that exhibit dielectric losses proportional to E_1^2 arising from ε'' and (ii) low-resistance samples that produce losses proportional to B_1^2.[35]

Most of the dielectric and conducting losses usually occur within the sample and the sample tube. This is especially important for materials that have a nonvanishing imaginary part, ε'', of the complex dielectric constant ($\varepsilon' - i\varepsilon''$) or high electron conductivity, σ, such as metals. Thus, it is important to position the sample in a region of the resonator at which the microwave electric field is minimal (Figure 4.8b). This is extremely important for samples that exhibit high dielectric losses (e.g. hydrous materials). For these, the most appropriate holder is a flat high-purity silica cell (Figure 4.8d) that can be accurately placed along the nodal plane of the E_1 field distribution. An excessive conduction loss in the case of bulk materials with good electron conductivity is another unfavourable factor that may prevent registration of the EPR spectra.

So-called 'point samples' (a few mm^3 in size) are usually sufficient in the X-band to obtain good signal:noise ratio, but linear geometry giving rise to higher filling factor, η, has a clear advantage, provided that dielectric and ohmic losses are small (Figure 4.8e). In the case of air-sensitive materials, cells with stopcocks that can be attached to a vacuum line may be used for sample handling (Figure 4.8f).

4.3   SPIN HAMILTONIAN AND SYMMETRY

The main interactions contributing to the EPR spectra of the magnetically diluted paramagnets are, as already discussed, the electronic Zeeman effect and the hyperfine coupling between the electron and the nuclear spins. In the case of $S \geq 1$, an additional fine structure interaction has to be accounted for. In some cases (pulsed EPR and systems with $I > 1/2$ quadrupole interactions) it may also be necessary to include the nuclear Zeeman effect. Compared to isotropic systems, in the solid state the EPR spectra exhibit additional complexity, arising from anisotropies of the magnetic interactions, determined by local symmetry at the paramagnetic centre. 'Anisotropy' here implies that the spectral features depend on the orientation of the paramagnetic centre with respect to the magnetic field; thus the corresponding spin Hamiltonian parameters have to be generally treated as second-rank tensor properties.

4.3.1   The g Tensor

In the case of solid materials (anisotropic systems), where g is no longer a simple scalar value, the EPR resonance condition has to be modified for inclusion of its angular dependence:

$$h\nu = \mu_B g(\theta, \varphi) B \qquad (4.21)$$

Thus, in order to describe fully the observed changes for all possible orientations, the scalar g (Equation 4.1) is substituted for in the spin Hamiltonian by the 3×3 symmetric g matrix. The electron Zeeman term can then be written as:

$$H_{EZ} = \hbar^{-1}\mu_B B^T g S \qquad (4.22)$$

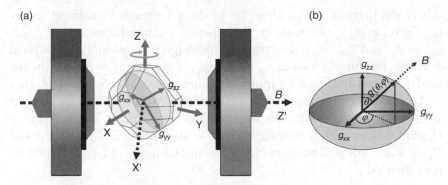

Figure 4.9 (a) Various coordinate systems used in EPR spectroscopy: $X'Y'Z'$ external (laboratory) coordinate system, XYZ crystallographic coordinates and xyz ($g_{xx}g_{yy}g_{zz}$) principal g tensor axes. (b) Angular dependence of $g(\theta,\varphi)$ for an orthorhombic g tensor, shown as a triaxial ellipsoid with the indicated principal axes.

In any arbitrary Cartesian coordinates XYZ, *e.g.* a crystallographic system as shown in Figure 4.9a, the g' tensor has a general non-diagonal form (Equation 4.23a):

$$g = \begin{pmatrix} g'_{xx} & g'_{xy} & g'_{xz} \\ g'_{yx} & g'_{yy} & g'_{yz} \\ g'_{zx} & g'_{zy} & g'_{zz} \end{pmatrix} \qquad (4.23a)$$

$$g = \begin{pmatrix} g_{xx} & 0 & 0 \\ 0 & g_{yy} & 0 \\ 0 & 0 & g_{zz} \end{pmatrix} \qquad (4.23b)$$

However, in its principal (intrinsic) axes systems (xyz), the corresponding g matrix is diagonal (Equation 4.23b). The components of the g' tensor are derived by recording several EPR spectra for various orientations of the investigated single crystal with respect to the external coordinate system ($X'Y'Z'$), as explained in Section 4.4.1. The (3×3) g' matrix is then diagonalised by finding a matrix $\mathbf{U}$ such that $g = \mathbf{U}^T \cdot g' \cdot \mathbf{U}$, allowing for the principal (diagonal) values of the g matrix and the direction cosines to be obtained, which mutually relate the external $X'Y'Z'$ and the principal (internal) xyz frames. The procedure for finding the $\mathbf{U}$ matrix is described elsewhere.[2]

In the principal axes system, the spin Hamiltonian for general rhombic symmetry assumes a simple form:

$$H_{EZ} = \hbar^{-1}\mu_B(g_{xx}B_xS_x + g_{yy}B_yS_y + g_{zz}B_zS_z) \qquad (4.24)$$

where the three diagonal elements of the g tensor have distinct values ($g_{xx} \neq g_{yy} \neq g_{zz}$), whereas in the case of an axial symmetry, $g_{xx} = g_{yy} = g_{\perp}$ and $g_{zz} = g_{\parallel}$. The trace of the diagonal elements is associated with the isotropic g value: $g_{iso} = 1/3(g_{xx} + g_{yy} + g_{zz})$, measured in the time-averaged spectra for rapidly tumbling species (see Section 4.4.2).

At any arbitrary orientation of the magnetic field, its components with respect to xyz axes can be described by the polar angles θ and φ (Figure 4.9b) in the following way: $B_z = B\cos\theta$, $B_y = B\sin\theta\sin\varphi$ and $B_x = B\sin\theta\cos\varphi$. The spin Hamiltonian (Equation 4.24) can then be rewritten as:

$$H_{EZ} = \hbar^{-1}\mu_B B(g_{zz}S_z\cos\theta + g_{yy}S_y\sin\theta\sin\varphi + g_{xx}S_x\sin\theta\cos\varphi) \quad (4.25)$$

The corresponding energies of the Zeeman levels can be derived for both spin states $\pm|1/2\rangle$ in the following way:

$$E_{\pm 1/2} = \pm\frac{1}{2}\mu_B B(g_{zz}^2\cos^2\theta + g_{yy}^2\sin^2\theta\sin^2\varphi + g_{xx}^2\sin^2\theta\cos^2\varphi)^{1/2} \quad (4.26)$$

noting that $S_x|\pm 1/2\rangle = 1/2|\mp 1/2\rangle$ and $S_y|\pm 1/2\rangle = -i/2|\mp 1/2\rangle$. In order for the resonance absorption to occur, the energetic condition (Equation 4.21) must be satisfied, where the angular dependence of an anisotropic g factor can be expressed as:

$$g^2(\theta,\varphi) = g_{zz}^2\cos^2\theta + g_{yy}^2\sin^2\theta\sin^2\varphi + g_{xx}^2\sin^2\theta\cos^2\varphi \quad (4.27)$$

In the case of an axial symmetry, it is reduced to:

$$g^2(\theta) = g_{\parallel}^2\cos^2\theta + g_{\perp}^2\sin^2\theta \quad (4.28)$$

The angular dependence of g values is often visualised in the form of an ellipsoid (Figure 4.9b).

Both the principal values and the directions (principal axes) of the g tensor depend on the electronic structure of the investigated paramagnet. The origin of its anisotropy is associated with a contribution of an angular orbital momentum to the magnetic moment, which is dictated by the interactions between the unpaired electron and its environment. Actually, due to the local electric fields present in the solid systems, the orbital degeneracy of the ground state is usually removed, and the angular momentum effectively quenched. However, its addition is partially restored by the spin–orbit coupling (SOC) mechanism that operates via an admixture of excited electronic states and the ground state.

The g_{ii} matrix elements depend essentially on the symmetry and the structure (separation) of the electronic energy levels of a paramagnetic centre. Within the crystal-field theory approximation for a non-degenerate ground state that is well separated from the excited states, the g_{ij} components can be calculated using the following equation:

$$g_{ij} = g_e \delta_{ij} - 2\lambda \Lambda_{ij} \tag{4.29}$$

The elements of the Λ matrix are obtained from the second-order perturbation theory formula:

$$\Lambda_{ij} = \Lambda_{ij} = \hbar^{-2} \sum_{n \neq 0} \frac{\langle 0|L_i|n \rangle \langle n|L_j|0 \rangle}{E_n - E_0} \tag{4.30}$$

In Equations 4.29 and 4.30, λ is the SOC constant (usually expressed in cm^{-1}, and tabulated elsewhere[2,3]), $|n\rangle$ are the spin-orbit coupled electronic states of the paramagnetic centre with energies E_n, $|0\rangle$ is the singly occupied molecular orbital (MO) with the ground-state energy E_0 and L_i ($i = x, y, z$) is the ith component of the orbital angular momentum operator. The integrals $\langle 0|L_i|n \rangle$ and $\langle n|L_j|0 \rangle$ can be computed simply for the unpaired electron centred on a single atom if the relevant MOs are written as a linear combination of the atomic orbitals (hybridisation). The $|L_i|p\rangle$ and $|L_i|d\rangle$ products are collated in Table 4.1.

Calculations of the Λ_{ij} elements can be simplified, as only those states $|n\rangle$ are admixed by SOC to the ground state $|0\rangle$ for which the direct product of the corresponding irreducible representations, $\Gamma_n \times \Gamma_L \times \Gamma_0$, includes a totally symmetric representation in a given point-symmetry group exhibited by the paramagnetic centre (L_i possesses the symmetry properties of rotations, R_i, about $i = x$-, y- and z-axes).

Table 4.1 Angular momentum operations L_i on the p and d atomic orbitals.[54]

Atomic orbital	Angular momentum		
	$\hbar^{-1}L_x$	$\hbar^{-1}L_y$	$\hbar^{-1}L_z$
$\|x\rangle$	0	$-i\|z\rangle$	$i\|y\rangle$
$\|y\rangle$	$i\|z\rangle$	0	$-i\|x\rangle$
$\|z\rangle$	$-i\|y\rangle$	$i\|x\rangle$	0
$\|x^2 - y^2\rangle$	$-i\|yz\rangle$	$-i\|xz\rangle$	$2i\|xy\rangle$
$\|z^2\rangle$	$-\sqrt{3}i\|yz\rangle$	$\sqrt{3}i\|xz\rangle$	0
$\|xy\rangle$	$i\|xz\rangle$	$-i\|yz\rangle$	$-2i\|x^2 - y^2\rangle$
$\|yz\rangle$	$i\|x^2 - y^2\rangle + \sqrt{3}i\|z^2\rangle$	$i\|xy\rangle$	$-i\|xz\rangle$
$\|xz\rangle$	$-i\|xy\rangle$	$i\|x^2 - y^2\rangle - \sqrt{3}i\|z^2\rangle$	$i\|yz\rangle$

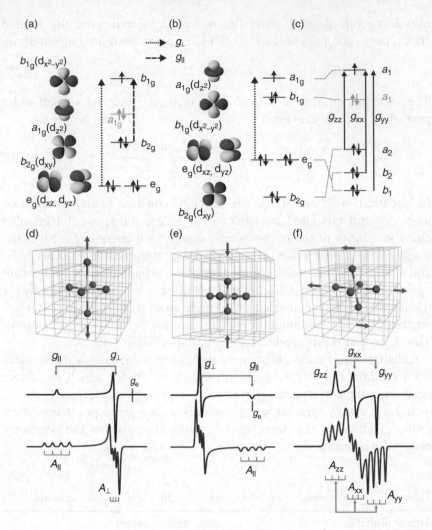

Figure 4.10 Local structure effects on the EPR signal of a $3d^9$ centre ($S = 1/2, I = 3/2$) in the D_{4h} and C_{2v} surroundings (elongated octahedron (a,d), compressed octahedron (b,e) and C_{2v}-distorted elongated octahedron (c,f)). The corresponding energy levels and the magnetic transitions are shown in (a,b,c), while the octahedral structures and the simulated EPR signals without and with hyperfine splitting are shown in (d,e,f).

For instance, in the case of a $3d^9$ centre (Cu^{2+}) in an elongated octahedral site of the local D_{4h} symmetry (Figure 4.10a,d), orbital angular operations belong to the irreducible representations $A_{2g}(L_z)$ and $E_g(L_x$ and $L_y)$. Taking into account a $d_{x^2-y^2}$ ground state (SOMO) of B_{1g} symmetry for calculation of the $g_{zz} = g_\parallel$ component, among three possible direct products $B_{1g} \times A_{2g} \times E_g = E_g, B_{1g} \times A_{2g} \times A_{1g} = B_{2g}$

and $B_{1g} \times A_{2g} \times B_{2g} = A_{1g}$, only the latter includes a totally symmetric representation, A_{1g}. Similarly, when calculating $g_{yy} = g_{xx} = g_\perp$ values it can be easily inferred that only for the $B_{1g} \times E_g \times E_g$ product is the SOC contribution possible. Thus, the expected shift of the $g_\parallel$ value is caused by coupling of the $d_{x^2-y^2}(B_{1g})$ orbital (SOMO) with the occupied $d_{xy}(B_{2g})$ orbital effected by the $L_z(A_{2g})$ operator. Using the data from Table 4.1, the corresponding matrix element $\langle d_{x^2-y^2}|L_z|d_{xy}\rangle\langle d_{xy}|L_z|d_{x^2-y^2}\rangle = (-2i\hbar)\langle d_{x^2-y^2}|d_{x^2-y^2}\rangle(2i\hbar)\langle d_{xy}|d_{xy}\rangle = 4\,\hbar^2$ can be evaluated; hence, from Equation 4.29, $g_\parallel = g_e + 8\lambda/[E(d_{x^2-y^2}) - E(d_{xy})]$. In other words, the magnitude of the g_{zz} component is determined by a paramagnetic current produced by electron circulation between the d_{xy} and $d_{x^2-y^2}$ orbitals that is induced by the magnetic field applied along the z-direction (Figure 4.10a). If the magnetic field is applied along the x-axis, the B_{1g} ground state is coupled by L_x with the degenerate d_{xz} and $d_{yz}(E_g)$ states, giving rise to the following matrix elements: $\langle d_{x^2-y^2}|L_x|d_{yz}\rangle\langle d_{yz}|L_x|d_{x^2-y^2}\rangle = i\hbar\langle d_{x^2-y^2}|id_{x^2-y^2} + \sqrt{3}id_{z^2}(-i\hbar)/\langle d_{yz}|d_{yz}\rangle = \hbar^2$ and $g_\perp = g_e + 2\,\lambda/[E(d_{x^2-y^2}) - E(d_{yz})]$. The same result is obtained for g_{yy}; thus, for such centres, $g_\parallel > g_\perp > g_e$. The corresponding powder EPR spectrum is shown in Figure 4.10d.

In the case of a compressed octahedron, the ground state changes to $3d_{z^2}$ of A_{1g} symmetry (Figure 4.10b,e). Since it exhibits a totally symmetric representation, the magnetically coupled orbitals must be of the same symmetry as the angular momentum operators. The L_x and L_y transform as E_g, so the $g_\perp$ value is associated with the coupling of SOMO with d_{xz} and $d_{yz}(E_g)$, while for the $g_\parallel$ direction no coupling is allowed by the symmetry, since none of the $3d$ orbitals exhibits the A_{2g} symmetry of the L_z operator. As a result, $g_\parallel = g_e$ (in this approximation) and $g_\perp = g_e + 6\lambda/[E(d_{z^2}) - E(d_{yz})]$, giving rise to a reversed EPR signal with $g_\perp > g_\parallel = g_e$ (Figure 4.10e). The splitting of the energy levels in the D_{4h} symmetry can be expressed in terms of crystal field parameters D_q, D_s and D_t, which can be determined from ultraviolet (UV)–visible spectroscopy, showing the complementarity of the two techniques.[3]

Similar calculations can be performed for any ground state and local symmetry at the paramagnetic centre. Equation 4.30 implies that for paramagnets whose number of unpaired electrons is smaller than that of the half-filled configuration (e.g. Ti^{3+}, V^{4+}), the Λ_{ij} values are negative and $g \ll g_e$, whereas the reverse ($g > g_e$) is true for more than half-filled configurations (Ni^+, Cu^{2+}). The deviations are larger for paramagnets with large SOC values and for those with small separation of the magnetically coupled electronic states.

In the case of a C_{2v}-distorted elongated octahedron (Figure 4.10c,f), the degeneracy of the E_g levels is removed, so the g_{xx} is no longer equal to g_{yy}. The d_{z^2} and $d_{x^2-y^2}$ orbitals, which both transform according to A_1 representation, may be admixed (hybridised) to form the SOMO $= c_1|d_{x^2-y^2}\rangle + c_2|d_{z^2}\rangle$. The resultant g tensor is rhombic, and its components are described by more complex expressions:

$$g_{zz} = g_e + 8c_1^2\lambda/[E(d_{x^2-y^2}) - E(d_{xy})]$$

$$g_{xx} = g_e + 2(c_1 + \sqrt{3}c_2)^2\lambda/[E(d_{x^2-y^2}) - E(d_{xz})] \qquad (4.31)$$

$$g_{yy} = g_e + 2(c_1 - \sqrt{3}c_2)^2\lambda/[E(d_{x^2-y^2}) - E(d_{yz})]$$

derived using the Λ matrix technique. Note that the $d_{x^2-y^2}/d_{z^2}$ orbital mixing alone can cause appreciable anisotropy in g_{xx} and g_{yy} values, which can clearly be seen in the EPR spectrum (Figure 4.10f).

In a more adequate approach, covalency of the bonds has to be taken into account, so the relevant MO can be written in a generic form as $|\psi_i\rangle = \alpha_i|M\rangle + \beta_i|L\rangle$, where α_i indicates the metal (M) contribution to the MO of the appropriate symmetry. The remaining part, gauged by β_i, is delocalised over the surrounding ligands (L), giving rise to a possible superhyperfine structure if only ligands with magnetic nuclei are involved and the spectra are sufficiently well resolved.

4.3.2   The Hyperfine A Tensor

The electron–nuclear hyperfine coupling observed in the condensed state arises mainly through two mechanisms: the previously discussed isotropic Fermi contact and an anisotropic dipolar interaction. As a result, the total hyperfine coupling between magnetic moments related to I and S must be described as a second-rank tensor, A, and we can write the corresponding spin Hamiltonian as:

$$H_{HFS} = \hbar^{-2}I^T A S \qquad (4.32)$$

In its principal axis systems, which generally do not need to be coincident with those of the g tensor, the A tensor is diagonal, and may be split into the isotropic and anisotropic parts, $A = a_{iso}E + T$, in the following way:

$$A = \begin{pmatrix} A_{xx} & 0 & 0 \\ 0 & A_{yy} & 0 \\ 0 & 0 & A_{zz} \end{pmatrix} = a_{iso}E + \begin{pmatrix} T_{xx} & 0 & 0 \\ 0 & T_{yy} & 0 \\ 0 & 0 & T_{zz} \end{pmatrix} \qquad (4.33)$$

The anisotropic part, T, corresponds to the dipolar interaction and is traceless. For this reason, in the case of paramagnetic species undergoing rapid tumbling, it can be averaged to zero, and the observed hyperfine splitting corresponds to the isotropic component alone: $a_{iso} = (A_{xx} + A_{yy} + A_{zz})/3$.

The anisotropic hyperfine Hamiltonian can be derived from classic dipolar equation averaged over spatial coordinates of the SOMO state:

$$H_{dipol} = -\hbar^{-2} g \mu_B g_n \mu_n \langle SOMO| \frac{I \cdot S}{r^3} - \frac{3(I \cdot r)(S \cdot r)}{r^5} |SOMO\rangle \quad (4.34)$$

where r is the vector joining the electron and the nucleus of interest. By expanding the scalar products and rearranging the appropriate terms into a matrix form, the following expression is obtained:

$$H_{dipol} = (-\hbar^{-2} g \mu_B g_n \mu_n)[S_x, S_y, S_z]$$

$$\cdot \begin{bmatrix} \left\langle \dfrac{r^2 - 3x^2}{r^5} \right\rangle & -\left\langle \dfrac{3xy}{r^5} \right\rangle & -\left\langle \dfrac{3xz}{r^5} \right\rangle \\ -\left\langle \dfrac{3xy}{r^5} \right\rangle & \left\langle \dfrac{r^2 - 3y^2}{r^5} \right\rangle & -\left\langle \dfrac{3yz}{r^5} \right\rangle \\ -\left\langle \dfrac{3xz}{r^5} \right\rangle & -\left\langle \dfrac{3yz}{r^5} \right\rangle & \left\langle \dfrac{r^2 - 3z^2}{r^5} \right\rangle \end{bmatrix} \cdot \begin{bmatrix} I_x \\ I_y \\ I_z \end{bmatrix}$$

$$= \hbar^{-2} S^T \cdot T \cdot I \quad (4.35)$$

After decoupling the radial and angular parts, the anisotropic tensor T assumes the form:

$$T_{ii} = (g_e \mu_B g_n \mu_n / r^3) \langle (3\cos^2\theta - 1) \rangle_{ii} = P \langle (3\cos^2\theta - 1) \rangle_{ii} \quad (4.36)$$

where $i = x, y, z$, $P = g_e \mu_B g_n \mu_n \langle r^{-3} \rangle$ and θ is the angle between r and B vectors. Because of their symmetry, for s orbitals the angular part averages to zero upon integration, while for p and d orbitals, T_{ii} assumes the following generic form:

$$\frac{2}{5} P_p \begin{pmatrix} 2 & & \\ & -1 & \\ & & -1 \end{pmatrix} \quad \text{and} \quad \pm \frac{2}{7} P_d \begin{pmatrix} 2 & & \\ & -1 & \\ & & -1 \end{pmatrix} \quad (4.37)$$

respectively. The positive sign applies for d_{z^2} and the negative for the other d orbitals.

Isotropic hyperfine coupling ($^A a_{iso} = 2/3 \mu_0 g \mu_B g_n \mu_n \rho^{\alpha-\beta}(R_A)$) takes place if a non-zero spin density, $\rho^{\alpha-\beta}(r) = \rho^\alpha(r) - \rho^\beta(r)$, is in contact with a given magnetic nucleus, A. The Fermi contact interaction may originate from the symmetry-allowed admixture of s orbitals to the SOMO (the so-called 'direct mechanism') or from polarisation of the inner-core s levels by the spin density localised on other orthogonal orbitals ('indirect' or 'polarisation' mechanism). The latter is usually accounted for by a simple expression of the type $Q\rho^{\alpha-\beta}(r)$, where Q indicates the polarisation constant:

$$a_{iso} = a_{iso}{}^{Fermi} + Q\rho^{\alpha-\beta}(r) \tag{4.38}$$

When the isotropic and anisotropic parts of the hyperfine tensor are determined and corrected for polarisation and spin–orbit effects (see below), approximate spin densities on the s, p and d orbitals constituting the $|SOMO\rangle = c_s|s\rangle + c_p|p\rangle + c_d|d\rangle$ can be calculated. While conducting such decomposition, the signs of the particular A tensor components (generally not available from experiment) should be assigned by judicious analysis of the magnetic interactions involved or by DFT calculations.[36] By using the tabulated atomic hyperfine constants for a given nucleus A, and assuming that the unpaired electron is entirely localised on a particular type of atomic orbital (s, p, d), the following is obtained:

$$c_s{}^2 = {}^A\rho_s = {}^Aa_{iso}(\text{experimental})/{}^Aa_{iso}(\text{atomic}) \tag{4.39a}$$

$$c_{p,d}^2 = {}^A\rho_{p,d} = {}^AT_{ii}(\text{experimental})/{}^AT_{ii}(\text{atomic}) \tag{4.39b}$$

Recommended calculated ${}^Aa_{iso}$ (atomic) and ${}^AT_{ii}$ (atomic) values for various nuclei and oxidation states can be found elsewhere.[37]

For a SOMO composed of a single p or d orbital, an axial hyperfine tensor is expected. Any departure from the axial symmetry that cannot be accounted for by the SOC correction is usually caused by hybridisation. For p orbitals constituting the SOMO:

$$|SOMO\rangle = c_x|px\rangle + c_y|p_y\rangle + c_z|p_z\rangle \tag{4.40}$$

neglecting the spin–orbit effects, the dipolar matrix elements are:[54]

$$T_{xx} = (2/5)P_p(2c_x^2 - c_y^2 - c_z^2)$$

$$T_{yy} = (2/5)P_p(-c_x^2 + 2c_y^2 - c_z^2) \tag{4.41}$$

$$T_{zz} = (2/5)P_p(-c_x^2 - c_y^2 + 2c_z^2)$$

$$T_{ij} = -3c_ic_j, (i \neq j)$$

An analogous set of equations for the SOMO composed of d orbitals:

$$|SOMO\rangle = c_{z^2}|z^2\rangle + c_{yz}|yz\rangle + c_{xz}|xz\rangle + c_{x^2-y^2}|x^2-y^2\rangle + c_{xy}|xy\rangle$$
$$(4.42)$$

can be written in the following way:[54]

$$T_{xx} = (2/7)P_d[-(c_{z^2})^2 - 2(c_{yz})^2 + (c_{xz})^2 + (c_{x^2-y^2})^2$$
$$+(c_{xy})^2 - 2\sqrt{3}(c_{z^2})(c_{x^2-y^2})]$$
$$T_{yy} = (2/7)P_d[-(c_{z^2})^2 + (c_{yz})^2 - 2(c_{xz})^2 + (c_{x^2-y^2})^2$$
$$+(c_{xy})^2 + 2\sqrt{3}(c_{z^2})(c_{x^2-y^2})]$$
$$T_{zz} = (2/7)P_d[2(c_{z^2})^2 + (c_{yz})^2 + (c_{xz})^2 - 2(c_{x^2-y^2})^2 - 2(c_{xy})^2] \quad (4.43)$$
$$T_{xy} = (2/7)P_d[-2\sqrt{3}(c_{z^2})(c_{xy}) + 3(c_{yz})(c_{xz})]$$
$$T_{yz} = (2/7)P_d[\sqrt{3}(c_{z^2})(c_{yz}) + 3(c_{xz})(c_{xy}) - 3(c_{yz})(c_{x^2-y^2})]$$
$$T_{xz} = (2/7)P_d[\sqrt{3}(c_{z^2})(c_{xz}) + 3(c_{yz})(c_{xy}) + 3(c_{xz})(c_{x^2-y^2})]$$

From these equations, it can be easily inferred that for certain combinations of the p and d orbitals involved in the SOMO, a non-diagonal hyperfine tensor $T_{ij} \neq 0$ will be obtained. This must be diagonalised in order to obtain the principal values (see Section 4.1) that can be compared with experiment. Furthermore, the SOC perturbs these results by adding terms to the diagonal matrix components of the order of $P(g_{ii} - g_e)$. Thus, a departure from, say, an axial symmetry may be due to the SOC effects if $A_\parallel = A_{zz}$ and $A_{xx} - A_{yy} \sim P(g_{xx} - g_{yy})$, or it may arise from d-orbital hybridisation when the rhombicity is larger. For instance, in the case of the previously discussed Cu^{2+} ions in elongated octahedral sites, if the SOMO is constituted by a sole $d_{x^2-y^2}$ orbital, the T tensor is axial, with $T_{zz} = -(4/7)P_d, T_{yy} = +(2/7)P_d$ and $T_{xx} = +(2/7)P_d$, whereas for the compressed octahedron with the d_{z^2} ground state, $T_{zz} = +(4/7)P_d, T_{yy} = -(2/7)P_d$ and $T_{zz} = -(2/7)P_d$. For the C_{2v} sites, with more involved SOMO $= c_{x^2-y^2}|d_{x^2-y^2}\rangle + c_{z^2}|d_{z^2}\rangle$, the anisotropic hyperfine tensor has the following components:

$$T_{xx} = (2/7)P_d[-(c_{z^2})^2 + (c_{x^2-y^2})^2 - 2\sqrt{3}(c_{z^2})(c_{x^2-y^2})]$$

$$T_{yy} = (2/7)P_d[-(c_{z^2})^2 + (c_{x^2-y^2})^2 + 2\sqrt{3}(c_{z^2})(c_{x^2-y^2})] \quad (4.44)$$

$$T_{zz} = (2/7)P_d[2(c_{z^2})^2 - 2(c_{x^2-y^2})^2]$$

Since the contributions of the d_{z^2} and $d_{x^2-y^2}$ AOs to T_{ii} are of different signs, hybridisation of those orbitals leads to a decrease of the hyperfine splitting (Figure 4.10f). A similar effect also takes place if p orbitals are allowed to mix with a $d_{x^2-y^2}$ SOMO. In that case, admixture of p_z decreases the dipolar splitting, $T_{zz} = -(4/7)(c_{x^2-y^2})^2 P_d + (4/5)c_z^2 P_p$, while admixture of p_x (or p_y) has an opposite effect, since $T_{zz} = -(4/7)(c_{x^2-y^2})^2 P_d - (2/5)c_x^2 P_p)$. Such a situation is observed for tetra-hedral and square planar centres.

In practice, an experimental A tensor $[A_{xx}, A_{yy}, A_{zz}]$ is determined from the experimental hyperfine structure (Figure 4.10f) and divided into a_{iso} and T components. Then, using equations such as Equation 4.43, the AO composition of the SOMO (c_i coefficients) is derived using the tabulated P values (equivalent to atomic T_{ii} values). An illustrative example of molecular interpretation of the g and A tensors is provided by analysis of the EPR spectrum of low-symmetry Cu^{2+} paramagnetic centres in $KZnClSO_4 \cdot 3H_2O$ matrix of kainite structure. The experimental g and A tensor components are $g_{xx} = 2.1535$, $g_{yy} = 2.0331$, $g_{zz} = 2.4247$ and $A_{xx}/10^{-4} = (-)31\ cm^{-1}, A_{yy}/10^{-4} = (+)63\ cm^{-1}, A_{zz}/10^{-4} = (-)103\ cm^{-1}$ (Figure 4.11a).[38] For the D_{2h} local point group symmetry of the hosting site (Figure 4.11b), the cop-per $3d$ orbitals can be mixed with $4s$ orbital and ligand-based L orbitals,

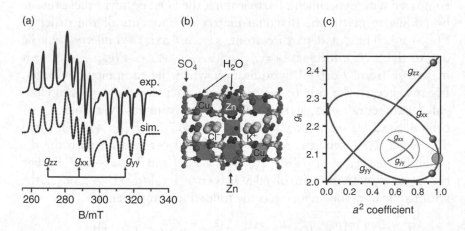

Figure 4.11 (a) X-band powder EPR spectrum of Cu^{2+} centres in $KZnClSO_4 \cdot 3H_2O$. (b) Crystal structure. (c) Influence of the $d_{x^2-y^2}$ contribution to SOMO (a^2 coefficient) on the g_{ii} components. Three balls represent experimental g_{ii} values. Reproduced with permission from Hoffmann *et al.* (2010) [38]. Copyright (2010) Elsevier Ltd.

resulting in the following MOs, which are responsible for the observed spin Hamiltonian parameters:

$$\Psi(A_g) = \alpha(ad_{x^2-y^2} + bd_{z^2} + c_s s) - \alpha' L_1$$

$$\Psi(A_g) = \alpha_1(ad_{z^2} - bd_{x^2-y^2} + c_s s) - \alpha_1' L_2$$

$$\Psi(B_{1g}) = \beta_1 d_{xy} - \beta_1' L_3 \qquad\qquad (4.45)$$

$$\Psi(B_{2g}) = \beta d_{xz} - \beta' L_4$$

$$\Psi(B_{3g}) = \beta' d_{yz} - \beta'' L_5$$

where $a^2 + b^2 + c_s^2 = 1$. For the A_g ground-state symmetry, the principal g and A values are given by:[39]

$$g_{zz} = g_e - 8\alpha^2 \beta_1^2 a^2 \frac{\lambda}{E_{xy}} \qquad\qquad (4.46a)$$

$$g_{yy} = g_e - 2\alpha^2 \beta^2 (a - \sqrt{3}b)^2 \frac{\lambda}{E_{xz}} \qquad\qquad (4.46b)$$

$$g_{xx} = g_e - 2\alpha^2 \beta^2 (a + \sqrt{3}b)^2 \frac{\lambda}{E_{yz}} \qquad\qquad (4.46c)$$

$$A_{zz} = \alpha^2 c_s^2 A_{4s} + P\left[-\alpha^2 \kappa + \Delta g_{zz} - \frac{4}{7}\alpha^2 \left(a^2 - b^2\right) \right.$$

$$\left. + \frac{1}{14}\frac{3a + \sqrt{3}b}{a - \sqrt{3}b}\Delta g_{yy} + \frac{1}{14}\frac{3a - \sqrt{3}b}{a + \sqrt{3}b}\Delta g_{xx} \right] \quad (4.47a)$$

$$A_{yy} = \alpha^2 c_s^2 A_{4s} + P\left[-\alpha^2 \kappa + \Delta g_{yy} + \frac{2}{7}\alpha^2 \left(a^2 - b^2 + 2\sqrt{3}ab\right) \right.$$

$$\left. - \frac{\sqrt{3}}{14}\frac{b}{a}\Delta g_{zz} - \frac{1}{14}\frac{3a - \sqrt{3}b}{a + \sqrt{3}b}\Delta g_{xx} \right] \quad (4.47b)$$

$$A_{xx} = \alpha^2 c_s^2 A_{4s} + P\left[-\alpha^2 \kappa + \Delta g_{xx} + \frac{2}{7}\alpha^2 \left(a^2 - b^2 - 2\sqrt{3}ab\right) \right.$$

$$\left. + \frac{\sqrt{3}}{14}\frac{b}{a}\Delta g_{zz} - \frac{1}{14}\frac{3a - \sqrt{3}b}{a + \sqrt{3}b}\Delta g_{yy} \right] \quad (4.47c)$$

where $\lambda = -829\,\text{cm}^{-1}$ is the SOC constant, $P = 0.036\,\text{cm}^{-1}$ and $\kappa = 0.43$ is the Fermi contact parameter of the free Cu^{2+} ion, E_i represents energy separation between A_g ground state and the $3d_{xy}$, $3d_{xz}$ and $3d_{yz}$ levels, and the $\alpha^2 c_s^2 A_{4s}$ term gives the direct contribution of the $4s$ orbital to the isotropic hyperfine coupling. Using the experimental g_{ii} and A_{ii} values, $a^2 = 0.947, \alpha^2 = 0.93$ and $\kappa = 0.284$ are derived from Equations 4.47a–c. At this stage, by using Equations 4.46a–c, separations between the energy levels can be calculated as a function of the covalency parameter, β_i. However, this problem may further be treated using the energies derived from the UV–visible spectrum ($E_{xy} = 11\,000\,\text{cm}^{-1}, E_{xz} = 11\,700\,\text{cm}^{-1}, E_{xz} = 13\,100\,\text{cm}^{-1}$),[40] which gives $\beta_1^2 = 0.80, \beta^2 = 0.80$ and $\beta'^2 = 0.67$. As a result, the following ground state of the Cu^{2+} centres, $|\text{SOMO}\rangle = 0.96(0.97|x^2 - y^2\rangle + 0.03|z^2\rangle$, has been determined. The influence of the degree of $|x^2 - y^2\rangle/|z^2\rangle$ hybridisation on the g tensor values is shown in Figure 4.11c. In the range of a^2 values characteristic of Cu^{2+} in kainite, admixture of $3d_{z^2}$ leads to a decrease of g_{zz} value, whereas g_{xx} increases and becomes much larger then g_{yy}.

4.3.3   The Fine Structure D Tensor

Several paramagnetic materials may have more than one unpaired electron and, consequently, $S > 1/2$. The most simple cases are related to $d^2 - d^8$ transition metal ions with $S \geq 1$ and no first-order contribution to the angular momentum ($\langle 0|L|0\rangle = const\langle 0|n\rangle = 0$, which occurs for the orbitally nondegenerate electronic ground state $|0\rangle$). In such cases, a new term must be added to the spin Hamiltonian to account for the unpaired electron–electron interactions. In noncubic systems this interaction leads to the removal of the M_S degeneracy even in the absence of external magnetic field. This phenomenon is called a zero-field splitting (ZFS), and is illustrated in Figure 4.12a for the Gd^{3+} ion ($S = 7/2, 5f^7$) in tetragonal $BaTiO_3$.[41] ZFS is vanishing in the cubic polymorph of $BaTiO_3$ (Figure 4.12b), however, indicating its high structure sensitivity on probing the phase transitions of solids. The essence of ZFS lies in two potential interactions: a weak dipolar electron–electron coupling and a mutual interaction between two spins mediated by the SOC, jointly described by tensor D. Experimentally, it is not possible to separate the spin–spin from the SOC contribution. The first contribution is more pronounced for paramagnets composed of light atoms in triplet states, while the SOC contribution is dominant for the transition metals.[42]

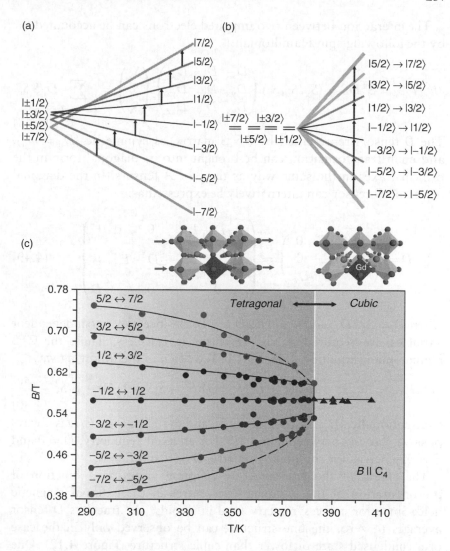

Figure 4.12 Zero-field splitting for the $Gd^{3+}(S = 7/2)$ ion in $BaTiO_3$ and the corresponding energy-levels splitting in magnetic field for (a) an axial symmetry showing seven fine structure transitions and (b) a cubic symmetry with a single transition. (c) Evolution of the positions of fine structure lines with temperature, showing a tetragonal to cubic phase transition. Single-crystal spectra are recorded along the C_4 axis. In the model of the $BaTiO_3$ structure, the direction of the tetragonal distortion is indicated with arrows. Reproduced from Rimai (1962) [41]. Copyright (1962) American Physical Society.

The interaction between two unpaired electrons can be accounted for by the following spin Hamiltonian:

$$\hbar^2 H_{ZFS} = S^T D S = (S_x, S_y, S_z) \begin{pmatrix} D_{xx} & D_{xy} & D_{xz} \\ D_{yx} & D_{yy} & D_{yz} \\ D_{zx} & D_{zy} & D_{zz} \end{pmatrix} \begin{pmatrix} S_x \\ S_y \\ S_z \end{pmatrix} = \sum_{i,j=x,y,z} D_{ij} S_i S_j$$

(4.48)

The D tensor (represented by 3×3 matrix) is symmetric ($D_{ij} = D_{ji}$), and upon transformation can be brought into its diagonal form in the principal axes in the same way as the g or A tensors. In the diagonal form, the D tensor can alternatively be expressed as:

$$D = \begin{pmatrix} D_{xx} & 0 & 0 \\ 0 & D_{yy} & 0 \\ 0 & 0 & D_{zz} \end{pmatrix} = \begin{pmatrix} -\frac{1}{3}D + E & 0 & 0 \\ 0 & -\frac{1}{3}D - E & 0 \\ 0 & 0 & \frac{2}{3}D \end{pmatrix}$$

(4.49)

where $D = 3/2 D_{zz}$ and $E = 1/2(D_{xx} - D_{yy})$, since only two independent variables are required as the D tensor is traceless. Thus, the ZFS Hamiltonian (Equation 4.48) may be rewritten in the following way:

$$\hbar^2 H_{ZFS} = D_{xx}S_x^2 + D_{yy}S_y^2 + D_{zz}S_z^2 = D(S_z^2 - S(S+1)/3) + E(S_x^2 - S_y^2)$$

(4.50)

Conventionally, $|D_{zz}| > |D_{yy}| > |D_{xx}|$, and in such a case, E/D is always positive, varying between 0 and 1/3. For an axial symmetry, $D \neq 0$ and $E = 0$, whereas for rhombic symmetry, $D \neq 0$ and $E \neq 0$.

The extent of the fine structure coupling varies as a function of the orientation of the paramagnetic centres in the external magnetic field. Since for cubic symmetry (and in liquids) the traceless D tensor averages to zero, the fine structure can be observed only in the case of a condensed state of lower than cubic structure (Figure 4.12). Due to some mixing between the magnetic states, the forbidden $\Delta M_S = \pm 2$ transitions become partially allowed and appear at low magnetic fields, usually with weak intensity.

In the already invoked example of Gd^{3+} ions doped in the tetragonal $BaTiO_3$ perovskite, the ZFS of the ^7S state produces four doubly degenerate spin states, $M_S = \pm 7/2, \pm 5/2, \pm 3/2$ and $\pm 1/2$. The Kramers degeneracy can be removed by the magnetic field, giving rise to the $2S = 7$ transitions $(-7/2 \rightarrow -5/2, -5/2 \rightarrow -3/2, -3/2 \rightarrow -1/2, -1/2 \rightarrow +1/2, +1/2 \rightarrow +3/2, +3/2 \rightarrow +5/2, +5/2 \rightarrow +7/2)$ shown in Figure 4.12a,c.

In contrast to the hyperfine structure, the components of the fine splitting have varying intensities $I \sim S(S+1) - M_S(M_S - 1)$, so that the intensity is greatest for the central line and smallest for the outermost lines.

The D_{ii} matrix elements can be calculated based on the Λ matrix elements from the following equation, where Λ_{ij} is defined by Equation 4.30:

$$D_{ij} = -\lambda^2 \Lambda_{ij} \tag{4.51}$$

As already mentioned, the ZFS is highly anisotropic and structure sensitive. This is reflected in both sign and absolute values of the D and E parameters. These can be related to the geometry of the local environment (coordination polyhedron) using an empirical superposition model (SPM), whose basic assumption is that a single ligand located at (R, θ, φ) relative to the transition-metal ion centre makes a contribution to the ZFS, taking the following form:

$$b_n^m = \sum_i b_n(R_i) K_n^m(\theta_i, \varphi_i) \tag{4.52}$$

where summation runs over all ligands. In this notation, $D = 3b_2^0$ and $E = 1/3b_2^2$. The angular K_n^m functions for various n and m values are tabulated elsewhere,[5,43] while the distance dependence of the radial functions, $b_n(R_i)$, is defined as:

$$b_n(R) = \bar{b}_n(R_0)(R_0/R)^{t_n} \tag{4.53}$$

where R_0 is a reference bond distance for a given ion and t_n is an empirical parameter. An application of SPM can be illustrated by Cr^{3+} centres $(S = 3/2)$ in an AgCl host.[44] From EPR studies, two tetragonal sites constituted by chromium Cr^{3+} associated with one or two vacancies of C_{4v} and D_{4h} local symmetry, respectively, have been found (Figure 4.13a,b). From the SPM, the D parameter for the Cr^{3+}–V_{Ag} centre can be expressed as:

$$D = \frac{3}{2}\bar{b}_2(R_0)\sum_i (R_0/R_i)^{t_n}(3\cos\theta_i - 1) \tag{4.54}$$

The Cr–Cl distance $R_0 \approx 0.2457\,\text{nm}$ and displacements of the Cl^- ions toward the chromium centre, $\Delta R \approx 0.018R_0$, induced by the presence of the adjacent vacancy have been determined by fitting the experimental values of $D = -2411 \times 10^{-4}\,\text{cm}^1$ and $-494 \times 10^{-4}\,\text{cm}^{-1}$ for the Cr^{3+}–V_{Ag} and V_{Ag}–Cr^{3+}–V_{Ag} centres, respectively, to those calculated from the SPM.

(a)   (b)

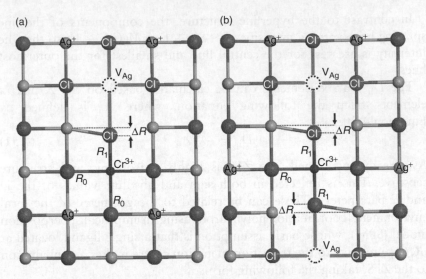

Figure 4.13 Schematic representation of tetragonal defect centres in Cr-doped AgCl crystals: (a) $Cr^{3+}-V_{Ag}(C_{4v})$ and (b) $V_{Ag}-Cr^{3+}-V_{Ag}(D_{4h})$ centres, where V_{Ag} represents a cationic vacancy in an AgCl lattice. Reproduced with permission from Wen-Chen and Shao-Yi (1998) [44]. Copyright (1998) Elsevier Ltd.

4.3.4 The Quadrupole Q Tensor

Any nucleus with a nuclear-spin quantum number $I \geq 1$ exhibits an electric quadrupole moment. As a result, the unpaired electron interacts with both the nuclear magnetic moment and the electric field gradient at the nucleus. The latter interaction is described by the quadrupole spin Hamiltonian:

$$H_{NQ} = \hbar^{-2} I^T Q I \qquad (4.55)$$

The quadrupole tensor is traceless ($\sum Q_{ii} = 0$) and the relationships between the diagonal matrix and the usual parameters $e^2 Qq/h$ and η are as follows:

$$Q = \begin{pmatrix} Q_1 & 0 & 0 \\ 0 & Q_2 & 0 \\ 0 & 0 & Q_3 \end{pmatrix} = \frac{e^2 Qq/h}{4I(2I-1)} \begin{pmatrix} -(1-\eta) & 0 & 0 \\ 0 & -(1+\eta) & 0 \\ 0 & 0 & 2 \end{pmatrix} \qquad (4.56)$$

where $e^2 Qq/h = 2I(2I-1)Q_3$ and $\eta = (Q_1 - Q_2)/Q_3$. Following the convention, the $e^2 Qq/h$ parameter can be positive or negative, whereas η varies between 0 and 1. The product eq is the largest-magnitude component of the electric field gradient at the nucleus, expressed in units of Vm^{-2}. Q is the electric quadrupole moment of the nucleus, typically given in units of barns (1 barn $= 10^{-28}$ m^2). The effect of the quadrupole interaction is usually complicated because it is accompanied by a much larger hyperfine interaction ($I \neq 0$). When the magnetic field is not collinear with the crystal axes, the quadrupole interaction changes the separation of the energy levels, which causes the spacing between the adjacent hyperfine lines to be greater at low- and high-field parts of the spectrum than in the middle. The resultant unequal spacing can be eliminated by recording spectra in higher microwave frequencies.[45] Another effect of the competition between the quadrupole electric field and the magnetic field is the appearance of additional $\Delta m_I = \pm 1$ and ± 2 lines, which are normally forbidden by the selection rule $\Delta m_I = 0$.

4.3.5 Electron–Electron Exchange Interactions J

The essence of the exchange interactions is the coupling of magnetic moments by formation of weak bonds between the interacting paramagnets, which may be spread over quite distant centres and a dipolar interaction through space. The spin Hamiltonian for interpretation of the magnetic properties of magnetically nondiluted materials, when all the magnetic centres have orbitally nondegenerate well-separated ground states, can be written as:[18,42]

$$H_{EX} = \hbar^{-2} \sum_{A,B} S_A \cdot J_{AB} \cdot S_B \qquad (4.57)$$

The H_{EX} Hamiltonian representing the exchange interactions describes a quadratic coupling between the magnetic moments localised on the centres A and B, bearing the spins S_A and S_B, respectively. This interaction is parameterised by the second-rank tensor, J_{AB}. The exchange coupling tensor J_{AB} can be decomposed in the following way:

$$J_{AB} = JE + D^{(S)}{}_{AB} + D^{(A)}{}_{AB} \qquad (4.58)$$

where E is the unit matrix, $D^{(S)}{}_{AB}$ is a symmetric traceless tensor $(J_{ij} - J_{ji})/2 - Tr(J)/3$, $D^{(A)}{}_{AB}$ is an antisymmetric tensor $(J_{ij} - J_{ji})/2$ and $J = Tr(J)/3$ is referred to as an isotropic exchange coupling constant. The resultant conventional form of the H_{EX} Hamiltonian is:

$$H_{EX} = \hbar^{-2} \sum_{AB} \lfloor -JS_A S_B + d_{AB}(S_A \times S_B) + S_A D^{(S)}{}_{AB} S_B \rfloor \qquad (4.59)$$

Thus, the overall exchange interactions are decomposed into the isotropic or Heisenberg–Dirac–van Vleck (HDvV) interaction (J), and the anisotropic parts of antisymmetric interaction (d_{AB}) and Dzyaloshinsky–Moriya interaction ($D^{(S)}_{AB}$).[46] The sign convention used herein is such that positive J corresponds to a ferromagnetic coupling (with spins aligned parallel), while negative J implies an anti-ferromagnetic state in which spins are aligned antiparallel in the ground state. All of these interactions arise from two main mechanisms: direct through-space magnetic interactions and electron–electron (through-bond) correlation effects. Although the through-space contribution is small, is may be dominant for distant centres, and in such cases it is usually modelled as a dipolar interaction between the localised magnetic moments and has the greatest effect on the anisotropic term d_{AB}. Dipolar interactions give rise to inhomogeneous broadening of the EPR spectra and Gaussian lineshapes. The d_{AB} tensor provides information on the distance between the individual paramagnetic centres. In the simplest cases, the relation $d_{AB}/cm^{-1} = 0.433 \times g^2/r^3$ can be used to convert the measured d_{AB} value to the distance r, expressed in angstroms. A more general account can be found elsewhere.[3]

The SOC that mixes excited states into the ground state contributes to the symmetric ($D^{(S)}$) and the antisymmetric (d_{AB}) dipolar terms of the H_{EX} Hamiltonian. Generally, it is much smaller that the HDvV interaction, and its influence on magnetic properties is significant only at low temperatures. The H_{EX} Hamiltonian should be complemented by individual ZFS and electron Zeeman terms for centres with spins $S > 1/2$.

For instance, in the case of $S = 1$, the effect of $J < 0$ is the separation of the energy levels into a singlet ground state $|0,0\rangle$ and triplet excited states $|1,0\rangle$, $|1,-1\rangle$ and $|1,1\rangle$. In the axial symmetry, ZFS splits the triplet state into two levels ($|1,0\rangle$ and $|1,\pm 1\rangle$), separated by the D value, as shown in Figure 4.14a. For $J < 0$, the population of the triplet state is governed by the Boltzmann distribution, $3/T[\exp(-J/kT)]$; the EPR signal can thus only be observed if the magnetic triplet state is thermally populated (*i.e.* for small J values). Copper-doped CeO_2 is probably one of the best examples of an $S = 1$ system with a clearly resolved EPR spectrum (Figure 4.14b), due to a pair of nearly equivalent Cu^{2+} ions with $g_{||} = 2.2079$, $g_{\perp} = 2.0403$, $|A_{||}| = 8.5$ mT, $|A_{\perp}| = 1.35$ mT, $D = 0.066\,cm^{-1}$, $d_{Cu-Cu} = -0.081\,cm^{-1}$, $D^{(S)} = 0.147\,cm^{-1}$ and $J = -52.5\,cm^{-1}$.[47] Alongside the allowed transitions ($\Delta M_s = 1$) observed in the mid field, weaker forbidden lines ($\Delta M_s = 2$) are also visible in the spectral low-field region.[48]

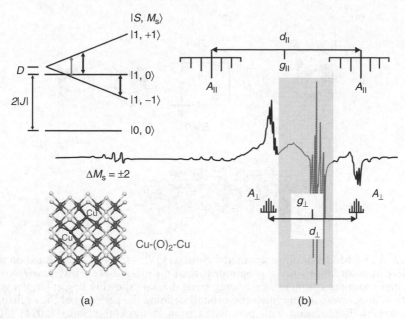

Figure 4.14 (a) Energy diagram for $S_A = S_B = 1/2$ dimeric species coupled *via* exchange interaction with an antiferromagnetic ground state. (b) X-band EPR spectrum and structural model of Cu^{2+} dimers in CeO_2 matrix. The dotted part of the spectrum represents an isolated monomeric Cu^{2+} site not discussed in the text. Spectrum reproduced with permission from Aboukaïs *et al.* (1992) [47]. Copyright (1992) Royal Society of Chemistry.

An example of the molecular interpretation of the exchange-coupled system is provided by dimeric vanadyl clusters, $OV^{IV}-O-V^{IV}O$($S_B = 1/2$ and $S_A = 1/2$), deposited on the surface of amorphous silica (Figure 4.15). In this case, a ferromagnetic coupling gives rise to an effective spin $S = 1$, whereas for an antiferromagnetic state, $S = 0$ (Figure 4.15b). The corresponding energy levels can be expressed as $E(S) = -JS(S + 1)$, and J can be calculated using quantum mechanical calculations within the broken-symmetry formalism.[49] DFT calculations have revealed that $J = -26\,cm^{-1}$, indicating that the antiferromagnetic ground state of the $OV^{IV}-O-V^{IV}O$ clusters is preferred.[50] The computed spin-density contour (Figure 4.15a) displays regions of positive and negative spin densities, accounting for the antiparallel orientation of the individual spins. After spatial averaging, the spin density vanishes, as the magnetic moment in the ground state is zero. As a consequence, surface dimers with the antiferromagnetic $OV^{IV}-O-V^{IV}O$ bridges are EPR silent, as has been experimentally proven.[51] More

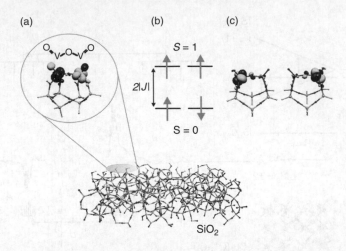

Figure 4.15 Model of dimeric vanadyl clusters, $OV^{IV}-O-V^{IV}O$, deposited on the surface of amorphous silica. (a) Spin density of the ground broken-symmetry state (antiferromagnetic coupling). (b) Energy-levels diagram showing singlet–triplet separation. (c) Corresponding magnetic orbitals defining the pathway of the exchange interactions. Reproduced with permission from Pietrzyk and Sojka (2011) [50]. Copyright (2011).

insight into the electronic nature of exchange coupling is provided by analysis of the corresponding magnetic orbitals, which are characterised by small yet nonzero overlap (Figure 4.15c). In this case, the only exchange pathway is made possible by symmetric $\pi-\pi$ overlap between the in-plane p orbitals of the oxo bridge and the $3d_{xy}$ vanadium orbital. Such a pathway indicates that the antiferromagnetic coupling is mediated by superexchange interactions *via* the μ–oxo bridge.

4.3.6 The Spin Hamiltonian

As already discussed, a quantum description of EPR is considerably simplified by introduction of the spin Hamiltonian obtained by averaging the full physical Hamiltonian over the space coordinates and spin coordinates of the paired electrons. The spin Hamiltonian is expressed in linear (*e.g.* interaction of the electron spin with the static field B) or bilinear (*e.g.* coupling between electron–electron and electron–nuclear spins) terms of spin-angular-momentum operators and contains a few phenomenological constants that can be related to the properties of the spin system. Since the spin Hamiltonian is invariant under symmetry operations of the point symmetry group of a paramagnetic centre, the site symmetry imposes restrictions on the constants (tensors) that factorise particular magnetic interactions. The spin Hamiltonian is a linear function of the magnetic field, taking a general form of $H_{sp}(B) = B^T G + G$ with the field-dependant (Zeeman operators) and field-independent (zero-field, hyperfine, quadrupole and exchange) operators. This energy operator is composed of a series of terms, as discussed above:

$$H_{sp} = H_{EZ} + H_{ZFS} + H_{HFS} + H_{NQ} + H_{NZ} + H_{EX} \qquad (4.60)$$

The most important are those related to the electron–Zeeman interaction, H_{EZ}, the hyperfine interaction, H_{HFS}, and the fine (zero-field) interactions, H_{ZFS}. These are parameterised by the g (Equation 4.23), A (Equation 4.33) and D (Equation 4.49) tensors, respectively. The energy of the electron Zeeman term is of the order of $0.3\,cm^{-1}$ for the X-band and increases with magnetic field. The zero-field interaction can vary within a much larger region ($0.01–50.0\,cm^{-1}$), while typical changes of the hyperfine interactions are between 10^{-4} and $0.01\,cm^{-1}$. The ZFS and HFS are field-independent quantities. These leading terms lay the ground for conventional CW-EPR and, in the case of large ZFS, HF-EPR measurements. The remaining interactions are of much lower energy: the quadrupolar interaction, H_{NQ} (Equation 4.56), is of the order of $10^{-3}\,cm^{-1}$, while the nuclear–Zeeman interaction, $H_{NZ} = -\hbar^{-1}\mu_n g_n BI$, is of the order of $10^{-2}\,cm^{-1}$ for X-band. Together with the smallest superhyperfine couplings ($<0.1–0.01$ mT), these can only be measured using pulsed EPR techniques.

Additionally, for polynuclear systems with exchange interaction, the spin Hamiltonian must be augmented with the exchange term, H_{EX} (Equation 4.59), which can span a wide range of energies (up to tens of cm^{-1} for isotropic exchange constant J).

The point symmetry at a given site of a paramagnetic centre imposes restrictions on the components of g and A (and other tensors), as well as the relative orientations of their principal axes,[3,52] providing a basis for classification of the EPR spectra. The symmetry determines, for example, whether any of the principal g, A or D values are required to be equal to each other and whether any of the principal g, A or D axes coincide. These criteria are summarised in Table 4.2, along with the corresponding nomenclature of the EPR spectra for magnetically diluted $S = 1/2$ species.[53] The importance of these relationships is that each type of EPR behaviour is associated with a restricted number of point group symmetries. For analysis of the hyperfine couplings, only those symmetry elements which the corresponding magnetic nucleus shares with the whole complex are important.[54] In axial or orthorhombic symmetries, both g and A matrices share the same principal axes, while

Table 4.2 Relationships between crystal systems, corresponding point group symmetries, EPR spectrum symmetry and the restrictions imposed on the components of the g and A tensors and the coincidence angles (α, β, γ).

Crystal system	Bravais lattice	Point symmetry	EPR symmetry	g and A tensor constraints	Coincidence of principal axes of g and A tensors
Triclinic		C_1, C_i	Triclinic	$g_{ij} \neq 0$ and $g_{xx} \neq g_{yy} \neq g_{zz}$; $A_{xx} \neq A_{yy} \neq A_{zz}$	All noncoincident, $\alpha \neq \beta \neq \gamma \neq 0°$
Monoclinic		C_2, C_s, C_{2h}	Monoclinic	$g_{xz} = g_{yz} = g_{zx} = g_{zy} = 0$ and $g_{xx} \neq g_{yy} \neq g_{zz}$; $A_{xx} \neq A_{yy} \neq A_{zz}$	One axis of g and one axis of A coincident, $\alpha \neq 0° \beta = \gamma = 0°$
Trigonal		C_3, S_6	Axial noncollinear	As for C_2 and $g_{xx} = g_{yy} \neq g_{zz}, g_{zz}, g_{xy} = -g_{yx}$; $A_{xx} = A_{yy} \neq A_{zz}$	Only g_{zz} and A_{zz} coincident, $\alpha \neq 0° \beta = \gamma = 0°$
		D_3, C_{3v}, D_{3d}	Axial	As for C_2 and $g_{xy} = g_{yx} = 0$ and $g_{xx} = g_{yy} \neq g_{zz}$; $A_{xx} = A_{yy} \neq A_{zz}$	All coincident $\alpha = \beta = \gamma = 0°$
Tetragonal		C_4, S_4, C_{4h}	Axial noncollinear	As for C_3	As for C_3
		$D_4, C_{4v}, D_{2d}, D_{4h}$	Axial	As for D_3	As for D_3
Hexagonal		C_6, C_{3h}, C_{6h}	Axial noncollinear	As for C_3	As for C_3
		$D_6, C_{6v}, D_{3h}, D_{6h}$	Axial	As for D_3	As for D_3
Orthorhombic		D_2, C_{2v}, D_{2h}	Rhombic	As for C_2 and $g_{xy} = g_{yx} = 0$	All coincident $\alpha = \beta = \gamma = 0°$
Cubic		T, T_h, O, T_d, O_h	Isotropic	As for D_3 and $g_{xx} = g_{yy} = g_{zz}$; $A_{xx} = A_{yy} = A_{zz}$	All coincident $\alpha = \beta = \gamma = 0°$

in monoclinic and triclinic symmetries the orientations of g and A axes may be different. It should be noted that inherent heterogeneity of the interfacial region decreases the permissible symmetry elements of surface species, often leading to low-symmetry EPR spectra.[55]

4.4 PRINCIPAL TYPES OF EPR SPECTRUM AND THEIR CHARACTERISTIC FEATURES

Inorganic materials may be synthesised as crystalline, nanostructured, partially disordered or amorphous (glassy) phases, and as such may give rise to well or poorly resolved, simple or complex EPR spectra. When a crystalline material is finely divided, additional line-broadening effects associated with distribution and modification of the EPR parameters can appear in the spectrum, owing not only to the effects of particle size distribution (usually in the range of 1–50 nm) but also to the distribution of dislocation strains, surface-related effects and local microheterogeneity in the nearest environment of the paramagnetic centres.[56−59]

Single-crystal investigations provide maximum information about the investigated paramagnetic centres and their spatial arrangement in the symmetry-related distinct sites in the crystal lattice. Much of this information, however, can also be extracted from more common but more challenging powder spectra, especially with the aid of advanced computer simulation and mutlifrequency measurements.[54] The use of powders may however generate a number of difficulties connected with poor resolution, overlapping signals, low-symmetry phenomena and g tensor anisotropy, too small to be observed at conventional frequencies.

4.4.1 Single-Crystal Spectra

Single-crystal spectra are recorded with a paramagnetic specimen mounted in a goniometer (Figure 4.8g) for precise measurement of the turning angles, so that the sample can be rotated in the resonant cavity with respect to the applied external field B about three orthogonal reference axes X, Y, Z. It is convenient for X, Y, Z to be related to the crystallographic axes (Figure 4.9a) or laboratory axes Z', X', Y'.[60] EPR spectra are recorded during stepwise rotations in the mutually

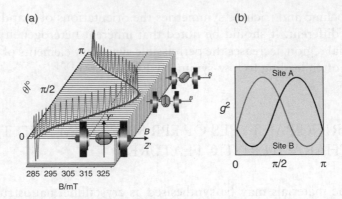

Figure 4.16 (a) Simulated single-crystal EPR spectra for axial symmetry ($S = 1/2, I = 0$) for various crystal orientations with respect to the magnetic field (the thick solid line is called a roadmap). (b) Angular dependence of the g value for two symmetry-related sites in a monoclinic crystal showing C_2 symmetry.

perpendicular $XZ-$, $YZ-$ and $XY-$ planes, and the actual g_{eff} values are measured from the current line positions (Figure 4.16a). In order to extract the $g_{ii}{}^2$ values, the obtained sinusoidal curves of the resonant field *versus* θ angle should be fitted by least squares to the theoretical formula:

$$g_{\text{eff}}^2 = g_{ii}^2 \cos^2\theta + g_{jj}^2 \sin^2\theta + 2g_{ij}^2 \sin\theta \cos\theta \qquad (4.61)$$

where θ is the angle between the direction of the static field B and the ith axis. In such a way, six independent variables g_{ij}^2 (since $g_{ij}^2 = g_{ji}^2$) can be determined ($i, j = X, Y, Z$, when rotations are about the Z-, X- and Y- axis)[5] and the full GG matrix can be constructed, as follows:

$$GG = \begin{bmatrix} g_{xx}^2 & g_{xy}^2 & g_{xz}^2 \\ g_{yx}^2 & g_{yy}^2 & g_{yz}^2 \\ g_{zx}^2 & g_{zy}^2 & g_{zz}^2 \end{bmatrix} \qquad (4.62a)$$

$$gg = \begin{bmatrix} g_{xx}^2 & 0 & 0 \\ 0 & g_{yy}^2 & 0 \\ 0 & 0 & g_{zz}^2 \end{bmatrix} \qquad (4.62b)$$

The principal $(gg)_{ii}$ values, along with the directional cosines relating the crystal axes X, Y, Z to the principal x, y, z axes of the gg tensor (Figure 4.9), are obtained by finding a C matrix ($C^T = C^{-1}$) that diagonalises the GG matrix, $gg = C \cdot GG \cdot C^T$. The principal values of the g tensor alone are obtained from $g_{ii} = (gg)_{ii}^{1/2}$. For the simple case of an axial symmetry, Equation 4.61 reduces to $g_{\text{eff}}^2 = g_\parallel^2 \cos^2\theta + g_\perp^2 \sin^2\theta$. The corresponding displacement of the EPR signal in the magnetic field upon angular rotation is shown in Figure 4.16a, marked with a thick solid line ('roadmap').

There are often situations in which magnetically equivalent centres of a crystal are located in different sites related by symmetry. In such a case, when B is placed along the Y- and X-axes, the two species will have the same angle with respect to the magnetic field and, therefore, the same effective g values. However, for arbitrary orientation of B, these two sites will obviously exhibit different effective g values, and consequently the roadmaps will be different (Figure 4.16b).

Equations similar to Equation 4.61 are used for more complex situations in which hyperfine coupling is present and the Zeeman term is dominant in the spin Hamiltonian. In such cases, g^2 should be replaced by $A^2 g^2$.[2,3] A single-crystal EPR study of $[O_2]^{3-}$ (Al) hole centres in stishovite ($P4_2/mmm$) – an aluminium-including high-pressure polymorph of SiO_2 – illustrates well the main features of a real spectrum (Figure 4.17) due to nonequivalent sites and clearly resolved hyperfine structure due to a ^{27}Al nucleus ($I = 5/2$).[61]

4.4.2 Static and Dynamic Disorder

From the point of view of EPR spectroscopy, two types of averaged signal – caused by partial or complete randomisation in space (static disorder) and in time (dynamic disorder) – may be distinguished.

Dynamic processes within and around the paramagnetic centres, such as tumbling and electron delocalisation, give rise to EPR linewidth changes or the shifting of line positions. Theoretical treatment of such time-dependent effects based on the stochastic Liouville equation is rather involved.[62,63] The timescales are gauged by the ratio of the correlation time of the dynamic processes to EPR signal spread ($\tau_c^{-1}/\Delta\omega$, where $\Delta\omega = \Delta A/\hbar$, $\Delta g\mu_B B/\hbar$). The slow motion regime covers the correlation times between 10^{-2} and $10^{-3} < (\tau_c^{-1}/\Delta\omega) \sim 1$, and gives rise to more complex EPR spectra that can be analysed by simulation using specialised software.[64,65] For the fast motion regime, 10^{-3}–$10^{-2} > (\tau_c^{-1}/\Delta\omega) \sim 1$, the EPR spectra approach the liquid-like limit, but random modulation of the energy levels may result in m_I-dependent line broadening, $\sigma(m_I) = \alpha + \beta m_I + \Omega m_I^2$, since the g- and A-tensor anisotropies are not completely averaged.

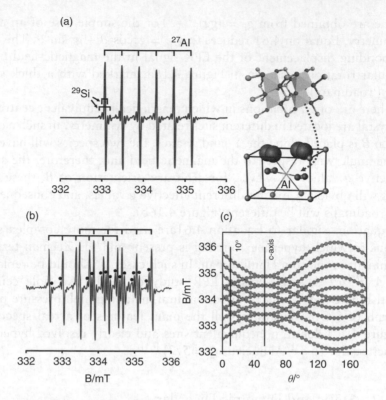

Figure 4.17 Single-crystal EPR spectra of electron-irradiated stishovite: (a) the spectrum measured with the magnetic field, *B*, parallel to the crystallographic axis, *c*; (b) at an angle of 10° (showing two types of Al-associated paramagnetic centre and forbidden transitions arising from quadrupole moment of Al); and (c) the angular dependence of the hyperfine lines in the (110) plane. Reproduced with permission from Pan *et al.* (2012) [61]. Copyright (2012) Springer Science + Business Media.

An EPR study of the superoxide centres encaged in nanoporous $12CaO \cdot 7Al_2O_3$ (mayenite) illustrates well how the lineshape is influenced by dynamic effects.[66] At 20 K, a clear orthorhombic signal with $g_{zz} > g_{yy} > g_{xx}$ is observed (Figure 4.18a). At ~100 K, g_{zz} decreases and g_{xx} increases (Figure 4.18b) due to the rocking motion along the molecular y-axis (Figure 4.18c). The EPR signal assumes an apparent axial shape ($g'_{zz} > g'_{xx} = g_{yy}$) with the following apparent g'_{ii} values:

$$g'^2_{xx} = g^2_{xx}\langle\cos^2\varphi_s\rangle + g^2_{zz}\langle\sin^2\varphi_s\rangle, g'^2_{zz} = g^2_{xx}\langle\sin^2\varphi_s\rangle + g^2_{zz}\langle\cos^2\varphi_s\rangle$$

(4.63)

Above 220 K, further precession of y about the crystal Y-axis (Figure 4.18c) causes the $g'_\parallel$ and $g'_\perp$ values to converge gradually, reaching

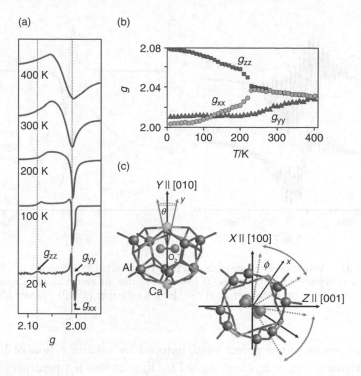

Figure 4.18 (a) Variable-temperature EPR spectra of $[O_2]^-$ species clathrated in mayenite, illustrating the effect of tumbling of the superoxide anions in nanocavities. (b) Temperature variation of the g_{zz}, g_{xx} and g_{yy} values. (c) Precession motion of an $[O_2]^-$ radical anion about the Y-axis and rocking motion in the XZ-plane. Reproduced with permission from Matsuishi *et al.* (2004) [66]. Copyright (2004) American Chemical Society.

$g_{av} = 2.03$ once 350–400 K is reached (Figure 4.18b). Other examples of motional averaging of an EPR signal are provided by Cu^{2+} in $KZnClSO_4 \cdot 3H_2O$[38] and Cu-doped silica gel[67] or $[VO]^{2+}$ in a $BaCl_2 \cdot 2H_2O$ matrix.[68]

The influence of electron delocalisation on the appearance of EPR spectra can be illustrated by a classic example of Li-doped V_2O_5 bronzes.[69] Between 4 and 77 K, a well-resolved 29-line hyperfine structure (Figure 4.19a) indicates that the unpaired electron is delocalised over four equivalent vanadium atoms ($I = 7/2, 100\%, 2nI + 1 = 2 \times 4 \times (7/2) + 1 = 29$). Above 77 K, the linewidth of the EPR signal gradually increases (Figure 4.19b), smearing out the hyperfine structure, so that at 140 K only a broad single line can be detected (Figure 4.19c). Such line broadening is associated with thermally activated electronic

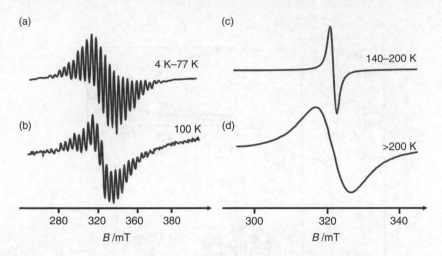

Figure 4.19   Temperature dependence of the EPR spectra of Li-doped V_2O_5 bronzes recorded at (a) 4–77 K, (b) 100 K, (c) 140–200 K and (d) above 200 K. Reproduced with permission from Livage *et al.* (1980) [69]. Copyright (1980) Elsevier Ltd.

jumping between the defect sites, induced by incorporation of lithium with frequency $\nu_h = \nu_o \exp(-E_a/kT)$. Thus, at low temperatures, when the frequency of jumping is much smaller than hyperfine splitting ($A_\parallel = 135\,\text{MHz}$), the hyperfine structure remains resolved, whereas coalescence of hyperfine lines is observed when ν_h becomes higher than the hyperfine coupling. Further change in the EPR signal shape upon temperature enhancement to 200 K is associated with the thermal excitation of electrons into the conduction band (produced by overlap of the empty $3d$ orbitals of V^{5+}), which results in pronounced broadening of the resonance line (Figure 4.19d).

Static spatial disorder (random orientation of crystallites) is a typical state for most polycrystalline materials. The shapes of the powder EPR signals are well understood, providing that statistical distribution is uniform (completely random). Although powders carry less spatial information than single crystals, there are many cases where powder spectra are sufficiently well resolved to provide chemical information rivalling that obtained from examination of dilute single crystals.

Partially orientated materials are more difficult to analyse, since a proper model of spatial orientation of the crystallite grains is usually lacking. However, in favourable cases in which they can be ordered in the form of monolayers with the crystallites orientated in one (axial) direction, an analysis of the resultant spectra is greatly

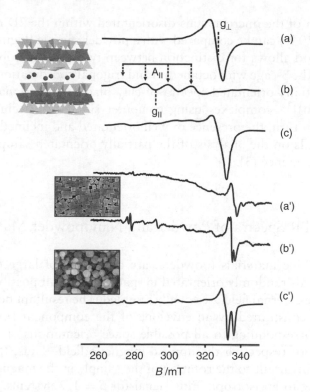

Figure 4.20 EPR spectra of partially orientated samples of (a,b,c) Cu-doped hectorite and (a',b',c') microcrystalline Na-A zeolite: (a,a') parallel orientation of Si–Al layers or Na–A microcrystals; (b,b') perpendicular orientation; (c,c') powder samples. Reproduced with permission from Pinnavaia (1981) [73] copyright (1982) Elsevier Ltd and from So *et al.* (2005) [71] copyright (2003) American Chemical Society.

simplified.[70,71] Various methods have been developed by which to assemble micrometre-sized crystals into a two-dimensional (2D) stratum on glass plates, opening the possibility for partial resolution of the orientation-dependent properties from microcrystalline materials.[72] As an example, Figure 4.20 shows EPR spectra of the partially orientated copper complexes sandwiched in alumina–silicate-layered materials (hectorite)[73] for parallel (Figure 4.20a), perpendicular (Figure 4.20b) and random orientations (Figure 4.20c). Similar behaviour can also be illustrated with an EPR study of copper-doped Na–A zeolite microcrystals deposited on a glass plate (Figure 20 a', b', c').[71] With the magnetic field perpendicular to the glass surface, a single-crystal-type EPR spectrum has been observed (Figure 4.20 a'), whereas for parallel orientation a spectrum of the microcrystals disorientated within the 2D monolayer (Figure 4.20 b′) can be compared with a powder sample (Figure 4.20 c′). This method allows for distinction between two types of copper centre located in the β-cage with octahedral and trigonal coordination. Analysis of the partially orientated spectra of clay platelets functionalised with [Cu(cyclam)]$^{2+}$ complexes using computer simulation techniques has revealed, in turn, the presence of well orientated and inclined plates.[70] More details on the analysis of the partially orientated samples can be found in reference [3].

4.4.3   EPR Spectra of Powder and Nanopowder Materials

Polycrystalline materials (powders) are composed of large number of microcrystals randomly orientated in space, so that all possible orientations in the external field are equally sampled. The resultant powder EPR pattern is constituted by an envelope of the component single-crystal spectra, corresponding to all possible space orientations of the microcrystals with respect to the applied magnetic field. Thus, the powder pattern is invariant to the rotation of the sample in the magnetic field.

Contrary to an isotropic EPR signal for $S = 1/2$ systems, caused by angular independence of the resonant magnetic field (Figure 4.21a), in the case of paramagnets with axial symmetry, the powder EPR spectrum (Figure 4.21b) is spread over the entire resonant field range ($\Delta B_r = B_\parallel - B_\perp$), depending on the orientation angle, θ, of the crystallites: $B_r(\theta) = (h\nu/\mu_B)(g_\parallel{}^2\cos^2\theta + g_\perp{}^2\sin^2\theta)^{-1/2}$. However, the simultaneous resonant absorption in this range does not lead to a flat envelope, since the orientation probability, $P(B) = 1/2\sin\theta/(dB/d\theta)$, is not uniformly distributed over this range. The resonant field changes rapidly with angle θ in the zx-plane, so that only the orientations close to the x- and z-axes contribute effectively to the total absorption. Characteristic features in the powder spectrum (associated with the $g_\parallel$ and $g_\perp$ values, Figure 4.21b) are associated with the turning points that appear when $\partial B/\partial\theta = 0$. Two solutions of this equation correspond to $\theta = 0$ and $\theta = \pi/2$ (*i.e.* to minimum and maximum in the $B(\theta)$ plot). Between these two extremes, a continuous absorption is observed, the derivative of which leads to the typical shape of an axial powder spectrum with

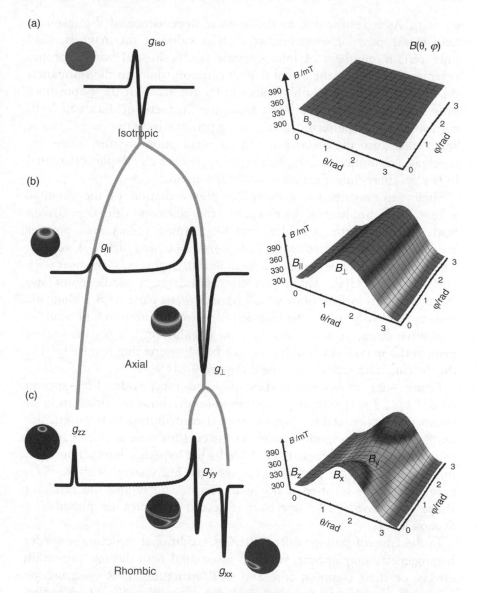

Figure 4.21 Simulated (a) isotropic, (b) axial and (c) rhombic powder EPR signals, along with the corresponding angular dependence of the resonant magnetic field $B(\theta, \varphi)$. Specific orientations of the crystallites with respect to the magnetic field that contribute to the given spectrum are depicted on the associated unit spheres.

$g_\perp < g_\parallel$. As a result, due to the angular dependence of the resonant field in the powder spectrum, at a given value of the magnetic field only certain numbers of microcrystals $(1/2N \sin\theta d\theta)$ with a proper orientation between the θ and $\theta + d\theta$ can contribute to the resonance. As implied by the unit-sphere plots in Figure 4.21b at the $g_\parallel$ position, only a few paramagnets aligned along the magnetic field axis $z(\theta = 0)$ contribute to the pattern. Upon moving progressively to $\theta = \pi/2$, more spins come into the resonance and the absorption intensity increases, reaching a maximum corresponding to $g_\perp$, for which all spins orientated in the xy- (perpendicular) plane are in resonance.

Similar arguments for a more complex situation of the rhombic g tensor are applicable. In this case, the resonant field depends on both the θ and the φ angles, and the turning points in a powder spectrum are associated with $\partial B(\theta, \varphi)/\partial\theta = 0$ and $\partial B(\theta, \varphi)/\partial\varphi = 0$. These equations have three solutions: $\theta = 0(z)$; $\theta = \pi/2$ and $\varphi = 0(x)$; and $\theta = \varphi = \pi/2(y)$. The y-axis corresponds to a saddle point and there are many orientations which have $B_r(\theta, \varphi)$ close to B_y. Magnetic resonance absorption in the first derivative representation will result in a positive going peak at the minimum B value (g_{zz}), a negative going peak at the maximum field (g_{xx}) and a baseline-crossing feature (g_{yy}) at the intermediate saddle-point field (Figure 4.21c).

Figure 4.21 shows the shapes observed for powder EPR spectra of $S = 1/2$, $I = 0$ systems. The unit spheres show the orientation of microcrystals (gauged by θ and φ angles) contributing to the particular features of the corresponding powder patterns for various anisotropies of the g tensor. These shapes are modified by the hyperfine interaction when nuclei with $I \neq 0$ are present in the systems. For systems with $S > 1/2$, a fine structure is often a dominant term, controlling the resultant shape of a powder EPR spectrum. Selected examples are provided in Section 4.5.1.

In the case of nanopowder materials, additional structural-disorder heterogeneity may appear, which is associated with the size and strain effects, or their common core and shell structures. EPR spectroscopy (especially at high frequencies) provides a useful probe of such phenomena, reflected in size-dependent g, A and ZFS values.[74] It has been observed, for instance, that g value varies in the range $1.960 < g^* < 1.968$ for n-type ZnO quantum dots with diameters between 3.0 and 7.0 nm.[75] Such changes, $g^* = g_e - 2/3[P^2\Delta/(E_g(E_g + \Delta))]$, can be rationalised in terms of the bandgap energy, E_g, the SOC of ZnO, Δ, and the interband matrix element, P. In the case of 5.0 nm CdSe quantum dots containing 0.6 wt% Mn^{2+}, the core and shell sites can be readily

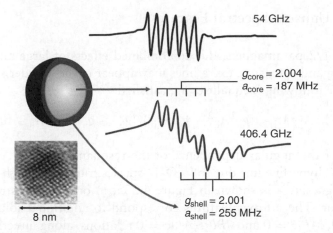

54 GHz

g_{core} = 2.004
a_{core} = 187 MHz

406.4 GHz

g_{shell} = 2.001
a_{shell} = 255 MHz

8 nm

Figure 4.22 EPR spectra of Mn-doped CdSe quantum dots recorded at 11 K and two microwave frequencies, showing a clear benefit of high field measurements in studies of a core–shell structure of such materials. Reproduced with permission from Wang *et al.* (2012) [76]. Copyright (2012) Elsevier Ltd.

distinguished on the basis of different ^{55}Mn hyperfine and g values (Figure 4.22).[76]

In polycrystalline, amorphous and nano materials, strain effects give rise to typical anomalous EPR line broadening. The g-, A- and D-strains, defined as a narrow random distribution (spread) of these parameters around their single values, reflect local microheterogeneity around the paramagnetic centres (static geometrical flexibility), associated with a slight variation of the bonding geometry (bond lengths and distances). In magnetically diluted systems, this phenomenon leads to apparent m_I and frequency dependence of the hyperfine (and fine) linewidths, which expressed in the frequency domain can be described in the following way:[77]

$$\sigma_v^2 = \left(\sum_{i=x,y,z} \left\{ \sigma_{R_i}^2 \left[\frac{\Delta g_i}{g_i} v_0(B) + \Delta A_i m_I \right]^2 g_i^2 l_i^2 \right\} \right) / g^2 \qquad (4.64)$$

where σ_{R_i} are the residual linewidths due to unresolved metal and/or ligand hyperfine splitting, homogeneous broadening and other sources and Δg_i and ΔA_i are the widths of the Gaussian distributions of the g and A values. The analysis of strains is not only a necessary condition for correct determination of the EPR parameters but also provides a valuable insight into the local heterogeneity of the sites.

4.4.4   Unusual Spectral Features

For $S = 1/2$ paramagnets, due to combined effects of large anisotropy of both g and A tensors, extra lines may appear in the powder spectrum when the following inequalities are satisfied:[78]

$$2A_{ii}^2 - h\nu A_{ii}/M_I < (g_{ii}^2 A_{ii}^2 - g_{jj}^2 A_{jj}^2)/(g_{ii}^2 - g_{jj}^2) < 2A_{jj}^2 - h\nu A_{jj}/M_I \quad (4.65)$$

A plot of the angular dependence of the resonant field $B_r(\theta, \varphi)$ of the $m_I = 3/2$ hyperfine line due to $[^{17}O_2]^-$ species trapped on high-surface-area magnesia[79] is shown in Figure 4.23a, in order to illustrate such behaviour. The off-axis features correspond to additional solutions of the $\partial B(\theta, \varphi)/\partial\theta = 0$ and $\partial B(\theta, \varphi)/\partial\varphi = 0$ relations, along directions that

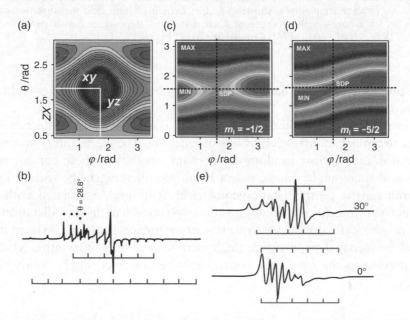

Figure 4.23   (a) Plot of the angular dependence of the resonant field $B_r(\theta, \varphi)$ for an $m_I = 3/2$ transition, showing a local minimum at $\theta = 28.8°$. (b) The corresponding powder EPR spectrum of $[^{17}O_2]^-$ radical anion trapped on an MgO surface, with the off-axis extra lines indicated by dots. (c,d) Resonant magnetic field of molybdenum(V) carbonyl species grafted on the silica surface with non-coincident g and A axes ($\beta = 30°$). The characteristic features in the first derivative powder EPR signal that are associated with the magnetic field minimum (MIN), maximum (MAX) and saddle point (SDP) for (c) $m_I = -1/2$ appear at different angles than those for (d) $m_I = 5/2$. (e) Corresponding EPR powder spectra, with oxygen hyperfine stick diagrams. Reproduced with permission from Pietrzyk *et al.* (2012) [6]. Copyright (2012) Wiley-VCH.

do not coincide with the principal magnetic axes (x, y, z). Indeed, apart from the normal field extremes along the principal directions $(\theta = 0, \varphi = 0$ and $\theta = 90°, \varphi = 0$ and $\theta = 90°, \varphi = 90°)$, there is a clear additional local minimum appearing at $\theta = 28.8°$ in the zx-plane, which gives rise to an extra absorption line. Such extra lines, being often more intense than the 'normal' ones (Figure 4.23b), may be used to analyse less resolved hyperfine features.[79]

Another possible complication of the powder EPR spectra may arise from pronounced noncoincidence of the principal axes of the g and A tensors. For single-crystal spectra, the noncoincidence of axes can be resolved definitively by analysis of the corresponding roadmaps for any symmetry. In the case of powders, however, it can be unravelled in practice for monoclinic spectra only, provided that $I > 1$ and both g and A tensors exhibit sufficiently large anisotropies, preferably of comparable magnitude $(\Delta g \mu_B B \approx \Delta A)$. For instance, analysis of the $B_r(\theta, \varphi, m_I)$ plots for the molybdenum carbonyls supported on silica $(S = 1/2, I = 5/2)$ reveals that for small m_I values $(\pm 1/2)$, the magnetic field extremes $(\partial B/\partial \theta = 0$ and $\partial B/\partial \varphi = 0)$, which are associated with characteristic features in the powder spectra, occur at the angles close to the principal axes of the g tensor (Figure 4.23c), while for large m_I values $(\pm 5/2)$ they appear at the angles close to the hyperfine principal axes (Figure 4.23d). A consequence of such competition is the appearance of hyperfine lines with strong variation in spacing with m_I (exceeding the second-order effects). Obviously, the $I = 1/2$ nuclei give rise to just two hyperfine lines and therefore no m_I variation of the spacing. Thus, only paramagnetic centres containing nuclei with $I > 1$ (such as ^{63,65}Cu with $I = 3/2$, ^{55}Mn with $I = 5/2$ and ^{59}Co with $I = 7/2$ for magnetic metal cores and ^{35,37}Cl with $I = 3/2$ or ^{17}O with $I = 5/2$ for ligands) are relevant in this context. Another useful criterion for axis noncoincidence is the absence of lines at positions anticipated from simple extrapolation beginning at either side of the spectrum or the appearance of lines that are not accounted for (Figure 4.23e). Failure to simulate an EPR spectrum with correct line positions and intensities provides a further indication of low symmetry. However, in practice, for small hyperfine splitting, the relative orientations of principal axes cannot be determined from powder EPR spectra, even if the relative anisotropies are large.[80] Another effect related to high angular anisotropy of the EPR lines in powder spectra is pronounced broadening of the fine structure lines, which together with an M_S dependence of the intensities (see Section 4.3.3) leads to their pronounced attenuation,[81] as illustrated in Figure 4.24 by Fe^{3+} ions $(S = 5/2)$ in $BaTiO_3$.[82]

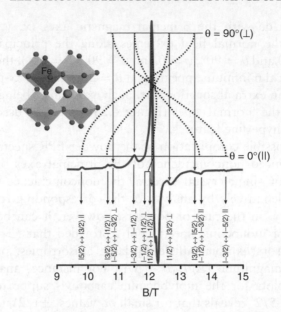

Figure 4.24 Calculated roadmaps and the corresponding Q-band EPR powder spectrum of Fe^{3+}- $(S = 5/2)$ doped $BaTiO_3$ (2 mol% Fe), showing an M_S-dependence of the fine structure lines. Reproduced from Böttcher *et al.* (2008) [82]. Copyright (2008) Institute of Physics.

An unusual temperature dependence of HF-EPR signals can be expected for nanomaterials that contain high-spin centres investigated at low temperatures, due to the combined results of Boltzmann population effects and M_S-dependent strain broadening. A spectacular example which ably illustrates this phenomenon is provided by manganese(II)-doped CdSe quantum dots, where the central $M_S = -1/2 \leftrightarrow M_S = 1/2$ transition with small angular variation and strain broadening is clearly resolved at 290 K (Figure 4.25), whereas the $\pm 3/2 \leftrightarrow \pm 1/2$, $\pm 5/2 \leftrightarrow \pm 3/2$ lines with strong angular dependence and strain broadening are smeared out, despite all states being effectively populated.[76] This contrasts with measurements at the lowest temperature (2 K), where only the $M_S = -5/2$ state is populated, giving rise to a single broad line.

4.4.5   Computer Simulation of Powder Spectra

In simple cases of coincident principal axes and the first-order limit, when EPR signals are not too complex and are sufficiently well resolved,

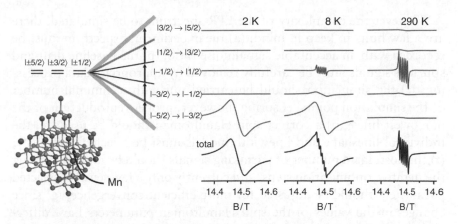

Figure 4.25 Zeeman energy levels and the simulated 406.4 GHz HF-EPR spectra for Mn-doped CdSe quantum dots. Broadening of the energy levels, indicated by varying thickness of the corresponding lines, is due to the ZFS strain. The overall spectra are obtained by adding the individual signals caused by the particular transitions. Reproduced with permission from Wang *et al.* (2012) [76]. Copyright (2012) Elsevier Ltd.

acceptable values of the EPR parameters can be derived directly from the experimental spectra by reading out the field positions of the characteristic turning points (Figures 4.10, 4.11, 4.14 and 4.21).[83] However, for more complicated powder EPR spectra and low symmetry, the number of the lines may vary considerably depending on the hyperfine anisotropy and multiplicity of magnetic nuclei. Since in such cases the line positions are not linear in any of the magnetic parameters, analysis of powder EPR spectra becomes a quite involved problem. This is especially valid when several paramagnetic species are present together, leading to complex multicomponent EPR spectra with overlapping partaking signals. In such cases, correct interpretation of the EPR spectra relies on computer simulation of the experimental profiles. The powder patterns computed using initial estimates of the spin Hamiltonian parameters are compared with experimental spectra. The parameters are then iteratively adjusted and the corresponding spectra reconstructed until the fit is satisfactory.

A more extensive account of the simulation algorithms and procedures can be found in review articles[84] and EPR textbooks.[3,7,15] A large number of computer programs of various levels of sophistication, such as EPRsim32,[85] POW[86] for $S = 1/2$ and EasySpin,[65] XSophe[87] and ZFSFIT[88] for $S \geq 1/2$, have been developed for this purpose in recent decades.

Whatever the complexity of the EPR spectrum to be simulated, there are a few hints to keep in mind: (a) the experimental spectrum must be recorded with an acceptable signal:noise ratio; (b) the baseline drift and spurious signals must be carefully removed by appropriate data processing; (c) the simulation should be carried out with a minimal number of the simulation points, ensuring, however, proper reproduction of the narrowest lines; (d) a correct spin Hamiltonian should be used; (e) the individual lineshape and linewidth models must be selected cautiously; (f) the postulated number of partaking signals should be minimal during the iterative optimisation steps, and currently only a few sensibly selected parameters should be adjusted for more efficient convergence; (g) since changes in the values of the spin Hamiltonian parameters have different effects on the signal shape, without appropriate scaling only some variables will be efficiently optimised; and (h) in order to estimate the uncertainty limits of the determined values, analysis of the parametric sensitivity of the simulated spectrum should be employed.

4.5   ADVANCED EMR TECHNIQUES

4.5.1   High-Field and Multifrequency EPR

HF-EPR techniques employ frequencies above 34 GHz (from V-band to far infrared). Although still rather rare and costly, they are becoming a well established method for studying high-spin materials with large ZFS,[89] resolution enhancement of the field-dependent terms, simplification of spectral shape with high hyperfine[90] and fine interactions[91] and enhancement of the orientation resolution for powder ENDOR or HYSCORE studies.[7,91]

A main advantage of using HF-EPR spectroscopy is associated with the investigation of high-spin systems containing clustered and isolated centres such as $[Mn_3]^{2+}$ ($S = 4$ and 6), Mn^{2+}, Fe^{3+}, $[Fe_3]^{3+}$ ($S = 5/2$), $[Mn_2]^{4+}$, Fe^{2+} ($S = 2$) or Co^{2+} ($S = 3/2$) with sizeable ZFS values. In such cases, when the probing frequencies are smaller than the ZFS, either the high-spin centres appear 'EPR-silent' (non-Kramers ions) or else only one of the many allowed transitions, $M_S = -1/2 \rightarrow M_S = 1/2$, can be observed (Kramers ions), making determination of the ZFS parameters unfeasible. Paramagnets with ZFS comparable with Zeeman term may often lead to extremely complicated spectra, whereas at higher frequencies such spectra become simpler and therefore easier to analyse.

This point is illustrated with the help of Mn^{2+} ions doped in $Zn_2V_2O_7$ in Figure 4.26a,b, where tremendous simplification of the angular variation of the EPR line positions upon passing from 9.6 to 249.9 GHz is evident, so that the D value can be read directly with a good precision. For more complex, low-symmetry centres, application of variable-frequency HF-EPR allows for straightforward determination of the fine structure D and E values. In this approach, the frequency is tuned quasicontinuously, and resonances are plotted as a function of transition energy, as exemplified in Figure 4.26c for an $[Ni(PH_3)_2Cl_2]$ $(S = 1)$ system.[92] Two zero-field resonances at 11.3 and 15.0 cm^{-1}, identified with $(D - E)$ and $(D + E)$ transitions, can be read directly from the plot. Furthermore, it is easy to differentiate between the allowed $(\Delta M_S = \pm 1)$ and forbidden $(\Delta M_S = \pm 2)$ transitions, since the latter have a much lower slope than the former. The corresponding single-frequency (611.2 GHz) HF-EPR spectrum is shown in Figure 4.26d.

When unresolved hyperfine structure determines primarily the linewidths, the condition $(\Delta g / g_{iso}) B_0 > \Delta B_{hfs}$ must be satisfied in order to observe g anisotropy (the anisotropic Zeeman interaction must exceed the inhomogeneous line broadening).[93] For instance, paramagnetic defect centres exhibit resonances at $g \sim 2.0$, and their typical Δg anisotropies are of the order of 10^{-3} to 10^{-4}. At the conventional X-band, the corresponding $(\Delta g / g_{iso}) B_0$ values are in the range 0.03–0.30 mT, whereas powder materials usually exhibit linewidths of around 1–3 mT. Resolved EPR features smaller than 0.1 mT have been reported for powder samples;[94] nevertheless, the line broadening and resolution generally pose a serious concern in studies of disordered nano- and amorphous materials. Yet, because the Zeeman interaction increases linearly with the external magnetic field, the field separation (ΔB_{12}) between two slightly different g_{ii} and g_{jj} components increases with frequency. Therefore, small anisotropy of the actually rhombic g tensor due to Cr(V) in $Na_3Cr(O_2)_4$ with $S = 1/2$ can be clearly revealed in HF-EPR spectra recorded at 375 GHz, whereas at 9.5 GHz only a broad, virtually structureless signal is observed (Figure 4.27a).[95] Resolution enhancement at high frequencies is possible only when the contribution of g strain to the linewidth is small.[96] Paramagnetic centres with considerable g-tensor anisotropy that is sensitive to local environment are more likely to exhibit pronounced g-strain broadening in heterogeneous systems. If such effects are already apparent at lower frequencies (X-band), probably little or no resolution enhancement will be gained at high frequencies. An example is given by X- and Q-band study of $Mo^V O_x$

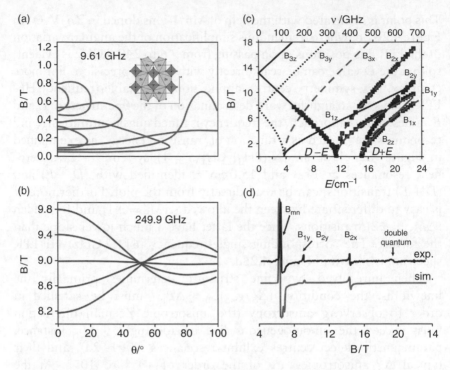

Figure 4.26 Angular variation of the simulated line positions for the allowed fine structure transitions of Mn^{2+} ($S = 5/2$) ions in a $Zn_2V_2O_7$ single crystal, showing dramatic simplification of the roadmap upon passing from (a) 9.61 GHz to (b) 249.9 GHz. (c) Tunable-frequency plots of the resonance field *versus* sub-THz photon quantum energy for a polycrystalline $Ni(PPh_3)_2Cl_2$ sample with $S = 1$. (d) The corresponding single-frequency HF-EPR spectrum. Experimental points are represented by squares. Simulations are shown by dashed lines for $B\|x$, dotted lines for $B\|y$ and solid lines for $B\|z$. Three zero-field resonances, two of which are directly detectable in the 95–700 GHz range, are labelled accordingly. Reproduced with permission from Krzystek *et al.* (2006) [92]. Copyright (2006) Elsevier Ltd.

clusters supported on silica, where upon passing to higher frequency the resolution of the spectrum is lost completely.[97] It should be noted that in the case of marked g strain, the resolution of hyperfine structure can be improved at lower L- or S-band frequencies.[96]

The multifrequency techniques may also be used for separation of overlapping signals. Indeed, differentiation between paramagnetic centres, A and B, characterised by slightly different g and $g + \Delta g$ values can be performed by recording EPR spectra at two different frequencies, since their separation, $\Delta B_{AB} = (h\nu/\mu_B)[\Delta g/g(g + \Delta g)]$, is made larger by increasing ν, as illustrated by complete separation of the hyperfine

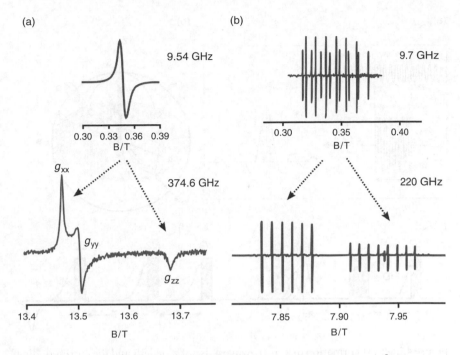

Figure 4.27 (a) Resolution enhancement of g-tensor anisotropy for Cr^{5+} ($S = 1/2$) by application of high-field EPR for magnetically undiluted $Na_3Cr(O_2)^4$ specimen (adapted from reference [95]). (b) Separation of the overlapping EPR signals due to Mn^{2+} and V^{2+} ions in MgO matrix by switching from X-band to 220 GHz measurement. Adapted from [30].

patterns of Mn ($I = 5/2$) and V ($I = 7/2$) cation-doped MgO hosts (Figure 4.27b).

4.5.2 Pulsed EPR Methods

Time-domain (pulsed) EPR spectroscopy is based upon the use of sequences of short, intense microwave pulses of given lengths and delay times at a constant magnetic field, B_0 (Section 4.2.4), which allows for excitation of a range of frequencies and gives rise to an EPR spectrum. The relation between the pulse length, τ_p, and the excitation bandwidth is shown in Figure 4.28a. The excitation profile of a pulse (frequency domain) is obtained by Fourier transformation of the pulse shape (time domain). For instance, an excitation bandwidth of 100 MHz

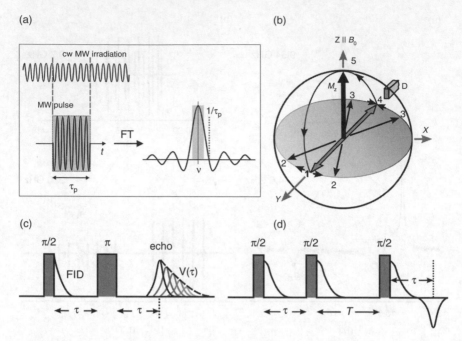

Figure 4.28   (a) Formation of a microwave pulse of τ_p length and the corresponding excitation band obtained after Fourier transformation. (b) Behaviour of the magnetisation, M_0, in the rotating frame upon a $\pi/2 \to \pi$ pulse sequence, leading to electron spin echo ((1) orientation of M after a strong $\pi/2$ pulse along the x-axis; (2) defocusing of spin packets with a rate of $1/T_m$; (3) orientation of spin packets after subsequent application of π pulse; (4) refocusing of spin packets and formation of an electron spin echo (ESE) signal recorded with detector D; (5) recovery of the longitudinal magnetisation with $1/T_1$). Pulse sequences for (c) primary (Hahn) echo and (d) stimulated echo.

corresponds to a 10 ns pulse, which is equivalent to only 3 mT for $g = 2$. Reduction of the pulse length leads to a broader excitation range. The result of applying a microwave pulse is rotation of the magnetisation, M, into the perpendicular xy-plane, where it can only be detected in a time of short duration compared to the relaxation times (Figure 4.28b). The rotation angle, α, depends on the pulse length, $\alpha = \omega_{MW}\tau_p = \gamma B_1 \tau_p$, where $B_1(\omega_{MW})$ is the field intensity of microwaves. In the simplest experiment, a single $\pi/2$ pulse transforms the original longitudinal magnetisation, M_z, into a transverse M_y magnetisation (Figure 4.28b). If the resonance condition, $\omega_{MW} = \gamma B_0$, is exactly fulfilled, the magnetisation stays along the y-axis following the $\pi/2$ pulse. However, if the Larmor frequency, $\omega = \gamma B_0$, is slightly mismatched, it initiates precession about the z-axis, with frequency $\Omega = \omega - \omega_{MW}$. Since the detector is aligned towards the $-y$-axis in the rotating coordinate framework, the frequency difference, Ω, is measured directly. The oscillating signal, which decays because of the transverse relaxation or inhomogeneous line broadening, constitutes the FID. By performing a Fourier transformation from time into a frequency domain, an EPR spectrum exhibiting a peak at Ω is obtained.

The inhomogeneous contribution to T_2 is due to a line-broadening mechanism that distributes the resonance frequencies for the individual spins, leading to an unresolved band. It happens because the individual spins in a sample are not all subjected to exactly the same local magnetic field. Thus, the observed line is the sum of the resonance transitions of multiple 'spin packets'. Each of the packets can be regarded as a pseudohomogeneous unit, slightly shifted in field from the others. The proper terminology for the relaxation time in solids is the 'phase-memory time', T_m, rather than T_2 (the latter is defined strictly by the Bloch equations). The common sources of inhomogeneous broadening are the inhomogeneous external magnetic field, unresolved hyperfine structure, anisotropic interactions in randomly orientated solids (g, A or D tensors) and dipolar interactions with adjacent paramagnetic centres, which are responsible for random local magnetic fields acting on a given unpaired electron.

The important advantages of performing pulsed rather than CW-EPR include simultaneous detection of all lines encompassed within the excitation bandwidth; in CW-EPR, only one spin packet is detected at a time. Thus, the full spectrum must be acquired by a relatively slow scanning of the magnetic field. Signal averaging is also made easy in FT-EPR as it requires only simple addition of the repetitive FID acquisitions. This advantage is particularly valid for systems with short T_1 relaxation times, since they allow for rapid pulse repetition and subsequent signal averaging. Obviously, such an advantage is lost in the case of very long T_1 times. Additionally, a single FT-EPR spectrum can be recorded in less than a microsecond range. This allows for instant sampling of paramagnets evolving in time for kinetic investigations. Furthermore, with FT-EPR techniques, the T_1 and T_m relaxation times can be measured directly from the response of special pulse-sequence experiments, rather than by extraction from convoluted lineshapes or saturation-behaviour measurements.

4.5.2.1   *Hahn Echo and Stimulated Echo*

The pulse sequence $\pi/2 \rightarrow \tau \rightarrow \pi \rightarrow \tau$ leads to the formation of a primary (Hahn) echo,[98] shown in Figure 4.28c. A spin echo experiment can be used to collect all those spins that have lost their phase due to the inhomogeneous phase memory loss after experience of a $\pi/2$ pulse. As the time, τ, increases, the fraction of the refocused spins decreases. This irreversible loss of phase memory corresponds to the homogeneous T_m relaxation in solids, which is a measure of the phase loss time for spins with identical frequency due to collisions with phonons.

In an $S = 1/2$ system in the rotating frame, the spin reversal is achieved by applying a π pulse at time τ after an initial $\pi/2$ pulse that flips the spins from the $+y$ to the $-y$ direction (Figure 4.28b, features (2) and (3)). The Hahn echo can be used to measure the phase memory time, T_m, of a sample by simply stepping the time interval τ between the $\pi/2$ and π pulses through an opposite range, and measuring the exponentially decaying intensity of echo with a rate of $2/T_m$ (Figure 4.28c). In addition, Hahn echo may also be applied to measure the frequencies of nuclear spin transition *via* ESEEM or HYSCORE methods (see Sections 4.5.2.2 and 4.5.2.3), and for the electron spin echo-detected EPR signals (ESE). In the case of pronounced inhomogeneous broadening, the latter technique is often the only way to detect an EPR signal.

With sufficiently narrow EPR spectra, excitation of the full range of the spectrum is possible within a microwave pulse. The ESE signal is obtained by subsequent Fourier transform of the second half of the primary echo, which is essentially the same as the FID, except that it is not preceded by a high-power microwave pulse, and therefore is not constrained by the dead-time limitations. However, the EPR spectra are usually too broad to be excited within the frequency range of a single microwave pulse. Yet, even if the spectrum exhibits a small span (2–3 mT), it may be difficult to observe FID due to too-broad, not-well-resolved lines, for which FID may decay very quickly and be lost in the spectrometer dead time. In this case, the magnetic field has to be swept as in a normal CW-EPR spectrum, and the spin echo intensity is acquired as a function of the field position. Apart from the inherent higher time resolution, the ESE technique may have other advantages over conventional CW-EPR. In the case of transition-metal ions, solid-state EPR spectra are often quite broad and hard to measure, essentially because the change of amplitude with the change of field is very small. Since ESE probes the z-magnetisation at each field position directly, such limitations may be readily avoided. Moreover, when the spin–lattice relaxation time is short enough, fast signal averaging can be applied to enhance the sensitivity in comparison to that available in CW-EPR.

Because the Hahn echo decays with (rather short) T_m, it is of limited utility for measurement of the weak hyperfine interactions. In such a case, a three-pulse (3-P) stimulated echo obtained by applying the pulse sequence $\pi/2 \rightarrow \tau \rightarrow \pi/2 \rightarrow T \rightarrow \pi/2 \rightarrow \tau \rightarrow$ echo (Figure 4.28d) is more appropriate, as it stores the spin coherence for a much longer time (T_1). During the first evolution time after the initial $\pi/2$ pulse, the electron spin coherence defocuses in the xy-plane. The second $\pi/2$ pulse tips the spins nearly along the z-direction (the components that remain in the xy-plane will be lost, since they dephase completely during time T, whereas the components aligned along the z-axis approach thermal equilibrium at a much slower rate ($1/T_1$). Upon application of the third $\pi/2$ pulse, the spins are flipped again into the xy-plane, where they refocus to produce the echo, which is well separated (delayed) from the idle detector dead time. The second and third $\pi/2$ pulses can be regarded as the π pulse in a Hahn sequence that has been separated in two parts at time T.

Extension of spin echo experiments with two-pulse (2-P) and 3-P sequences leads to ESEEM and HYSCORE techniques that are extensively used for probing of weak and unresolved hyperfine and quadrupole interactions with distant ligands.

4.5.2.2 ESEEM Spectroscopy

ESEEM experiments are performed by recording the echo intensity generated by a sequence of microwave pulses, spaced by evolution times. The generated echo decay ($V(\tau, T)$) is not simply exponential, but is featured by superimposed modulation associated with the hyperfine frequencies of nuclei adjacent to the unpaired electron.[99] Thus, judicious analysis of the modulation pattern can provide useful information about neighbouring nuclei, particularly those weakly coupled to the paramagnetic centre due to remote location and/or a small $g_n I$ factor. The significance of the ESEEM effect lies in its ability to indirectly measure nuclear frequencies (NMR frequencies in Figure 4.5b) due to its outstanding EPR sensitivity. The disadvantage of ESEEM is that the modulation depth, k, is often only a few per cent of the echo and depends strongly on the hyperfine coupling and its anisotropy.

The standard one-dimensional ESEEM experiment consists of 2-P and 3-P sequences, and is based on the primary and the stimulated echo, respectively. The modulation of the echo results from changes in amplitude due to precession of the nuclear moments, which alters the local effective magnetic field experienced by the unpaired electron. This causes the phase of the electron magnetic moment to lag or lead, giving rise to modulation of the echo envelope.

For investigation of single-crystal materials, 3-P experiments are preferred, since modulation effects can be observed for longer times than they can with a 2-P echo.[5] The 3-P ESEEM sequence may be written as: $\pi/2 \rightarrow \tau \rightarrow \pi/2 \rightarrow t_1 \rightarrow \pi/2 \rightarrow \tau \rightarrow$ echo (Figure 4.29a). The second $\pi/2$ pulse transfers the electron spin coherence into nuclear spin coherence. The decay of this modulation is determined by the nuclear phase memory time, or else the electron T_1 time (if the latter is shorter than the former), which yields narrower lines after Fourier transformation and, therefore, a better resolution of the NMR transitions. The third $\pi/2$ pulse transfers the nuclear spin coherence back to observable electron spin coherence.

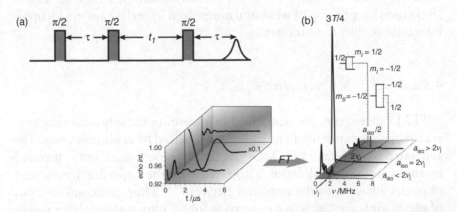

Figure 4.29 (a) Pulse sequence used for 3-P ESEEM experiment. (b) Simulated powder ESEEM spectra for an $S = 1/2$ spin coupled to a single $I = 1/2$ nucleus for weak ($a_{iso} \ll 2\nu_I$), matching-coupling ($a_{iso} \sim 2\nu_I$) and strong ($a_{iso} > 2\nu_I$) conditions. Frequency-domain spectra are obtained by Fourier transformation of the time-domain signals. Reproduced with permission from Deligiannakis *et al.* (2000) [102]. Copyright (2000) Elsevier Ltd.

The quantitative description based on the density matrix formalism for $S = I = 1/2$ yields the following modulation of the 3-P ESEEM:[10]

$$V(T, \tau) = 1 - \frac{k}{4}\{[1 - \cos(\omega_\beta \tau)][1 - \cos(\omega_\alpha(\tau + T))]$$

$$+ [1 - \cos(\omega_\alpha \tau)][1 - \cos(\omega_\beta(\tau + T))]\} \quad (4.66)$$

where ω_α and ω_β denote nuclear frequencies for α and β electron spin projections (m_S), respectively. The modulation depth parameter, k, is defined as:

$$k = \sin^2 2\eta = \left(\frac{\omega_I T'}{\omega_\alpha \omega_\beta}\right)^2 \quad (4.67)$$

where 2η is the angle between two local fields at the nucleus for $m_S = \pm 1/2$ manifolds, ω_I is a Larmor frequency for nuclear spin I and T' is related to the dipolar hyperfine tensor.[10]

The relative sizes of the hyperfine and nuclear Zeeman interactions, together with the quadrupole interaction, determine the shapes of ESEEM spectra.[100,101] In general, the shape of an ESEEM profile does not allow for direct determination of the isotropic and anisotropic hyperfine interactions from the positions of the maxima of the spectrum without numerical simulations. Figure 4.29b shows typical simulated time- and frequency-domain 3-P ESEEM spectra (generated with EasySpin software[65]) for three limiting coupling regimes. In the case of an axial hyperfine tensor, near the $4\nu_I = 2a_{iso} + T$ condition (known as the 'matching condition'[101]) the ESEEM spectrum is characterised by two frequencies: a dominant narrow feature at $\nu_\alpha \sim 3T/4$ and a broad, weak one at $\nu_\beta \sim 2\nu_I$. For hyperfine couplings that appreciably deviate from the matching conditions, the intensity and resolution of the spectral features decrease significantly. For a weak coupling, $a_{iso} < 2\nu_I$, condition, the main peaks are centred approximately at ν_I, whereas for strong couplings, $a_{iso} > 2\nu_I$, the spectral features are centred at $a_{iso}/2$.[102]

A disadvantage of the 3-P ESEEM can immediately be recognised from Equation 4.66. For certain τ values, one may encounter so-called 'blind spots', where the ω_β modulation is completely suppressed when $(1 - \cos \omega_\alpha \tau) = 0$ and *vice versa*. This implies that, in order to be sure that all frequencies are detected, ESEEM measurements should be performed for at least two different τ values. Furthermore, the evolution of the nuclear coherence apparently starts earlier (at $-\tau$) than the recorded stimulated echo. This increases the dead time for the experiment, since the third pulse cannot be placed in front of the second. This is especially valid for the disordered materials, where the anisotropic line broadening effect may lead to faster decay of the ESEEM signals. One may use more modern blind spot- and dead time-free ESEEM pulse sequences.[10]

4.5.2.3   HYSCORE Spectroscopy

3-P ESEEM spectroscopy can be extended into its four-pulse (4-P) variant by inserting a strong π pulse between the second and third $\pi/2$ pulses, resulting in the following pulse sequence: $\pi/2 \to \tau \to \pi/2 \to t_1 \to \pi \to t_1 \to \pi/2 \to \tau \to$ echo. Owing to this extra π pulse, the nuclear coherence created by the first two $\pi/2$ pulses, which evolves during the first time period t_1, is interchanged between the two m_s manifolds of the unpaired electron. The local magnetic fields experienced by the nuclei thus change and the inhomogeneous hyperfine packet is refocused during the second evolution period following the π pulse. As a result, a nuclear coherence transfer echo is formed, which is detected as an amplitude modulation of the stimulated echo formed at the time, τ, after the third $\pi/2$ pulse.

The 2D version of the 4-P ESEEM experiment, obtained by incrementing two different t_1 and t_2 evolution periods independently (Figure 4.30a), is the HYSCORE.[103] The name 'HYSCORE' reflects transfer of the nuclear coherence by the central π pulse between the m_S manifolds. Fourier transformation of the two time domains gives a 2D spectrum with the frequency axes v_1 and v_2 (Figure 4.30b). The evolution of the nuclear coherence in the two time intervals, t_1 and t_2, with different NMR frequencies, v_{12} and v_{34}, produces cross peaks at the frequencies (v_1, v_2), (v_2, v_1) and $(v_1, -v_2)$, $(v_2, -v_1)$ in the $(+,+)$ and $(+,-)$ quadrants of the 2D spectrum, respectively. These peaks are placed symmetrically about the diagonal $v_1 = v_2$ in the $(+,+)$ quadrant (for weak couplings). If the transfer between the $m_S = \pm1/2$ manifolds is not complete (i.e. when the π pulse is not strong enough), additional peaks (v_1, v_1) and (v_2, v_2) appear on the diagonal. As in 3-P ESEEM for certain combinations of v_1, v_2, and τ values, cross peaks can disappear, resulting in the blind spots. As the HYSCORE experiment involves four pulses, up to 13 unwanted echoes can result when the excitation is incomplete; however, these can all be eliminated by suitable phase cycles.[104]

The HYSCORE technique exhibits several important advantages over the ESEEM method when used for solids and disordered systems. The extension into a second dimension simplifies greatly the analysis

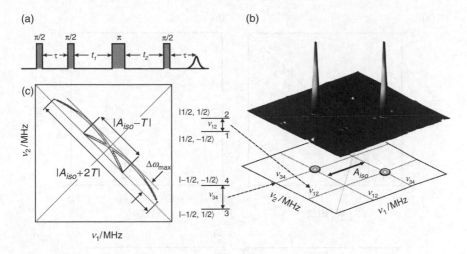

Figure 4.30 (a) Pulse sequence used for HYSCORE experiment and simulated spectra for $S = I = 1/2$ with an (b) isotropic ($a_{iso} = 20$ MHz) and (c) axial ($a_{iso} = 1.8$ MHz, $T = -8$ MHz) hyperfine tensor.

of complicated spectra, since the cross peaks can only be observed between NMR frequencies of the same nucleus. Due to the refocusing effect of the π pulse, the HYSCORE technique is especially useful for investigation of anisotropically broadened ESEEM lines. In the case of disordered systems (powders), the broad ESEEM lines will emerge as correlation ridges. From the positions and the shapes of these ridges, information can be extracted about the magnitude and anisotropy of the hyperfine couplings. As an example, Figure 4.30c shows typical simulated HYSCORE spectra for a weakly coupled $S = I = 1/2$ system with an axial hyperfine tensor. The dipolar coupling leads to curvature of the ridges away from the antidiagonal line ($\nu_2 = 2\nu_1 - \nu_1$). The maximum shift, $\Delta\omega_{max}$, from the antidiagonal is related to the dipolar hyperfine coupling, T, in the following way:[105]

$$\Delta\omega_{max} = \frac{9}{32}\frac{T^2}{|\omega_I|} \tag{4.68}$$

The value of the isotropic coupling constant, a_{iso}, can be estimated from the separation of the ridges, as illustrated in Figure 4.30c. The endpoints of the ridges, which correspond to the canonical orientations $\theta = 0°$ and $90°$, are located on the antidiagonal.

The weak and strong (nuclear Zeeman *versus* hyperfine interaction) couplings can be easily distinguished in HYSCORE spectra. For the

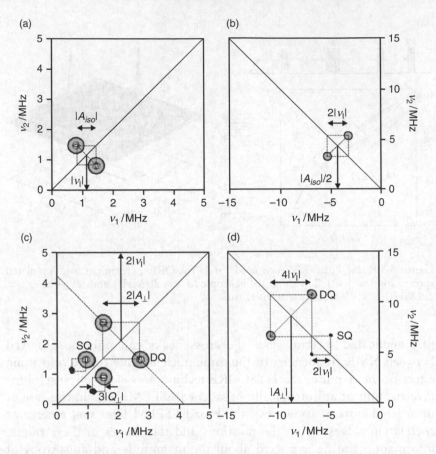

Figure 4.31 Simulated HYSCORE spectra for (a) weak coupling ($a_{iso} = 0.8\,\text{MHz}$, $T = -0.2\,\text{MHz}$, $\nu_I = 1.04\,\text{MHz}$), (b) strong coupling ($a_{iso} = 9\,\text{MHz}$, $T = -0.5\,\text{MHz}$, $\nu_I = 1.04\,\text{MHz}$), (c) weak coupling and quadrupolar interaction ($Q_1 = Q_2 = -0.15\,\text{MHz}$, $Q_3 = 0.3\,\text{MHz}$) and (d) strong coupling and quadrupolar interaction ($Q_1 = Q_2 = -0.5\,\text{MHz}$, $Q_3 = 1.0\,\text{MHz}$).

weak-coupling situation ($a_{iso} < 2\nu_I$), the peaks appear in the $(+, +)$ quadrant, and are thus well separated from the strong-coupling features ($a_{iso} > 2\nu_I$) observed in the $(+, -)$ quadrant.

For nuclei with $I \geq 1$ (*e.g.* ^{14}N, ^{63}Cu) and appreciable quadrupole interaction, analytical formulae describing the frequencies and shapes of the cross peaks have been derived elsewhere (*e.g.* for $I = 1$, see reference [106]). In the case of one ^{14}N nucleus, 18 correlation ridges are predicted, but not all of them are resolved in the experimental HYSCORE spectra. In general, frequencies characterised by strong dispersion due to anisotropy are difficult to resolve in HYSCORE. In the case of pronounced hyperfine coupling anisotropy, the ^{14}N cross peaks involving the double quantum (DQ) lines are usually less dispersed and dominate the HYSCORE spectrum. The DQ transitions appear with the frequencies $v_\beta{}^{DQ}$ ($|m_S = -1/2, m_I = 1\rangle \rightarrow |m_S = -1/2, m_I = -1\rangle$) and $v_\alpha{}^{DQ} |m_S = 1/2, m_I = -1\rangle \rightarrow |m_S = 1/2, m_I = 1\rangle$, given by:

$$v_{\alpha,\beta}^{DQ} = 2\sqrt{\left(v_I \pm \frac{a}{2}\right)^2 + \left(\frac{e^2 qQ}{4\hbar}\right)^2 (3 + \eta^2)} \qquad (4.69)$$

where a is the hyperfine coupling constant at the particular observer position and quadrupole parameters are defined in Equation 4.56. The DQ cross peaks will be observed in the $(+, +)$ quadrant for the weak-coupling case and in the $(-, +)$ quadrant for the strong-coupling case. Characteristic simulated spectra for one nucleus with $I = 1$ but zero quadruple interaction are displayed in Figure 4.31a,b (weak and strong case, respectively), while the effect of nonzero quadrupole interaction is shown in Figure 4.31c,d. More involved quantitative discussion of the HYSCORE spectra and a detailed treatment of the formal machinery can be found in specialised textbooks.[9,10,99]

With the development of commercial spectrometers, various experimental techniques of FT-EPR became more frequently applied for materials characterisation. Applications include studies of vacancies in perovskites,[107] oxide nanomaterials,[108] framework substitution in porous materials,[109] carbon-based materials[110] and catalytic materials,[111] among numerous others. The utility of the pulsed techniques may be illustrated using mayenite, $12CaO \cdot 7Al_2O_3$, as an example.[108] The distribution of the counterbalancing $[OH]^-$ anions within the positively charged framework, seen through their interaction with the isomorphously substituted copper(II), is achieved based on the measured proton hyperfine dipolar coupling obtained from HYSCORE spectra. Location of the $[O_2]^-$ radical anions present inside the mayenite nanocages was revealed by ESEEM and HYSCORE measurements of the quadrupole tensor of the adjacent framework aluminium ions ($I = 5/2$). Proposed locations of the extra-framework anionic species, along with the corresponding CW-EPR, HYSCORE and ESEEM spectra, are shown in Figure 4.32.

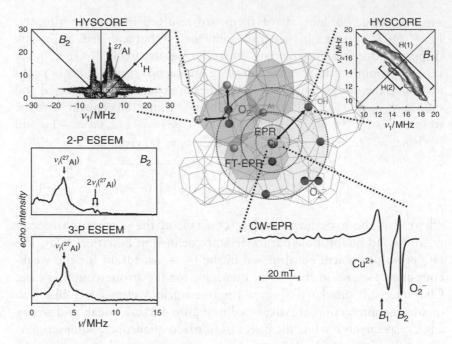

Figure 4.32 Spatial extent of the information accessible for CW-EPR and FT-EPR techniques, illustrated using mayenite, $12CaO \cdot 7Al_2O_3$, doped with framework Cu^{2+} ions and extra-framework $[O_2]^-$ and $[OH]^-$ ions. Also shown are the corresponding CW-EPR spectrum, the 2D HYSCORE spectra of protons and ^{27}Al nuclei and the one-dimensional ESSEM spectra of ^{27}Al nuclei. B_1 and B_2 are the observer position of the magnetic field used for FT-EPR measurements. Reproduced with permission from Matsuishi *et al.* (2004) [66] copyright (2004) American Chemical Society and Maurelli *et al.* (2010) [108] copyright (2010) Royal Society of Chemistry.

REFERENCES

[1] J. R. Pilbrow and M. R. Lowerey, *Rep. Prog. Phys.*, **43**, 433 (1980).

[2] J. A. Weil, J. R. Bolton and J. E. Wertz, *Electron Paramagnetic Resonance: Elementary Theory and Practical Applications*, John Wiley & Sons, Ltd, New York, NY, 1994.

[3] F. E. Mabbs and D. Collison, *Electron Paramagnetic Resonance of d Transition Metal Compounds*, Elsevier, Amsterdam, 1992.

[4] M. Brustolon and E. Giamello (Eds), *Electron Paramagnetic Resonance: A Practitioner's Toolkit*, John Wiley & Sons, Ltd, Hoboken, 2009.

[5] J. R. Pilbrow, *Transition Ion Electron Paramagnetic Resonance*, Oxford University Press, Oxford, 1990.

[6] P. Pietrzyk, Z. Sojka and E. Giamello, in *Characterization of Solid Materials and Heterogeneous Catalysis – From Structure to Surface Reactivity*, M. Che and J. C. Védrine (Eds), Wiley-VCH, Weinheim, 2012.

[7] S. K. Misra (Ed), in *Multifreqiency Electron Paramagnetic Resonance – Theory and Applications*, Wiley-VCH, Weinheim, 2011.

[8] M. Ikeya, *New Applications of Electron Spin Resonance: Dating, Dosimetry and Microscopy*, World Scientific, Singapore, 1993.

[9] L. Kevan and M.K. Bowman (Eds), in *Modern Pulsed and Continuous-wave Electron Spin Resonance*, John Wiley & Sons, Ltd, New York, NY, 1990.

[10] A. Schweiger and G. Jeschke, *Principles of Pulse Electron Paramagnetic Resonance*, Oxford University Press, Oxford, 2001.

[11] C. S. Sunandana, *Bull. Mater. Sci.*, **21**, 1 (1998).

[12] M. Che and E. Giamello, in *Physical Techniques for Solid Materials*, B. Imelik and J. C. Vedrine (Eds), Springer, New York, NY, 1994.

[13] Z. Sojka, *Catal. Rev. Sci. Eng.*, **37**, 461 (1995).

[14] A. Schweiger, *Angew. Chem., Int. Ed.*, **30**, 265 (1991).

[15] A. Bencini and D. Gatteschi, in *Inorganic Structure and Spectroscopy*, Vol. I, E. I. Solomon and A. B. P. Lever (Eds), John Wiley & Sons, Ltd, New York, NY, 1999.

[16] G. Jeschke, *Curr. Opin. Solid State Mater. Sci.*, **7**, 181 (2003).

[17] D. M. Murphy and C. C. Rowlands, *Curr. Opin. Solid State Mater. Sci.*, **5**, 97 (2001).

[18] R. Boča, *A Handbook of Magnetochemical Formulae*, Elsevier, Amsterdam, 2012.

[19] J.-P. Grivet, *J. Magn. Reson.*, **125**, 102 (1997).

[20] M. Chiesa, G. Amato, L. Boarino, E. Garrone, F. Geobaldo and E. Giamello, *Angew. Chem., Int. Ed.*, **42**, 5033 (2003).

[21] G. Clavel, M.-G. Willinger, D. Zitoun and N. Pinna, *Adv. Funct. Mater.*, **17**, 3159 (2007).

[22] S. Simon R. Pop, V. Simon and M. Coldea, *J. Non-Cryst. Solids*, **331**, 1 (2003).

[23] I. Djerdj, G. Garnweitner, D. Arčon, M. Pregelj, Z. Jagličic and M. Niederbergera, *J. Mater. Chem.*, **18**, 5208 (2008).

[24] K. Dyrek and Z. Sojka, *J. Chem. Soc., Faraday Trans. I*, **78**, 3177 (1982).

[25] A. O. Ankiewicz, M.C. Carmo, N.A. Sobolev, W. Gehlhoff, E.M. Kaidashev, A. Ranm, M. Lorenz and M. Grundmann, *J. Appl. Phys.*, **101**, 024 324 (2007).

[26] G. Mitrikas, *Phys. Chem. Chem. Phys.*, **14**, 3782 (2012).

[27] S. R. Seyedmonir and R. F. Howe, *J. Chem. Soc., Faraday Trans. I*, **80**, 2269 (1984).

[28] M. Chiesa, E. Giamello, C. Di Valentin, G. Pacchioni, Z. Sojka and S. Van Doorslaer, *J. Am. Chem. Soc.*, **127**, 16 935 (2005).

[29] A. Weidinger, M. Waiblinger, B. Pietzak and T. Almeida Murphy, *Appl. Phys. A*, **66**, 287 (1998).

[30] http://www.kyospin.com/KSMQ7Applications.htm#g1 (last accessed 28 May 2013).

[31] Z. Sojka and G. Stopa, *J. Chem. Edu.*, **70**, 675 (1993).

[32] B. A. Goodman and J. B. Raynor, *Adv. Inorg. Chem. Radiochem.*, **13**, 135 (1970).

[33] A. Baldansuren, H. Dilger, R.-A. Eichel, J. A. van Bokhoven and E. Roduner, *J. Phys. Chem. C*, **113**, 19 623 (2009).

[34] W. Gehlhoff and W. Ulrici, *Phys. Status Solid. B*, **102**, 11 (1980).

[35] C. P. Poole Jr, *Electron Spin Resonance: A Comprehensive Treatise on Experimental Techniques*, John Wiley & Sons, Ltd, New York, NY, 1983.

[36] P. Pietrzyk, K. Podolska and Z. Sojka, in *Electron Paramagnetic Resonance*, Vol. 23, B. C. Gilbert, D. M. Murphy and V. Chechik (Eds), RSC Publishing, Cambridge, 2013.

[37] P. H. Rieger, *J. Magn. Reson.*, **124**, 140 (1997).

[38] S. K. Hoffmann, J. Goslar and K. Tadyszak, *J. Magn. Reson.*, **205**, 293 (2010).

[39] D. Attanasio, *J. Magn. Reson.*, **26**, 81 (1977).

[40] B. D. P. Raju, N. O. Gopal, K. V. Narasimhulu, C. S. Sunandana and J. L. Rao, *J. Phys. Chem. Solid*, **66**, 753 (2005).

[41] L. Rimai and G. A. DeMars, *Phys. Rev.*, **127**, 702 (1962).

[42] A. Bencini and D. Gatteschi, *Electron Paramagnetic Resonance of Exchange-coupled Systems*, Springer, Berlin, 1990.

[43] C. Rudowicz, *J. Phys. C: Solid State Phys.*, **20**, 6033 (1987).

[44] Z. Wen-Chen and W. Shao-Yi, *Physica B*, **252**, 173 (1998).

[45] R. S. Drago, *Physical Methods in Chemistry*, W. B. Saunders Company, Philadelphia, PA, 1977.

[46] T. Moriya, in *Magnetism, Vol. 1*, G.T. Rado and H. Suhl (Eds), Academic Press, New York, NY, 1963.

[47] A. Aboukaïs, A. Bennani, C. F. Aissi, G. Wrobel, M. Guelton and J. C. Vedrine, *J. Chem. Soc., Faraday Trans. I*, **88**, 615 (1992).

[48] S. S. Eaton, K. M. More, B. M. Sawant and G. R. Eaton, *J. Am. Chem. Soc.*, **105**, 6560 (1983).

[49] T. Soda, Y. Kitagawa, T. Onishi, Y. Takano, Y. Shigeta, H. Nagao, Y. Yoshioka and K. Yamaguchi, *Chem. Phys. Lett.*, **319**, 223 (2000).

[50] P. Pietrzyk and Z. Sojka, *Appl. Magn. Reson.*, **40**, 471 (2011).

[51] A. Adamski, T. Spałek and Z. Sojka, *Res. Chem. Intermed.*, **29**, 793 (2003).

[52] J. R. Pilbrow and M. R. Lowerey, *Rep. Prog. Phys.*, **43**, 433 (1980).

[53] F. E. Mabbs, *Chem. Soc. Rev.*, **22**, 313 (1993).

[54] P. Rieger, *Electron Spin Resonance: Analysis and Interpretation.*, RSC Publishing, Cambridge, 2007.

[55] M. Chiesa, E. Giamello and M. Che, *Chem. Rev.*, **110**, 1320 (2010).

[56] K. M. Whitaker, S. T. Ochsenbein, V. Z. Polinger and D. R. Gamelin, *J. Phys. Chem. C*, **112**, 14 331 (2008).

[57] H. Zhou, A. Hofstaetter, D. M. Hofmann and B. K. Meyer, *Microelectronic Engineering*, **66**, 59 (2003).

[58] Z. Sojka and M. Che, *Appl. Magn. Reson.*, **20**, 433C (2001).

[59] S. K. Misra, *Appl. Magn. Reson.*, **10**, 193 (1996).

[60] A. Bencini and D. Gatteschi, in *Transition Metal Chemistry*, G. A. Melson and B. N. Figgis (Eds), Marcel Dekker, New York, NY, 1982.

[61] Y. Pan, M. Mao, Z. Li, S. M. Botis, R. I. Mashkovtsev and A. Shatskiy, *Phys. Chem. Minerals*, **39**, 627 (2012).

[62] N. M. Atherton, *Electron Spin Resonance*, John Wiley & Sons, Ltd, New York, NY, 1973.

[63] J. H. Freed, *Ann. Rev. Phys. Chem.*, **23**, 265 (1972).

[64] D. E. Budil, S. Lee, S. Saxena and J. H. Freed, *J. Magn. Reson. A*, **120**, 155 (1996).

[65] S. Stoll and A. Schweiger, *J. Magn. Reson.*, **178**, 42 (2006).

[66] S. Matsuishi, K. Hayashi, M. Hirano, I. Tanaka and H. Hosono, *J. Chem. Phys. B*, **108**, 18 557 (2004).

[67] S. Mukherjee and A. K. Pal, *J. Non-Cryst. Solids*, **341**, 170 (2004).

[68] M. Suhara and K. Iikawa, *J. Phys. Chem. Solids*, **42**, 129 (1981).

[69] J. Livage, A. Pasturel, C. Sanchez and J. Vedel, *Solid State Ionics*, **1**, 491 (1980).

[70] H. So, H. Jung, J.-H. Choy and R. L. Belford, *J. Phys. Chem. B*, **109**, 3324 (2005).

[71] H. So, K. Ha, Y.-J. Lee, K. B. Yoon and R. L. Belford, *J. Phys. Chem. B*, **107**, 8281 (2003).

[72] A. Kulak, Y.-J. Lee, Y. S. Park and K. B. Yoon, *Angew. Chem., Int. Ed.*, **39**, 950 (2000).

[73] T. J. Pinnavaia, in *Advanced Techniques for Clay Mineral Analysis*, J. J. Fripiat (Ed), Elsevier/North-Holland, New York, NY, 1981.

[74] P. G. Baranov, S. B. Orlinskii, C. de Mello Donegá and J. Schmidt, *Appl. Magn. Reson.*, **39**, 151 (2010).

[75] K. M. Whitaker, S.T. Ochsenbein, V. Z. Polinger and D.R. Gamelin, *J. Phys. Chem. C*, **112**, 14 331 (2008).

[76] Z. Wang, W. Zheng, J. van Tol, N. S. Dalal and G. F. Strouse, *Chem. Phys. Lett.*, **524**, 73 (2012).

[77] J. R. Pilbrow, *J. Magn. Reson.*, **93**, 12 (1991).

[78] I. V. Ovchinnikov and V. N. Konstantinov, *J. Magn. Reson.*, **32**, 179 (1979).

[79] M. Chiesa, E. Giamello, M. C. Paganini, Z. Sojka and D. Murphy, *J. Chem. Phys.*, **116**, 4266 (2002).

[80] Z. Sojka, A. Adamski and M. Che, *J. Mol. Catal.*, **112**, 469 (1996).

[81] D. E. De Vos, B. M. Weckhuysen and T. Bein, *J. Am. Chem. Soc.*, **118**, 9615 (1996).

[82] R. Böttcher, H. T. Langhammer, T. Müller and H.-P. Abicht, *J. Phys. Condens. Matter*, **20**, 505 209 (2008).

[83] M. Che and Y. Ben Taarit, *Adv. Coll. Inter. Sci.*, **23**, 235 (1985).

[84] J. A. Weil, *Mol. Phys. Rep.*, **26**, 11 (1999), and references therein.

[85] T. Spałek, P. Pietrzyk and Z. Sojka, *J. Chem. Inf. Model.*, **45**, 18 (2005).

[86] P. H. Rieger, *J. Magn. Reson.*, **50**, 485 (1982).

[87] M. Griffin, A. Muys, C. Noble, D. Wang, C. Eldershaw, K. E. Gates, K. Burrage and G. R. Hanson, *Mol. Phys. Rep.*, **26**, 60 (1999).

[88] G. Morin and D. Bonnin, *J. Magn. Reson.*, **136**, 176 (1999).

[89] S. Datta, E. Bolin, R. Inglis, J. Milios, E. K. Brechin and S. Hill, *Polyhedron*, **28**, 1788 (2009).

[90] C. Dablemont, C. G. Hammaker, R. Thouvenot, Z. Sojka, M. Che, E. A. Maatta and A. Proust, *Chem. Eur. J.*, **12**, 9150 (2006).

[91] W. R. Hagen, *Coord. Chem. Rev.*, **190–192**, 209 (1999).

[92] J. Krzystek, S. A. Zvyagin, A. Ozarowski, S. Trofimenko and J. Telser, *J. Magn. Reson.*, **178**, 174 (2006).

[93] D. Stehlik and K. Möbius, *Annu. Rev. Chem.*, **48**, 745 (1997).

[94] E. Giamello, M. C. Paganini, D. M. Murphy, A. M. Ferrari and G. Pacchioni, *J. Phys. Chem. B*, **101**, 971 (1997).

[95] B. Cage, A. K. Hassan, L. Pardi, J. Krzystek, L. C. Brunel and N. S. Dalal, *J. Magn. Reson.*, **124**, 495 (1997).

[96] W. Froncisz and J. S. Hyde, *J. Chem. Phys.*, **73**, 3123 (1980).

[97] M. Che, J. C. McAteer and A. J. Tench, *J. Chem. Soc., Faraday Trans. I*, **74**, 2378 (1978).

[98] E. L. Hahn, *Phys. Rev.*, **80**, 580 (1950).

[99] S. A. Dikanov and Y. D. Tsvetkov, *Electron Spin Echo Envelope Modulation (ESEEM) Spectroscopy*, CRC Press, Boca Raton, FL, 1992.

[100] E. J. Reijerse and C. P. Keijers, *J. Magn. Reson.*, **71**, 83 (1987).

[101] A. Lai, H. L. Flanagan and D. J. Singel, *J. Chem. Phys.*, **89**, 7161 (1988).

[102] Y. Deligiannakis, M. Louloudi and N. Hadjiliadis, *Coord. Chem. Rev.*, **204**, 1 (2000).

[103] P. Höfer, A. Grupp, H. Nebenführ and M. Mehring, *Chem. Phys. Lett.*, **132**, 279 (1986).

[104] C. Gemperle, G. Aebli, A. Schweiger and R. R. Ernst, *J. Magn. Reson.*, **88**, 241 (1990).

[105] A. Pöppl and L. Kevan, *J. Phys. Chem.*, **100**, 3387 (1996).

[106] S. A. Dikanov, L. Xun, A. B. Karpiel, A. M. Tyryshkin and M. K. Bowman, *J. Am. Chem. Soc.*, **118**, 8408 (1996).

[107] R.-A. Eichel, *Phys. Chem. Chem. Phys.*, **13**, 368 (2011).

[108] S. Maurelli, M. Ruszak, S. Witkowski, P. Pietrzyk, M. Chiesa and Z. Sojka, *Phys. Chem. Chem. Phys.*, **12**, 10 933 (2010).

[109] S. Maurelli, M. Vishnuvarthan, M. Chiesa, G. Berlier and S. Van Doorslaer, *J. Am. Chem. Soc.*, **133**, 7340 (2011).

[110] B. Corzilius, K.-P. Dinse and K. Hata, *Phys. Chem. Chem. Phys.*, **9**, 6063 (2007).

[111] S. Van Doorslaer and D. M. Murphy, *Top. Curr. Chem.*, **321**, 1 (2012).

5

Analysis of Functional Materials by X-ray Photoelectron Spectroscopy

Karen Wilson[a] and Adam F. Lee[b, c]

[a] European Bioenergy Research Institute, School of Engineering and Applied Science, Aston University, Aston Triangle, Birmingham, UK
[b] Department of Chemistry, University of Warwick, Coventry, UK
[c] School of Chemistry, Monash University, Melbourne, Australia

5.1 INTRODUCTION

X-ray photoelectron spectroscopy (XPS) plays a key role in the surface characterisation and design of improved functional materials.[1] The mechanical, chemical, optical and electronic properties of condensed matter underpin a vast array of functional materials whose behaviour is critical to modern society. For example, the surface properties of polymers can be tuned for use in packaging, nonstick coatings and medical applications; coatings can be used to improve frictional resistance (tribology) between moving components, *e.g.* in engines; surface-catalysed chemical transformations are used to create new materials, foods, medicines and energy sources; the electronics industry is dependent on surface modification and characterisation techniques for the development of semiconductor components and nanoscale devices; and surface processes in the atmosphere play an important role in ozone depletion

Local Structural Characterisation, First Edition. Edited by Duncan W. Bruce, Dermot O'Hare and Richard I. Walton.
© 2014 John Wiley & Sons, Ltd. Published 2014 by John Wiley & Sons, Ltd.

cycles. Thus, whether we are looking at a car's body shell, a biological implant, a catalyst or a solid-state electronic device, surface reactivity is crucial to performance.

XPS is a vacuum-based technique, developed in the mid 1960s by Siegbahn and coworkers, which employs the photoelectric effect to characterise the atomic composition of surfaces.[2] Based on this work, Siegbahn was awarded the Nobel Prize for Physics in 1981 for his contribution to the development of high-resolution electron spectroscopy.[3] XPS can play an important role in guiding the design of new materials, tailored to meet increasingly stringent constraints on performance devices, by providing insight into their surface compositions and the fundamental interactions between the surfaces and the environment. This chapter will outline the principles and application of XPS as a versatile, chemically specific analytical tool in determining the electronic structures and (usually surface) compositions of constituent elements within diverse functional materials.

5.1.1   The Basic Principles of XPS

XPS employs soft X-rays to photoexcite core electrons, which are then emitted from their parent atoms with a kinetic energy characteristic of their initial atomic energy level and element (Figure 5.1).

The kinetic energy, E_k, of the emitted photoelectron is calculated as follows:

$$E_k = h\nu - E_B + \phi \tag{5.1}$$

where E_B is the binding energy of the atomic energy level and ϕ is the combined workfunction of the sample and the electron analyser.

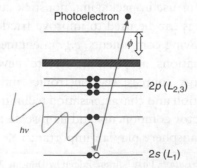

Figure 5.1   The process of X-ray core-level photoemission.

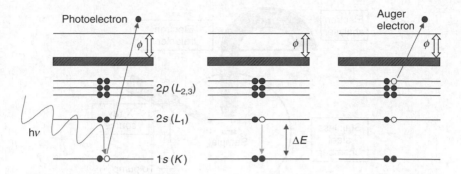

Figure 5.2   The process of X-ray-stimulated Auger electron emission.

The core hole resulting from the photoemission process can be filled by relaxation of an electron from a higher energy level, according to Figure 5.2. When this occurs, the energy difference, ΔE, can be emitted *via* either fluorescence or transfer to another electron, which is itself subsequently emitted from the parent atom; this is referred to as the 'Auger electron'. The Auger electron energy can be approximated by Equation 5.2, but it should be noted that $E_{(L2,3)}$ is raised above that within the neutral parent atom pre-excitation, due to decreased Coulomb repulsion following creation of the L_1 hole. The Auger electron is indexed based upon the energy levels involved in the emission process; the example in Figure 5.2 corresponds to a $KL_1L_{2,3}$ Auger electron.

$$E_{(Auger)} \sim E_{(K)} - E_{(L1)} - E_{(L2,3)} \qquad (5.2)$$

Photoelectrons are readily absorbed or scattered by air, so the photoemission process is almost always performed under ultra-high vacuum (UHV) using the experimental arrangement shown in Figure 5.3. Energy-selecting electron analysers employ tuneable magnetic fields, typically in a concentric hemispherical analyser (CHA) configuration. A voltage is applied across an inner and an outer hemisphere, selectively steering electrons through the gap between the hemispheres. Electrons travelling too quickly collide with the outer hemisphere, while those travelling too slowly are attracted to the inner hemisphere, such that only electrons in a narrow energy region succeed in travelling around the hemispherical arc to the detector. The resolving power, $E_0/\Delta E$, of the analyser is related to the entrance slit width, s, and the radius, R, of the hemisphere as $2R/s$. To maintain good resolving power at high energy, and thus a constant energy resolution across a spectrum, a series of electron lenses are placed before the CHA, which can be

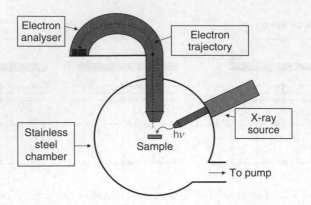

Figure 5.3 Arrangement of a typical XPS spectrometer, using a CHA to detect photoelectrons.

used to retard the incoming electrons to a fixed 'pass energy'. In such a constant-analyser energy mode, if the pass energy is 50 eV, the lenses will be required to slow electrons of 1000 eV to 950 eV in order that they can be detected.

The energy resolution in XPS is approximated by:

$$\Delta E = (\Delta E_x{}^2 + \Delta E_y{}^2 + \Delta E_z{}^2)^{1/2} \qquad (5.3)$$

where ΔE_x is the full-width half-maximum (FWHM) of the X-ray line, ΔE_y is the width of the analyser energy window and ΔE_z is the natural line width of the core level from which the photoelectron is emitted.

Limiting energy resolution is often controlled by the linewidth of the X-ray source, with laboratory instruments typically employing Mg and Al K_α anodes with intrinsic linewidths of 0.75 and 0.85 eV, respectively. Monochromation (Figure 5.4) can reduce the linewidth of the Al K_α X-ray to 0.28 eV by removing the satellite peaks and the Bremsstrahlung emission. Laboratory anodes made from other elements (*e.g.* Ag) can be used when higher excitation energies are required; however, their characteristic X-ray lines are often broader than monochromated Al K_α. Further improvements in photoelectron energy resolution *via* tuning of the incident X-ray energy can be achieved by using synchrotron radiation, which also permits the photoelectron kinetic energy to be adjusted in order to confer greater surface sensitivity in material analysis.

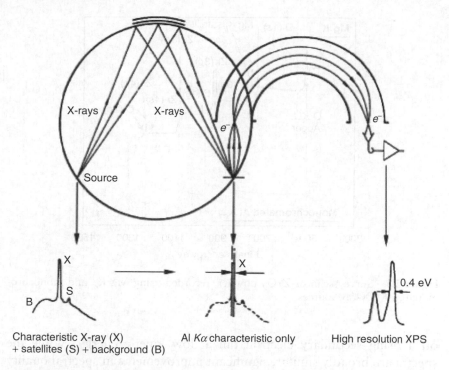

X-rays X-rays

e^- e^-

Source

X X 0.4 eV

B S

Characteristic X-ray (X) Al $K\alpha$ characteristic only High resolution XPS
+ satellites (S) + background (B)

Figure 5.4 Operation of an Al K_α monochromator for XPS. Bragg diffraction from a bent crystal and the Rowland circle generates monochromatic X-rays *via* removal of satellites and background. Image courtesy of Chris Blomfield, Kratos Analytical. Copyright (2013) Kratos Analytical.

5.1.2 Quantification of X-ray Photoelectron Spectra

Every element besides hydrogen and helium can be probed by XPS. Sample compositional analysis is readily performed over a period of minutes by acquiring a survey scan over the broadest energy range accessible by the available X-ray source. Each element has a unique set of well-defined energy levels, which allows their identification through matching of the photoelectron binding energies in the survey spectra to those of tabulated atomic energy levels.

By way of illustration, a survey scan from ZrO_2 powder is shown in Figure 5.5, recorded using both Mg K_α and monochromated Al K_α X-rays. All of the accessible energy levels for each element are visible, spanning O $1s-2p$ and Zr $3s-4p$ energy levels, which are superimposed

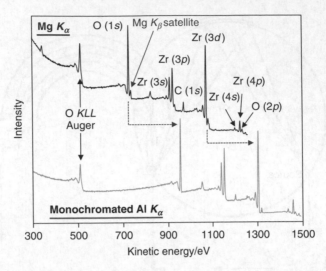

Figure 5.5 Survey scan of ZrO_2 powder, recorded using Mg K_α and monochromated Al K_α X-ray sources.

on a rising secondary electron tail at low kinetic energy. While the spectra are broadly similar, significant improvements in spectral quality are achieved with monochromatic X-rays; specifically, the secondary electron tail is decreased and the satellite associated with the Mg K_β line is removed. Increasing the photon energy raises the kinetic energy of the photoelectron, whereas the Auger transition (highlighted for the O KLL, where a $1s$ electron and a $2s$ or $2p$ electron is involved in the Auger emission process) arising from an internal electron rearrangement process remains unchanged. This differential response of photoelectron peak to changing photon energy provides a simple means of discriminating Auger from X-ray photoelectrons (XPs). Early XPS studies of such insulating oxide materials proved problematic due to sample charging following photoelectron emission, which results in peak shifts and line-broadening effects. These difficulties of peak-shape analysis and energy referencing have since been overcome with the development of improved charge compensation devices (or flood guns), which are fitted to modern XPS instruments. With good charge compensation, analysis of insulating materials and deconvolution of complex spectra containing multiple components can now be reliably performed.

Elemental compositions are readily obtained following background subtraction, by correcting the resulting integrated photoelectron peak areas using element-specific relative sensitivity factors (RSFs). RSFs take into account such factors as the inelastic mean free path (IMFP), the atomic absorption cross-section and the degeneracy of the particular core level involved. Such sensitivity factors have been tabulated by Schofield[4] and are used to correct XP peak intensities during quantification. Accurate quantification must, however, involve spectral calibration, in order to take into account the density of the material and instrumental factors based around the sensitivity of the analyser. The effect of these parameters can be seen by comparing the Zr 3s, 3p and 3d peaks, which all have different relative intensities even though they originate from the same number of Zr atoms.

High-resolution spectra reveal more information about the chemical nature of the element being probed, but also fine structure arising from spin–orbit coupling (SOC), which gives rise to two peaks for all p, d and f orbitals. The notation used to describe the spin–orbit split states is derived from L–S coupling where $j = (l \pm s)$, where l and s are the orbital and the spin angular momentum, respectively. The magnitudes of the spin–orbital splitting and of the peak area ratios of the two components for an element in different compounds are treated as constant in XPS analysis, with the latter deduced from the degeneracy of j states, given by $(2j + 1)$. This spin–orbit splitting is shown in Figure 5.6 for Zr 3d transitions, for which the $3d_{3/2} : 3d_{5/2}$ photoelectrons exist in a 2 : 3 ratio and are 2.4 eV apart. A compilation of such information is available at the NIST X-ray Photoelectron Spectroscopy Database.[5]

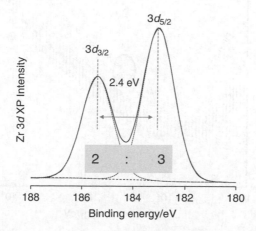

Figure 5.6   High-resolution Zr 3d XP spectrum for ZrO_2 powder.

5.1.3   The Origin of Surface Sensitivity

The surface sensitivity of XPS is derived from the fact that photoelectrons emitted over the range 50–2000 eV have a high probability of inelastic scattering. Electrons can lose energy *via* a number of mechanisms, including plasmon scattering, core electron ionisation in adjacent atoms and single-particle electron excitation by valence electrons. The mean free path (MFP) of an electron is generally described by the universal escape depth curve,[6] which has a broad minimum at a kinetic energy of around 70–100 eV (Figure 5.7).[7]

The universal nature of the curve in this energy range originates from the fact that inelastic scattering of electrons mostly involves excitation of conduction electrons, which have a similar density in all elements.[9] Note, however, that at lower energies other scattering mechanisms such as phonon scattering become dominant. It should be borne in mind that this universal curve concept is an approximation and that there can be appreciable variations in the IMFP for materials with significant density differences, as illustrated by the variation in MFP of the curves shown in

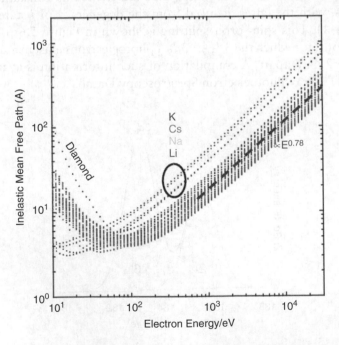

Figure 5.7   Calculated IMFPs for a range of elements spanning Li–Bi, as a function of electron kinetic energy. Reproduced with permission from Fadley (2010) [8]. Copyright (2010) Elsevier Ltd.

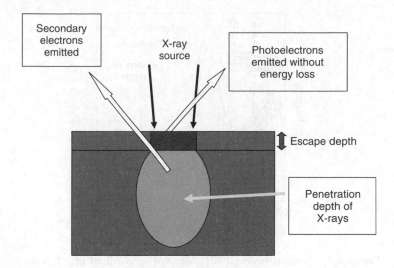

Figure 5.8 Only electrons in top surface layers contribute to the photoelectron peak; electrons emitted from the bulk are inelastically scattered and contribute to the secondary electron tail seen at low kinetic energy in Figure 5.5.

Figure 5.7 for the range of elements examined. Significant deviation is observed for the alkali metals, which have lower atomic densities than *e.g.* Al, Cu, Ag and Au. Indeed, K is reported to have an IMFP three to four times higher than that of the more dense material, Au.[7]

Thus, by using soft X-ray sources, generated in laboratory instrumentation by Mg K_α (1253.6 eV) or Al K_α (1486.3 eV) anodes, low-energy photoelectrons with short IMFP can be generated.[10] IMFP values of 0.5–2.0 nm are typically observed over the electron kinetic energy range 10–1300 eV, so the range of emerging electrons is characteristic of the atomic composition of the uppermost layers of the sample (Figure 5.8). There is growing interest in the application of hard X-ray photoemission, which operates in the multi-keV regime[11] and would extend the bulk sensitivity of photoemission beyond what is achievable *via* conventional XPS. By taking advantage of high-flux X-rays from third-generation synchrotron radiation facilities, 'buried interfaces' can be probed.

5.1.4   Angular Resolved XPS

The surface specificity of XPS means that it is also a powerful tool by which to probe information about the film thickness or

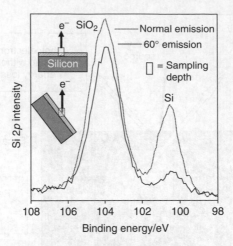

Figure 5.9   Angular-resolved XP spectra of an oxidised silicon wafer.

surface-terminating layer of a material. This is particularly relevant to atomic-layer coatings and oxidation processes, as demonstrated by Figure 5.9, which shows two spectra of an oxidised silcon wafer, recorded at normal and grazing emission relative to the analyser. When spectra are recorded at normal emission, the analysis is at its least surface sensitive and probes the furthest into the bulk of the sample. In the case of the oxidised Si surface, if the oxide layer is sufficiently thin, both surface oxide and subsurface silicon are detected.[12] Surface sensitivity can be increased by recording spectra at greater grazing emission, with the signal from sub-surface Si atoms attenuated when the sample is orientated such that the electrons have to travel a longer distance in order to emerge.

By calculating the attenuation of the underlying silicon component, an estimation of oxide film thickness (d) can be made as follows:

$$I_d = I_0 \exp\left(\frac{-d}{\lambda \cos\theta}\right) \tag{5.4}$$

where I_d is the intensity of the underlying substrate signal, I_0 is the intensity of the clean surface, λ is the IMFP of the photoelectron and θ is the angle between the analyser and the surface normal.[4]

By measuring the signals for two components in *e.g.* an alloy as a function of angle, it is also possible to discrimate a uniform alloy from a system with a surface-capping layer, as illustrated in Figure 5.10. While it is convenient to rotate a wafer or metallic foil, for powdered samples (*e.g.*

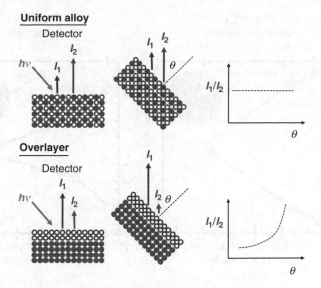

Figure 5.10 Use of angular resolved XPS to discriminate uniform alloys from surface overlayers.

core–shell nanoparticles) this is not experimentally possible. However, by taking advantage of the different escape depths of different photoelectrons, it is possible to estimate shell thickesses (d) if different X-ray energies are used to excite the photoemission process and thus vary λ. Using commercial Mg K_α or Al K_α X-rays offers a means of achieving this with a laboratory XPS system. The continuously tuneable photon energies avaialable at synchrotron sources, however, allow a more detailed depth profile of such capping layers to be created. Figure 5.11 shows how IMFP and a measured atomic fraction of ~15 nm-diameter Pd : Rh or Pt : Pd in core–shell nanoparticles vary with electron kinetic energy (or incident photon energy). Photoelectrons with a kinetic energy of ~1152 cV originate from both the shell, the intermediate layer and the central core of each nanoparticle, whereas at lower kinetic energy electrons emerge principally from the shell due to the shorter MFP of these photoelectrons.

5.1.5   Chemical Shift Information from XPS

Photoelectron binding energies are particularly sensitive to the oxidation state or local electron density of the atom being probed. In order to

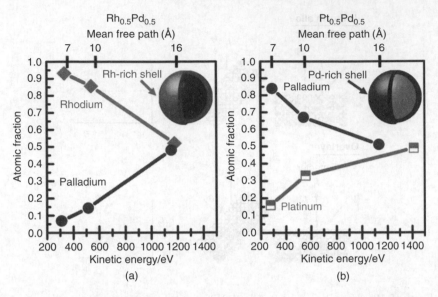

Figure 5.11 Depth profiling of core–shell nanoparticles as a function of photoelectron energy. Reproduced with permission from Tao *et al.* (2008) [13]. Copyright (2008) American Association for the Advancement of Science.

assess the chemical shift, all binding energies must be quoted in reference to an energy calibration. Typically, such calibration is performed relative to adventitious carbon at 284.8 eV or to the Fermi edge. More precise referencing can be achieved according to a gold standard, for which the Au $4f_{7/2}$ is expected at 84 eV.

The effect of oxidation of elemental silicon, shown in Figure 5.9, demonstrates that the transition from elemental Si to oxidised SiO_2 results in a 3.9 eV increase in the Si $2p$ electron binding energy.[14] Such binding energy shifts can result from a change in the electron density of the initial state of the atom, or from changes in core–hole screening mechanisms, which are referred to respectively as *initial-* and *final-state* effects. In the case of silicon oxidation, an initial-state change arises from the effect of covalent attachment of electronegative oxygen atoms to Si in the surface layer, which increases the binding energy of the Si core electrons. Detailed studies of the oxidation of clean Si(111) and Si(100) actually reveal that surface oxidation proceeds *via* substoichiometric oxides.[15] The formation of Si^+, Si^{2+}, Si^{3+} and Si^{4+} results in new spectral features in the Si $2p$ region, which have increasing chemical shifts of 0.95, 1.75, 2.48 and 3.9 eV, respectively.[14]

Based on simple initial-state arguments, the binding energy of a core electron of an anion or cation increases with the electronegativity of the

Table 5.1 Selected binding energies of core XP lines, illustrating the effect of electronegativity and coordination number on the magnitude of chemical shift.

Compound/Ion	Line	Binding energy/eV
K_2SO_4	S $2p$	168.9
Li_2SO_4	S $2p$	169.4
K_2CO_3	C $1s$	289.3
Li_2CO_3	C $1s$	289.9
$[NO_3]^-$	N $1s$	408–407
$[NO_2]^-$	N $1s$	404.6–403.6
$[NH_4]^+$	N $1s$	401–402
$[SO_4]^{2-}$	S $2p$	170–169
$[SO_3]^{2-}$	S $2p$	168–167

counterion. For example, both S $2p$ and C $1s$ XP regions for $[SO_4]^{2-}$ and $[CO_3]^{2-}$ anions increase on changing the cation from K to more electronegative Li (Table 5.1).

Binding energies also increase with the number of electronegative groups surrounding the central atom, as illustrated in Table 5.1.[16] Studies of transition-metal phosphides reveal that the P $2p$ binding energy increases as the difference in electronegativity between the metal and the phosphorus atom decreases from Cr–P to Co–P.[17] S $2p$ photoelectrons have a particularly wide chemical-shift range, spanning 160–170 eV from S^{2-} to S^{6+} in sulfate. Figure 5.12 shows a typical spectrum obtained for a thiol-modified silica, illustrating that oxidation with H_2O_2 can completely oxidise the sulfur species to $-SO_3H$.

More complex spectra and chemical shifts are often observed for elements in which several *final states* are possible and for rare earth or transition metals in which excitation of electrons to the valence conduction band occurs in parallel to the photoemission process.[19–21] Ligand–metal charge-transfer effects change the electron configuration of the final excited state, giving rise to satellites that are at high binding energy relative to the main photoemission peak, as demonstrated for Cu $2p$ spectra of CuO, shown in Figure 5.13. The occurrence of these multielectron excitation or '*shake-up*' processes is, however, dependent on the presence of vacant d states in the valence band; in contrast, Cu_2O, having a d-electron configuration of d^{10}, does not exhibit any satellite transitions.[22] A study of a range of Cu complexes – [Cu(*S*-3-aminohexahydroazepine)$_2$] $(BF_4)_2$, CuO, $CuF_2 \cdot 2H_2O$, $CuCl_2$, $CuBr_2$ and copper(II) phthalocyanine – also reveals that the Cu $2p$ satellite intensity increases with the covalent character of the Cu–ligand bond.[23]

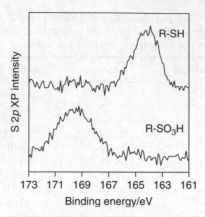

Figure 5.12 S 2p XP spectra for thiol and sulfonic acid functionalised silicas. Reproduced with permission from Wilson and Lee (2009) [18]. Copyright (2009) Wiley-VCH.

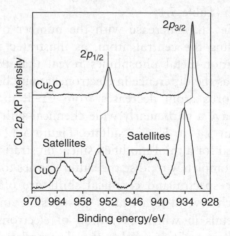

Figure 5.13 Cu 2p XP spectra of CuO and Cu_2O. Reproduced with permission from Wilson and Lee (2010) [27]. Copyright (2010) Royal Society of Chemistry.

The intensity of satellite fine structure is also dependent upon ligand field splitting, the coordination number and the high/low spin characteristics of the metal centre,[19] allowing the oxidation states and coordination environments of compounds of transition metals such as Ni,[24] Co,[25] Fe[26] and Cu[22] to be distinguished readily.

Identification of oxidation states is occasionally hampered by energy referencing issues, and can be particularly problematic when binding energy shifts are small, as between Cu^0 and Cu^+ (Table 5.2). This can

Table 5.2 Cu $2p_{3/2}$ binding energies and Auger
parameters for Cu, Cu_2O and CuO.

Material	$2p_{3/2}$ binding energy/eV	Auger parameter/eV
Cu	932.7	1855.2
Cu_2O	932.6	1849.4
CuO	933.8	1855.7

be overcome by determining the Auger parameter associated with the
emitted core electron, which is defined as follows:[28]

Auger parameter (α) = kinetic energy (Auger e^- from photoelectron)

$$+ \text{ binding energy (photoelectron)} \qquad (5.5)$$

The Auger parameter is independent of the energy calibration of the
spectrometer and is an unequivocal probe for changes in the electronic
properties of the atom.

An overview of the application of high-resolution XPS to understanding of the nature of the active site in functional materials and in
heterogeneous catalysts will be considered in the following sections.

5.2 IMAGING XPS

Advances in detector electronics have opened the way for development of
photoelectron microscopes and instruments with XPS imaging capabilities. Figure 5.14 shows the lens arrangement used to project an image on
the detector, and has two operation modes: microspot spectroscopy and
imaging spectromicroscopy. Significant improvements in the efficiency
of scanning photoelectron microscopy (SPEM) have been achieved by
implementation of multichannel electron detectors, where each channel measures electrons with a specific kinetic energy window.[29] This
facilitates simultaneous acquisition of images by each channel, allowing
reconstruction of the spectrum corresponding to the covered energy
window from a selected micro-area. Images acquired from a CrN-doped
TiAlN-coated material (typically applied to improve the performance
of cutting tools, moulds and dies) are shown, illustrating the elemental
distribution at the surface of the material after oxidation at $>900\,°C$ in

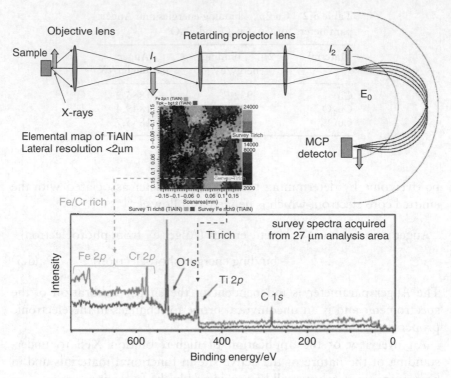

Figure 5.14 Analyser arrangement for imaging XPS. Inset shows an image illusrating the spatial distribution of Ti and Fe contaminant in the surface of CrN-doped TiAlN. Selected area image shows the elemental composition from a 27 μm analysis area. Image courtesy of Chris Blomfield, Kratos Analytical. Copyright (2013) Kratos Analytical.

air. XPS chemical mapping and small spot XPS are able to identify the formation of domains rich in Ti or Fe/Cr.

High-intensity X-ray sources available at third-generation synchrotron sources offer significant enhancement in imaging capabilities using scanning X-ray photoemission electron microscopy (XPEEM).[30] Figure 5.15 shows the arrangement of an XPEEM based on the spectromicroscopy beamline at the Elettra synchrotron in Italy. The specimen is first irradiated with X-rays, then, by using a combination of electrostatic and magnetic lenses, the magnified image of the sample is projected onto a phosphorus screen and finally collected by a charge-coupled device (CCD) camera. XPEEM offers an energy resolution of 0.3 eV and lateral resolution of a few tens of nanometres.

The resolving power of XPEEM is demonstrated in Figure 5.16, which shows images from Ag-coated carbon nanotubes. The high spatial

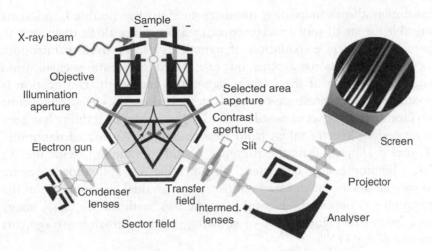

Figure 5.15 Set-up of the low-energy electron microscopy/X-ray photoelectron emission microscopy (LEEM/XPEEM) instrument. Reproduced with permission from http://www.pc5.ch.tum.de/index.php?id=795 (last accessed 31 May 2013) (after Fig. 2 from Design of a Spectroscopic Low-energy Electron Microscope by Lee H. Veneklasen, *Ultramicroscopy*, 36, 76–90, Elsevier Ltd). Copyright (2013) Professor Dr Sebastian Günther.

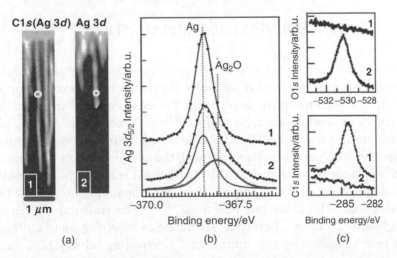

Figure 5.16 (a) C 1s image of C nanotubes covered with a thin Ag film, taken before exposure to oxygen plasma. (b) Ag $3d_{5/2}$ image taken after gasification and consumption of C nanotubes. (c) Ag $3d_{5/2}$, C 1s and O 1s spectra taken on the indicated spots in the images, before oxygen exposure in (1) and after gasification in (2). Reproduced with permission from Barinov *et al.* (2009) [30]. Copyright (2009) Elsevier Ltd.

resolution allows individual nanostructures to be visualised and areas suitable for small spot spectroscopic measurements along their axes to be selected. Surface oxidation of nanotubes can be used to introduce oxygenated functional groups, but at elevated temperatures gasification and consumption of the carbon nanotubes can result. Gasification is promoted by the presence of metals deposited on the carbon nanotube surface, so the impact of postdeposited Ag on nanotube stability has been investigated. Spectra taken from fresh individual Ag-coated nanotubes (Figure 5.16c) confirm the presence of only C and Ag, with the Ag $3d_{5/2}$ binding energy consistent with metallic Ag. Following exposure to oxygen plasma at elevated temperatures, gradual consumption of the nanotubes is observed, starting from the tips, with the Ag $3d_{5/2}$ image and spectra in Figure 5.16b–c showing that the remaining fragments consist of Ag_2O.[30]

Future improvements to optics and detectors, including aberration correction and faster detection systems, will improve both the lateral resolution and the time resolution, enabling photoelectron microscopy in the sub-10 nm and ps regime.

5.3   TIME-RESOLVED HIGH-RESOLUTION XPS

Advances in surface science instrumentation to enable time-resolved spectroscopic measurements (at ambient pressure) offer exciting opportunities to quantitatively investigate the composition, structure and dynamics of working catalyst surfaces.[31,32] Time-resolved X-ray photoelectron spectroscopy (TR-XPS) of reacting adsorbates on sub-minute timescales has become possible through the combination of high photon flux and high resolution available at third-generation synchrotrons[33] and improved electron analyser detection capabilities (such as two-dimensional (2D) delay line detectors).[34,35] Fundamental mechanistic insight into a variety of catalytic technologies involving small molecules has been obtained by this approach,[36] providing information on the threshold temperature for reaction and the nature and surface concentration of key intermediates. The following sections will present some highlights of where XPS has proven instrumental in providing insight into surface-catalysed reaction mechanisms.

5.3.1   Selective Catalytic Alcohol Oxidation

Catalytic selective oxidations are an elegant class of chemical transformation widely employed throughout academia and the chemical
sector to valorise platform chemicals such as olefins and alcohols. The
direct, aerobic selective oxidation of functionalised hydrocarbons *via*
platinum-group metal catalysts has been a matter of intense research
and debate over the past decade. Such heterogeneously catalysed routes
to valuable fine and agrochemical intermediates and products offer
great potential process, safety and environmental benefits over current industrial syntheses employing stoichiometric reagents. However,
full-scale commercialisation of these alternative catalytic clean technologies still awaits detailed knowledge of the optimal reaction conditions,
activation protocols and deactivation pathways. This information in
turn requires better insight into both surface reaction mechanisms
and the nature of the active catalyst site. Thus, an improved understanding the surface chemistry of alcohols is important for a number
of clean catalytic technologies where they are employed as renew-
able energy or chemical sources, such as direct methanol–ethanol fuel
cells,[37] H_2 generation *via* selective reforming[38] and bioethanol fuel
additives[39] (wherein toxic oxygenate emissions are of concern[40]).
Ethanol is also used to model the degradation of higher-molecular-
weight oxygenates found in biomass.[41] In all cases, the strength of
interaction between alcohol and metal catalyst is central to controlling reaction pathways and improving these catalytic processes. For
example, ethanol reformation over bimetallic/Pt(111) systems has shown
that the H_2 yield increases as the *d*-band centre moves towards the
Fermi level.[42] Alcohol chemisorption over transition metal surfaces
is generally accepted to occur initially *via* oxygen lone-pair donation,
resulting in an $\eta^1(O)$ conformation.[43] A TR-XPS study of methanol
over clean Ir(111)[44] found only molecular chemisorption <170 K, with
adsorbed methanol decomposing directly to surface CO and H on heating. A minor C–O dissociation pathway was also observed, resulting in
trace CH_x fragments which sequentially dehydrogenated to carbon by
500 K. Preadsorbed oxygen helped stabilise a methoxy surface intermediate between 160 and 220 K, which decomposed to formate (HCO_2-)
and CO (Figure 5.17). Formate in turn decomposed ~350 K evolving
CO_2; similar processes operate at lower temperatures over Pd(111)

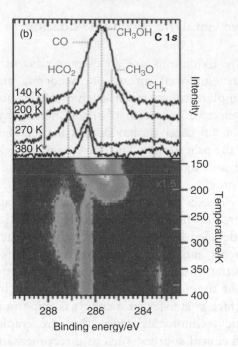

Figure 5.17 C 1s TR-XPS of CH_3OH reacting over Ir(111). Reproduced with permission from Weststrate *et al.* (2007) [44]. Copyright (2007) American Chemical Society.

and Pt(111). The adsorbed CO and H reacted, with residual oxygen desorbing as water and CO_2 ($\sim$400 K), limiting site blocking by residual CO.

Methanol has also been followed by TP-XPS over Ni(100)[45] to elucidate its reaction pathway over Raney nickel catalysts, whose catalytically active centres possess mainly low-indexed (100) planes and are potential electrodes for low-temperature fuel cells. Like over Ir(111), methanol dehydrogenates at >160 K, forming a methoxy species; however, this decomposes to CO without generating surface formate, and nickel catalysts may therefore be more susceptible to CO poisoning during operation at ambient temperature.

The surface chemistry of ethanol has also been investigated by TP-XPS over Pt(111),[46] where the appearance of ethoxy species following adsorption below 170 K is noted. C 1s TP-XPS shows sequential dehydrogenation of this ethoxy intermediate through acetaldehyde to a stable surface acetyl species over the range 170–240 K (Figure 5.18).

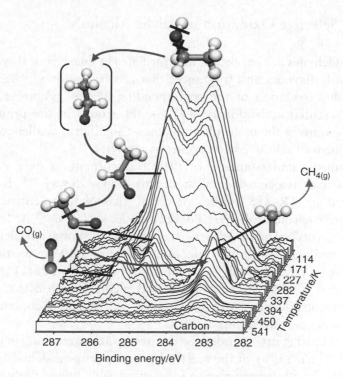

Figure 5.18 Reaction pathway of ethanol over Pt(111) followed by TR-XPS. Reproduced with permission from Lee *et al.* (2004) [46]. Copyright (2004) Elsevier Ltd.

Acetaldehyde dehydrogenation to acetyl is energetically favoured over Pt(111), with an activation barrier of only $1\,kJ\,mol^{-1}$,[47] with the resulting surface acetyl decarbonylating above 200 K, to form CH_4 and CO, which desorb at 310 and 450 K, respectively. The reaction pathway elucidated from TP-XPS was deduced as follows (the symbol (a) denotes an adsorbed species):

$$CH_3CH_2OH(a) \rightarrow CH_3CH_2O(a) + H(a) \qquad \sim180\,K\,(1)$$

$$CH_3CH_2O(a) \rightarrow CH_3CH=O(a) + H(a) \qquad >180\,K\,(2)$$

$$CH_3CH=O(a) \rightarrow CH_3C=O(a) + H(a) \qquad >180\,K\,(3)$$

$$CH_3CO(a) \rightarrow CH_3(a) + CO(a) \qquad >200\,K\,(4)$$

$$CH_3(a) + H(a) \rightarrow CH_4(g) \qquad >310\,K\,(5)$$

$$CO(a) \rightarrow CO(g) \qquad >450\,K\,(6)$$

5.3.2 Selective Oxidation of Allylic Alcohols

Allylic aldehydes are important intermediates in fine chemical synthesis, particularly flavours and fragrances that can be readily synthesised by the aerobic oxidation of the corresponding alcohol (Figure 5.19).[48] The preferential activation of $-H_2C-OH$ groups in the presence of olefinic groups in the molecule also poses a significant challenge for the development of selective heterogeneous catalysts.

Mechanistic understanding of these transformations over Pd-based catalysts and associated catalyst deactivation pathways[49] has been elucidated by TR-XPS studies of crotyl alcohol (a precursor to the preservative sorbic acid) over Pd(111) model surfaces.[50] At low temperature, crotyl alcohol adsorbs with the C=C bond parallel to the surface, analogously to allyl alcohol.[51] Figure 5.20 shows the C 1s TR-XPS of a reacting crotyl alcohol adlayer over clean Pd(111), which reveals that ~90% of the chemisorbed alcohol oxidises to surface-bound crotonaldehyde (with C=C and C=O functions coplanar with the surface) upon warming to 200 K.

Further heating drives undesired decarbonylation, resulting in propene and CO. The majority of the reactively formed propene dehydrogenates *via* propylidyne intermediates to form irreversibly bound carbonaceous deposits. The adsorbed CO (which persists to 430 K) and residual carbon may be responsible for the deactivation of metallic Pd clusters *via* site blocking. Surprisingly, pre-adsorbed oxygen suppresses crotonaldehyde decarbonylation, promoting its release from the surface (rather than combustion) and thus increasing reaction selectivity. These findings suggest that the selective oxidation of allylic alcohols over supported Pd catalysts requires careful temperature control and a high surface oxygen coverage and/or Pd dispersion.[52] Such predictions are borne out by our parallel *in operando* X-ray absorption investigations of dispersed Pd/C[53] and Pd/Al$_2$O$_3$[54] catalysts, which indeed show that surface palladium oxide is the active site and that deactivation follows the chemical reduction of PdO$_x$ to Pd^0.

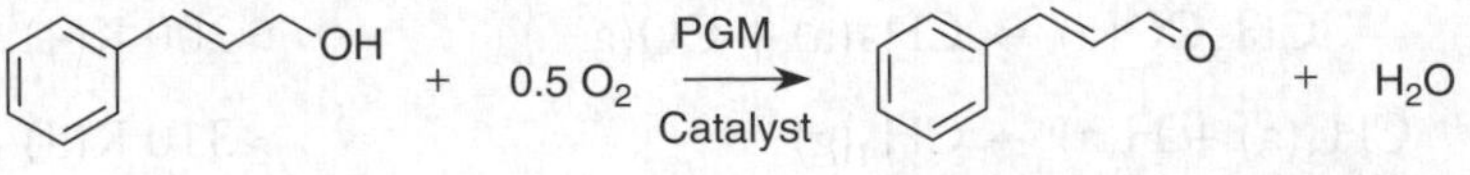

Figure 5.19 Aerobic oxidation of cinnamyl alcohol (PGM = platinum group metal).

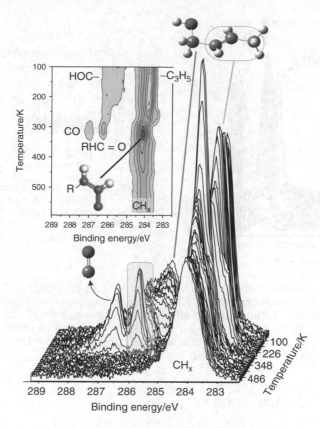

Figure 5.20 C 1s TR-XPS of a reacting crotyl alcohol adlayer over clean Pd(111). Reproduced with permission from Lee *et al.* (2007) [50]. Copyright (2007) American Chemical Society.

In an exciting breakthrough in this field, bimetallic AuPd catalysts were found to exhibit exceptional activity and selectivity towards the oxidative dehydrogenation of diverse alcohols over those achievable with Pd alone.[55] In order to understand the promotional effect of Au on the selox chemistry reaction of crotyl alcohol over Au/Pd(111), alloy systems were explored by TR-XPS (Figure 5.21).[56]

These measurements revealed that gold adlayers facilitate desorption of reactively formed crotonaldehyde, while also suppressing aldehyde decomposition, thus increasing product selectivity. The optimum alloy composition for crotyl alcohol conversion, with minimal decomposition of the resultant crotonaldehyde product, appears between AuPd$_4$ and AuPd$_7$.

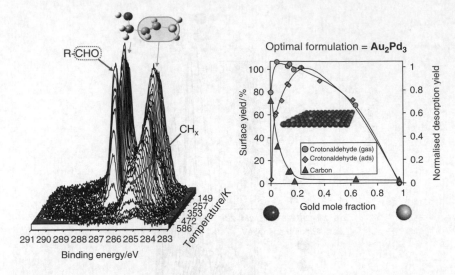

Figure 5.21 Summary of crotyl alcohol chemistry on Au-modified Pd(111) surfaces. (a) Temperature-programmed C 1s XP spectra of a saturated crotyl alcohol adlayer over an ultrathin 3.9 monolayer Au/Pd(111) film annealed to 973 K. (b) Strong promotional effect of alloyed Au on the selective oxidation of crotyl alcohol to crotonaldehyde over Pd(111). Reproduced with permission from Lee *et al.* (2009) [56]. Copyright (2009) Elsevier Ltd.

5.3.3 C–X Activation

Organohalides are widely employed as solvents, refrigerants and fungicides, in addition to being valuable building blocks in organic synthesis, *e.g.* for pharmaceuticals manufacture. Catalysts that enable facile C–X bond activation are necessary both for the destruction of chlorinated[57] or brominated compounds, to prevent their release into aquatic or atmospheric environments, and for chemical synthesis such as cross-coupling of alkyl/aryl halides under mild conditions.[58]

Palladium-catalysed cross-couplings (such as Heck arylation) and Suzuki–Miyaura reactions represent key technologies within academia and industry for the formation of C–C bonds. These are characterised by the coupling of an electrophilic species, *e.g.* organic halides, with a nucleophile, such as an alkene or organoboron, catalysed by Pd^0 in the form of either homogeneous organometallic Pd complexes or colloidal and supported Pd particles. The surface chemistry of aryl halides over Pd surfaces is thus of great relevance to this important class of reaction. Aryl halides are often investigated as a means of seeding surfaces with phenyl derivatives,[59–70] but couping to form biphenyl derivatives *via* the Ullman reaction can also be observed over Pd(111), Pd(110), Ag(111),[71] Cu(111)[72] and Au(111)[73] surfaces. Over Pd, the majority of phenyl fragments react with hydrogen to form benzene, but a smaller number are observed to couple to form biphenyl. While the (110) surface of Pd is found to be more active towards C–halogen bond cleavage, the (111) face is more efficient for ring coupling. In contrast, over Pt surfaces,[74–76] aryl halides are found to only form adsorbed phenyl moieties and halide atoms. The reaction of bromobenzene over Pt(111) has been studied by Fast-XPS and NEXAFS, which reveal that bromobenzene adopts a tilted geometry, with the ring plane at $60 \pm 5°$ to the surface.[77] Debromination occurs at ~300 K to yield a stable (phenyl) surface intermediate and atomic bromine, with further heating resulting in desorption of reactively formed H_2, C_6H_6 and HBr. However, there was no evidence for either biphenyl or Br_2 formation (Figure 5.22).

The surface chemistry of organochlorine species is also relevant to our understanding of the effect that anthropogenic sources of halohydrocarbons have on ozone depletion cycles. Industrial organochlorine species released at ground level are responsible for >70% of stratospheric chlorine, hence efficient catalysts for the ground-level stripping of halides from waste halohydrocarbons would have profound environmental benefits. In an effort to understand platinum group metal (PGM) hydrodechlorination catalysts, TP-XPS was used to investigate the thermal chemistry of 1,1,1-trichloroethane (TCA), a common commercial solvent until 2005, over Pt(111).[78] TCA adsorbs nondissociatively at 100 K, bonding through the three Cl atoms in a vertical geometry akin to ethylidyne. Between 120 and 200 K, around 60% of chemisorbed TCA undergoes rapid, stepwise C–Cl bond cleavage *via* metastable CH_3CCl_2 and CH_3CCl surface intermediates, liberating atomic chlorine and ethylidyne (Figure 5.23). The majority of ethylidyne decomposes at >350 K, resulting in CH_x moieties whose subsequent dehydrogenation deposits atomic carbon and removes most of the adsorbed chlorine as evolved HCl at between 350 and 420 K. The remaining ethylidyne (15%) reacts with H adatoms and desorbs as ethane. Although Pt(111) is very effective for low-temperature TCA dehydrochlorination under UHV, we identified the residual surface carbon and chlorine as potential poisons under continuous catalytic operation. Subsequent light-off measurements of commercial Pt/Al_2O_3 catalysts verified this hypothesis, and insight from TP-XPS enabled reaction optimisation with co-fed hydrogen to achieve efficient TCA hydrodechlorination to HCl (readily scrubbed from an effluent stream) and C_2H_6 at only 400 K.

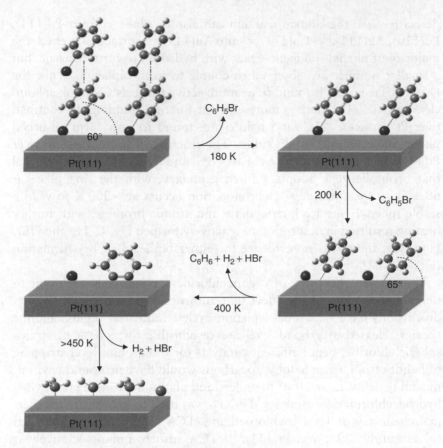

Figure 5.22 Hydrodebromination of bromobenzene over Pt(111). Reproduced with permission from Lee *et al.* (2007) [77]. Copyright (2007) American Chemical Society.

5.4 HIGH- OR AMBIENT-PRESSURE XPS

XPS is a very powerful technique for studying surface-catalysed reactions, particularly for the evolution of adsorbed surface species as a function of either coverage or temperature. High-resolution soft X-ray facilities available at third- and fourth-generation synchrotron facilities have enabled detailed kinetic investigation of surface chemical transformations to be undertaken. However, current investigations, particularly with relevance to heterogeneous catalysis, are limited by the requirement to work under UHV conditions, where the coverage of key reactively formed intermediates is often low.

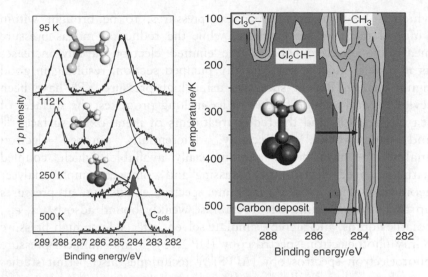

Figure 5.23 C 1s TP-XPS of TCA dehydrohalogenation over Pt(111), highlighting stepwise C–Cl scission and ethylidyne intermediate. Reproduced with permission from Lee *et al.* (2010) [79]. Copyright (2010) Royal Society of Chemistry.

The major technical difficulty associated with performing XPS at elevated pressure ($>1 \times 10^{-6}$ Torr) is associated with the short MFP of low-energy electrons employed in XPS. Attempts to study the effects of material processing in realistic environments currently involves the use of high-pressure cells in which samples can be exposed to reactive environments. This is generally informative in studies where irreversible changes in materials properties such as surface oxidation or reduction processes can be observed. However, the isolation of weakly chemisorbed reaction intermediates is particularly challenging as these may well be unstable when analysis is subsequently performed under UHV conditions.

Pioneering work in analyser and vacuum chamber design has paved the way for high-pressure XPS measurements to be undertaken. While spectral aquisition at elevated pressure entails a number of technical difficulties, including the operation of detectors and X-ray sources and the high pressure to be overcome, methods aimed at improving spectral sensitivity are the most challenging. When pressures of ~1 mbar are reached, the MFP of an electron is ~1 mm, and acquisition of an XP spectrum under such conditions requires sample analyser distances to be on this order of magnitude. Such analytical capabilities have been facilitated by the use of analysers fitted with small-aperture nose cones with differentially pumped lens sections. This allows the sample, which is held under millibar gas pressures, to be brought within 1 mm of the analyser entrance, while the reduction in gas pressure means the effective MFP of the emitted electron rapidly increases as it travels into the differentially pumped section, resulting in good signal quality. Recent successes using high-pressure XPS have been reviewed, and include studies of ice-melting processes, the surfaces of sea-salt aerosols and the redox treatments of metal-oxide surfaces[80] and nanoparticles.[81] Recently, differentially pumped electron analysers[82] have become commercially available, which, coupled with improved electrostatic focussing and compact sample analyser geometries (Figure 5.24), facilitate spectral acquisition at pressures up to 1 mbar,[83] sufficient to stabilise weakly bound adsorbates, *e.g.* volatile liquids, and thereby simulate solvent effects. Such high-pressure X-ray photoelectron spectroscopy (HP-XPS)[84] and ambient-pressure photoelectron spectroscopy (APPS)[85] techniques also permit studies over thermodynamically unstable catalyst surface structures/phases.[82] These developments mirror the general trend within the heterogeneous catalysis community towards *in operando* studies of catalytic phenomena under 'real-world' operating conditions. Future improvements in chamber design and increased flux at modern synchtron facilities should lead to the potential for higher-pressure, real-time measurements to be performed. In addition to catalytic applications, such capabilities open up opportunities for studies over ice–water interactions relevant to atmospheric chemistry and of biological systems in which maintenance of a background pressure of water reduces cell denaturation.

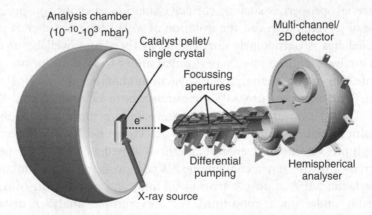

Figure 5.24 Typical instrumental set-up for high-pressure, time-resolved XPS measurements of heterogeneous catalysts. Reproduced with permission from Lee *et al.* (2010) [79]. Copyright (2010) Royal Society of Chemistry.

The following examples illustrate the power of HP-XPS in materials characterisation.

5.4.1 AP-XPS Studies of the Surface Chemistry of Oxidised Metal Surfaces

The major commercial route to ethylene oxide is the catalytic selective oxidation of ethene over $Ag/\alpha\text{-}Al_2O_3$. The underlying reaction mechanism has attracted significant industrial[86] and academic[87] interest, with the nature of the active oxygen species (molecular or atomic oxygen) and hydrocarbon intermediates (oxametallacycle[88,89] or π-bonded ethene[90]) being hotly debated. It is now widely accepted that atomic oxygen is the active oxidant, with more electron-deficient (electrophilic) species favouring epoxidation and more basic (nucleophilic) species promoting combustion.[91] The ability to generate and stabilise oxidised PGM surfaces requires high background pressures of O_2 to be maintained while spectroscopic measurements are performed. Such HP-XPS was first reported in the late 1970s, when interaction of oxygen on a Ag surface was studied at pressures of up to 2 mTorr,[82] revealing the presence of three unique oxygen environments attributed to atomic, surface and subsurface oxygen. However, these high-pressure measurements were quite demanding on the instumentation and it is only with recent advances in spectrometer design that new interest in HP-XPS has been ignited.

More recent HP-XPS[92] has shown that although both nucleophilic and electrophilic oxygen reside on the catalytically active surface, the yield of ethylene oxide only correlates with the concentration of electrophilic oxygen; basic oxygen adatoms (which spectroscopically resemble bulk Ag_2O[93]) promote ethene combustion to CO_2 and H_2O (Figure 5.25). The same study also showed that silver is inactive towards oxidation at $\sim$420 K due to site blocking by carbonates and carbonaceous residues.

AP-XPS studies have also demonstrated the dynamic nature of Pd surface oxidation in near-ambient conditions with varying O_2 pressures.[94,95] Figure 5.26 shows how the Pd $3d$ and the corresponding O $1s$ high-pressure XP spectra evolve following oxygen exposure, leading to the formation of different oxide phases. Analysis of the phase diagram reveals that bulk oxidation is a two-step process, starting with the metastable growth of the surface oxide into the bulk and continuing

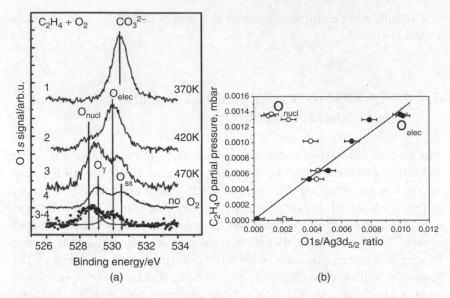

Figure 5.25 HP-XPS, showing (a) evolving surface oxygen species during ethene epoxidation over Ag foil and (b) the correlation between ethylene oxide yield and electrophilic $O_{(a)}$. Reproduced with permission from Bukhtiyarov *et al.* (2006) [92]. Copyright (2006) Elsevier Ltd.

with a first-order transformation to PdO. Likewise, AP-XPS studies of Pt(111) oxidation show that oxide growth commences *via* a PtO-like surface oxide which coexists with chemisorbed oxygen, with an ultra-thin α-PtO_2 trilayer that may nucleate at steps or defects and which is a precursor to bulk oxidation.[96]

Methane oxidation over PdO surfaces has also been investigated, with a strong dependence of methane combustion activity on PdO content being observed. Over the temperature range 530–650 K, PdO seeds are thought to grow within the Pd_5O_4 surface oxide, with the reaction rate increasing exponentially with temperature as PdO forms. Above 650 K, the oxide phase decomposes and catalytic activity decreases. At still higher temperatures, >700 K, reaction over the metal is observed, with combustion occurring on the almost clean metal surface and the near-surface region saturated with a dissolved oxygen species.[97]

Methanol oxidation over oxidised Cu surfaces has also been investigated by AP-XPS.[98] Cu is an effective catalyst for the oxidative dehydrogenation of alcohols to their respective aldehydes. The oxidation of methanol can proceed *via* either partial oxidation to formaldehyde or total oxidation. AP-XPS demonstrates that under experimental conditions of 400 °C and 0.6 mbar, the active catalyst for selective

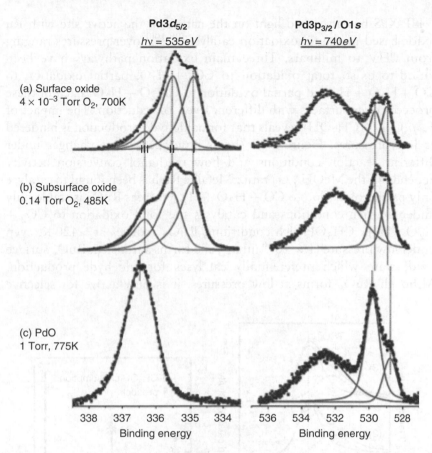

Figure 5.26 Pd $3d_{5/2}$ (left) and O $1s$/Pd $3p_{3/2}$ (right) XPS regions, corresponding to different stages in the oxidation of Pd(111): (a) $\sqrt{6} \times \sqrt{6}$ surface oxide; (b) 'subsurface oxide'; and (c) bulk PdO. Reproduced with permission from Ketteler *et al.* (2005) [94]. Copyright (2005) American Chemical Society.

oxidation to formaldehyde comprises a metallic Cu surface with a subsurface oxygen species. Pure metal is not an active catalyst for the methanol oxidation reaction, but subsurface oxygen atoms are suggested to perturb the activity of the metallic surface by introducing strain in the Cu surface lattice. The surface and near-surface regions vary strongly as a function of the oxygen–methanol ratio, with the concentration of subsurface oxygen species correlating with formaldehyde yield. Bulk oxide formation has been found to be detrimental to formaldehyde production. Thus, the catalytic activity of metals is not determined by their surface properties alone: the composition of the subsurface region must also be taken into account.

HP-XPS has also shed light on the nature of the active site with Ru oxide-based methanol oxidation catalysts[99,100] over pressures ranging from UHV to millibars. Three main oxidation pathways have been found to exist: total oxidation to $CO_2 + H_2O$, partial oxidation to $CO + H_2O + H_2$ and partial oxidation to $CH_2O + H_2O$. Each of these proceeds over surfaces with different levels of reduction. The impact of T, $p(O_2)$, $p(CH_3OH)$ reveals that formaldehyde production is hindered at low pressures. Figure 5.27 shows how the surface changes under different reaction conditions and how methanol conversion activity depends on the $MeOH : O_2$ ratio. Metallic Ru has been found to catalyse only partial oxidation to $CO + H_2O + H_2$, whereas RuO_2 is stable only under O_2-rich conditions and catalyses the total oxidation to $CO_2 + H_2O$. Under CH_3OH-rich conditions, RuO_2 reduces at >420 K, even under low pressure (1×10^{-6} mbar), and forms a transient RuO_x surface oxide state which preferentially catalyses formaldehyde production. Although RuO_x forms at low pressures, it is only active for selective

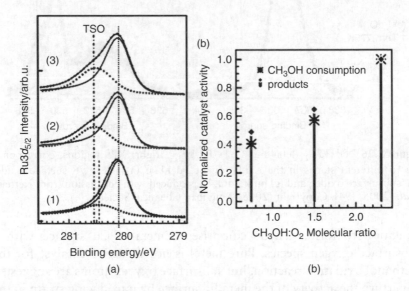

Figure 5.27 (a) Deconvoluted Ru $3d_{5/2}$ spectra of the steady-state systems measured in the temperature range 570–620 K, representing the regimes of oxygen potential favouring oxidation to (1) $CO + H_2 + H_2O$, (2) $CH_2O + H_2O$ and (3) $CO_2 + H_2O$. The dotted and dashed lines indicate the Ru_{bulk}, $Ru–O_{ad}$ and TSO (RuO_x) components. (b) Catalyst activity at 580–600 K for the oxidised precatalyst state as a function of the $CH_3OH : O_2$ mixing ratio, normalised against the product yield and the methanol consumption measured for a $CH_3OH : O_2$ mixing ratio of 2.3. Total pressure: 2.4×10^{-1} mbar. Reproduced with permission from Blume *et al.* (2007) [99]. Copyright (2007) Elsevier Ltd.

oxidation at 10^{-1} mbar. This pressure-dependent behaviour has been ascribed to oxygen stabilising the lifetime of the CH_2O intermediate at the surface.

5.4.2 Selective Hydrogenation

Hydrogenation is considered to be one of the most important catalytic routes in organic synthesis, with the hydrogenations of olefins, ketones, carboxylic acids and nitriles among the most widely employed processes in the fine chemicals industry. Heterogeneous catalytic hydrogenation by gaseous H_2 using supported Pd, Ni or Pt catalysts is one of the cleanest methods by which to reduce a chemical bond. An improved understanding of the surface chemistry of multifunctional hydrocarbon hydrogenation is of great importance to the development of clean catalytic technologies.

Palladium is particularly effective in the selective hydrogenation of alkynes to alkenes,[101] is an important commercial process for purifying olefin feedstocks and undergoes pressure- and temperature-dependent hydrogen dissolution, forming α- and β-hydride phases, which play an important role in hydrogenation.[102] Under UHV, only H adatoms can be stabilised; however, high-equilibrium hydrogen pressures can also stabilise α-PdH (4 mbar) and β-PdH ($\sim 7 - 10$ mbar). The β-phase can be generated by exposing Pd to 1 bar H_2 at 338 K then cooling it and reducing the pressure to 4 mbar. The role of subsurface and bulk hydrides in trans-2-pentene hydrogenation over Pd foil and Pd(111) single-crystal model catalysts has been investigated by HP-XPS.[103] Pentane is efficiently produced over the foil via an bound alkene surface species, whereas Pd(111) is virtually inactive. These striking differences are linked to the presence of surface and subsurface carbon, with higher PdC concentrations promoting activity. Pd α-hydride formation is postulated as a prerequisite for efficient trans-2-pentene hydrogenation, with surface hydrogen rate limiting at >500 K due to hydride decomposition and weak H_2 adsorption.

The importance of surface carbon is further highlighted in the selective hydrogenation of 1-pentyne over Pd(111) and foil surfaces. Hydrogenation has been found to occur via a σ-bound alkene surface species, with clean Pd foil favouring total hydrogenation to pentane, but as the reaction proceeds, selectivity switching to $\sim 98\%$ pentene.[104]

While Pd foil has been found to exhibit good activity for pentane formation, Pd(111) is virtually inactive. *In situ*, depth-profiling XPS indicates that the active surface comprises a Pd–C phase around three layers deep, which serves to inhibit hydride formation and the emergence of unselective bulk-dissolved hydrogen to the surface. Carbon is believed to both alter the *d*-band centre and expand the Pd lattice, which may result in the generation of sites that are more active for hydrogenation.

In a further elegant series of experiments combining prompt gamma activation analysis and *in situ* XPS studies on both practical Pd black and model Pd foil samples,[105,106] Teschner *et al.* demonstrated that selective hydrogenation of acetylene, propyne and 1-pentyne follows the genesis of surface PdC arising from fragmentation of alkyne reactants during the early stage of reaction (Figure 5.28).

Although Pd(111) is less active for 1-pentyne hydrogenation than for foil surfaces, it exclusively yields 1-pentene. This higher alkene

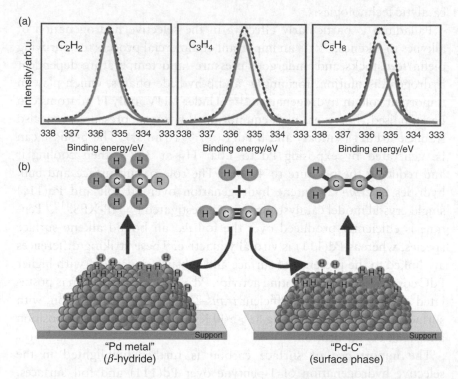

Figure 5.28 C 1*s* HP-XPS spectra (a), revealing the formation of total or selective ethyne hydrogenation products over the surface palladium carbide and hydride surface shown schematically in (b). Reprinted from Teschner *et al.* (2008) [105], with permission from AAAS.

selectivity is at odds with measurements of dispersed catalysts possessing large (111)-orientated Pd particles, which preferentially favour alkanes[107,108] and may reflect the difficulty in current HP-XPS configurations of achieving sufficiently high pressures to stabilise the β-hydride phase, which favours total hydrogenation. Selective alkyne hydrogenation thus reflects a subtle interplay between palladium surface carbide and hydride formation.

5.4.3   HP-XPS Studies of Core–Shell Nanoparticulate Materials

HP-XPS is a powerful technique by which to study dynamic changes in catalyst surface composition under reactive environments and to develop a more accurate understanding of the nature of the active site. Application to the study of surface restructuring of bimetallic Rh–Pd and Pt–Pd core–shell nanoparticles[13] under cycling CO/NO environments (50 wt% Pd in both cases) reveals significant changes in surface composition when the reaction conditions are changed from oxidising to reducing (Figure 5.29). Under oxidising conditions, Rh segregation is observed over an Rh–Pd nanoparticle, whereas under reducing conditions migration of Pd to the surface is observed.

Studies of Ag–Cu bimetallic catalysts[109] in ethylene epoxidation also show that these particles undergo surface structural changes under reactive environments. Preferential segregation of Cu is observed during annealing under vacuum or under an atmosphere of O_2 or an ethylene/ O_2 mixture. HP-XPS experiments have revealed the formation of surface copper oxide on the outer surfaces of the particles, and that a previously assumed Cu–Ag surface alloy is not the active catalytic surface for ethane epoxidation.

5.5   APPLICATIONS TO INORGANIC MATERIALS

5.5.1   Bimetallic Nanoparticles

The thermal evolution of titania-supported Au shell–Pd core bimetallic[110] nanoparticles, prepared *via* colloidal routes, has been

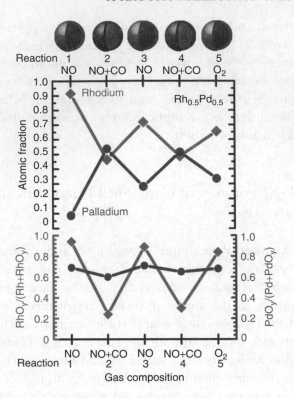

Figure 5.29 Surface termination of Rh–Pd core–shell nanoparticles under NO/CO environments. Reproduced with permission from Tao *et al.* (2008) [13]. Copyright (2008) AAAS.

investigated by *in situ* XPS. Annealing *in vacuo* or under inert atmosphere rapidly pyrolyses the citrate ligands, but induces only limited Au/Pd intermixing and particle growth at <300 °C (Figure 5.30). Higher temperatures promote more dramatic alloying, accompanied by significant sintering and surface roughening. These changes are mirrored by the nanoparticle-catalysed liquid-phase selective aerobic oxidation of crotyl alcohol to crotonaldehyde; palladium surface segregation enhances both activity and selectivity, with the most active surface alloy attainable containing ~40 atom% Au.

A synchrotron XPS-based depth-profiling measurement was used to study the atomic arrangement of Pt–Pd nanoparticles[111] under thermal treatment. From a knowledge of the photoemission cross-section and the MFP of the photoelectrons for the selected photon energy excitation, the distribution of Pt and Pd could be determined. When $Pt_{0.7}Pd_{0.3}$ nanoparticles were heated under vacuum, rearrangement of Pt and Pd

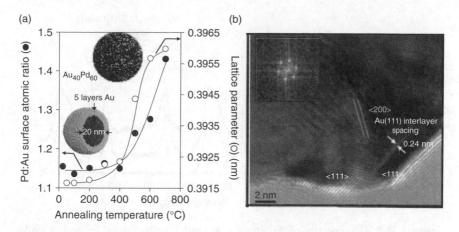

Figure 5.30 (a) Surface atomic Pd : Au ratio of Au shell–Pd core/TiO_2 NPs from *in situ* XPS and bulk lattice parameter from powder XRD as a function of annealing temperature. (b) Atomic-resolution HRTEM images of as-prepared (Fourier transform inset) Au shell–Pd core/TiO_2 NPs. Reproduced with permission from Lee *et al.* (2010) [110]. Copyright (2010) Elsevier Ltd.

atoms inside the nanoparticles led to a core–shell structure. On the basis of the XPS 'depth profiling', the core radius and the shell thickness of the annealed nanoparticles could be estimated. The presence of Pd atoms at the surface of Pt–Pd nanoparticles is expected from a theoretical point of view, due to the lower surface free energy, γ, of Pd and the higher cohesive energy of Pt. The observation of Pd segregation under vacuum annealing suggests there is a lower heat of segregation for Pd in $Pt_{0.7}Pd_{0.3}$ nanoparticles when compared to the corresponding bulk alloys.

Differential charging of core–shell Au@SiO_2 nanoparticles has been investigated by deliberately applying external voltage stress to the sample while recording XPS spectra of the nanoparticles.[112] This is proposed as a simple noncontact method by which to determine the location and the chemical identity of the medium in which the charges reside within the nanostructures. Charging can be identified from a shift in the measured binding energy of the corresponding XPS peak. While bare Au nanoparticles exhibit no measurable binding energy shift in the Au $4f$ peaks, the Au@SiO_2 system shows shifts in both the Au $4f$ and the Si $2p$ peaks, which are attributed to charging under the applied potential. Positive charges arise because core holes are created by the very fast ($<10^{-12}$ s) photoelectron emission process and subsequent filling of these holes by outer electrons. The holes end up in the valence band

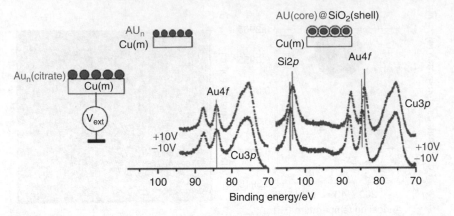

Figure 5.31 XPS spectra of Si 2*p*, Au 4*f* and Cu 2*p* peaks, recorded under application of a −10 and +10 V DC voltage stress to Au or Au(core)@SiO₂ (shell) nanoparticles dispersed on a Cu sample holder. Reproduced with permission from Tunc *et al.* (2005) [112]. Copyright (2005) American Chemical Society.

and are stabilised in the large band gap of the silicon dioxide. Later, these holes are neutralised, on the timescale on which they are measured (0.01–10.0 s), by the low-energy electrons falling onto the sample, the flux of which can be controlled by the applied external potentiality, thereby allowing the charging and discharging of the nanocapacitors to be controlled.

Using the shift in the Au 4*f* peaks (Figure 5.31), the capacitance of the 15 nm gold (core)/6 nm silica (shell) nanoparticle/nanocapacitor is estimated as 60 aF.[112] This simple method of controlling the charging, by application of an external voltage stress during XPS analysis, enables the charges developed on surface structures to be detected and quantified in a noncontact fashion.

5.5.2   XPS Studies of Heteropolytungstate Clusters

Tightening legislation on the generation of toxic waste during the synthesis of many fine and speciality chemicals is driving industry to implement alternative clean technologies.[113] Solid acids are environmentally benign alternatives to homogeneous acids such as H_2SO_4, $AlCl_3$, $ZnCl_2$ and BF_3, traditionally employed in organic transformations, which offer the advantages of facile product separation and opportunities to operate in continuous flow reactors. Among the heterogeneous acid catalysts

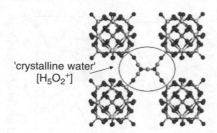

'crystalline water'
$[H_5O_2^+]$

Figure 5.32 Interaction of phosphotungstic acid ($H_3PW_{12}O_{40}$) clusters *via* $[H_5O_2]^+$. Central atom = P, light grey = W and dark grey = 0.

under investigation, heteropoly acids (HPAs) have attracted significant attention, since they possess high Brønsted acid strengths (approaching the superacidic region) and tuneable acidity.[114,115]

HPAs are polyoxometalate inorganic cage structures with the general formula $H_nM^{m+}X_{12}O_{40}$, where M is the central atom (typically Si or P) and X the heteroatom (typically W or Mo). The acidity in HPA crystallites arises due to crystalline water residing between adjacent clusters, as shown in Figure 5.32.

However, HPAs have inherently low surface areas ($1-5$ m^2 g^{-1}), so are often employed *via* support on a porous carrier material such as carbon, silica or alumina.[116,117] During the immobilisation of HPA on oxide surfaces, it is important to achieve high dispersions in order to maximise the accessibility of the acid sites. The evolution of $H_3PW_{12}O_{40}$ (HPW) adlayers over a porous, hydroxylated silica can be probed effectively by XPS, allowing identifation of monolayer and multilayer species.

Following deposition of HPW onto silica, W $4f$ XPS (Figure 5.33 inset) reveals that the spectra can be decomposed into two sets of spin–orbit split $4f_{7/2}$ and $4f_{5/2}$ doublet components in a constant $3:1$ ratio.[118,119] Interaction of the HPW cluster with a surface will break the symmetry of the molecule, so it seems reasonable to assume that the high-binding-energy components are attributable to the three W=O units in contact with the surface. The more intense, low-binding-energy component is at the same energy as bulk HPW, so is caused by the nine WO_x units pointing away from the interface. The observed chemical shift for the terminal W=O groups coordinating to the surface ('facing down') can be understood in terms of their experiencing more efficient core–hole screening than the remainder, reducing final-state contributions and thus lowering their W $4f$ binding energy relative to those W=O groups 'facing away', which reside in a bulk-like environment.

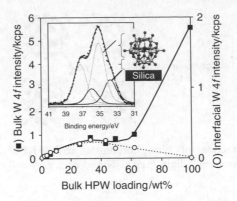

Figure 5.33 Inset shows W $4f$ spectra of monolayer HPW species. Main image shows the evolution of interfacial and bulk W $4f$ signal. Reproduced with permission from Wilson and Lee (2010) [27]. Copyright (2010) Royal Society of Chemistry.

Analysis of the component areas as a function HPW loading shows that these components are retained for compositions of up to 40 wt%, after which the intensity of the main low-binding-energy components increases. These observations suggest that the growth of the HPW clusters proceeds in a layerwise manner, with the first monolayer completed by ~40 wt%, after which multilayer growth occurs *via* the formation of three-dimensional (3D) crystallites and the interfacial signal is attenuated.

The importance of these observations becomes apparent when HPWs are employed as heterogeneous catalysts in polar and nonpolar chemistries. Nonpolar reactants are found to react only over the surface of the HPW units, whereas polar reactants can penetrate into the bulk of the structure, according to Figure 5.34. Hence, for optimum performance in reactions with nonpolar organic molecules, such as alkylation or terpene isomerisation, a monolayer dispersion of the catalytically active HPW phase is required. Stability of the monolayer requires silica supports with high hydroxyl coverage, as these stabilise the inteaction of W=O groups with the surface.

The high solubilities of HPAs in polar media makes such supported variants unsuitable for the catalysis of liquid-phase reactions involving substrates such as alcohols, ethers or esters.[120] Alkali exchange of HPAs to form partially substituted salts (*e.g.* $Cs_xH_{(1-x)}PW_{12}O_{40}$) (Figure 5.35) induces a remarkable change in their morphology and solubility in water.[121] For example, $M_xH_{(1-x)}PW_{12}O_{40}$ salts with large monovalent ions such as Cs^+, $[NH_4]^+$ and Ag^+ are insoluble in water and have higher

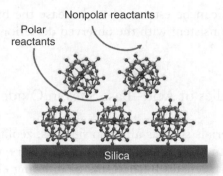

Figure 5.34 Surface- and bulk-catalysed reactions of polar and non-polar reactants over supported HPAs.

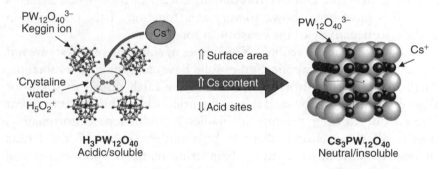

Figure 5.35 Cs^+ exchange of crystalline water between Keggin cages to form $Cs_xH_{(3-x)}PW_{12}O_{40}$ moieties that are insoluble in polar media. Complete exchange yields neutral $Cs_3PW_{12}O_{40}$ salt. Reproduced with permission from Wilson and Lee (2010) [27]. Copyright (2010) Royal Society of Chemistry.

surface areas than undoped $H_3PW_{12}O_{40}$.[122] The precise interaction of the cation with the Keggin anions (i.e., the depronated HPA) is unclear (Figure 5.35), but the most plausible theory suggests that $H_3PW_{12}O_{40}$ is deposited on core ultrafine crystallites of $Cs_3PW_{12}O_{40}$.[123] Detailed XPS studies of the evolution of $Cs_xH_{(1-x)}PW_{12}O_{40}$ species reveal that there is a significant deviation between surface and bulk Cs−W ratios at low Cs content. This deviation has been proposed to be in line with a structural model, where the Cs signal from a $Cs_3PW_{12}O_{40}$ core is attenuated by surface $H_3PW_{12}O_{40}$ clusters. Indeed, by taking the inelastic escape depth for the Cs $3d$ photoelectron to be 0.7 nm and the thickness of an $H_3PW_{12}O_{40}$ cluster to be ~1 nm, an external layer of Keggin clusters can be estimated to decrease the bulk Cs signal by ~75%, which is consistent with the observed deviation.[124]

5.5.3 XPS Studies of Acid–Base Sites in Oxide Catalysts

Mixed-oxide materials such as alumino silicates, zeolites and hydrotalcites possess acid and base sites in their surface layers, which arise from polarisation of surface sites by elements of differing electronegativities or cation–anion vacancies.[125,126] The change in electron density in the surface layer associated with the acid (decreased electron density at cation) or base (increased electron density at anion) site formation should result in binding energy shifts that are detectable by XPS. Indeed, the electronic structure and electron binding energy of framework elements in zeolites have been shown to vary as a function of the composition and electronegativity of the constituent ions.[127]

In contrast to acidic zeolites, XPS studies of solid base catalysts are not widely reported. The application of solid base catalysts such as alkaline-earth oxides,[128] hydrotalcites,[129] alkaline-earth doped mesoporous silicas[130] and resins[131] and even dolomitic rock[132] has attracted great interest due to their promise in biodiesel production. Unfortunately, despite all these efforts, there is little understanding of the effects of basicity on catalytic activity, hampering material optimisation and commercial exploitation of these materials.

MgO is an inexpensive and readily available solid base catalyst, but its basic properties are very sensitive to their preparation route.[133] To date, there is no simple way to characterise surface basicity, which is intimately linked to catalyst reactivity in solid base-catalysed transesterifications. In order to improve understanding of factors influencing solid base catalysts for biodiesel synthesis, noninvasive methods of determining basicity are required. As Lewis base strength is related to the O^{2-} donor ability, it should also correlate with electron density and $O\,2p$ binding energy. However, as outlined in Section 5.1, small shifts in XP binding energies must be treated with caution due to energy referencing accuracy, particularly when comparing results from different research groups. Studies of metal oxides have shown that the oxygen Auger parameter increases with the polarisability of metal–oxide bonds.[134] Auger parameter measurements should therefore provide a measure of the oxide ion electron density, which is independent of energy referencing problems, and a simple spectroscopic method for the quantitative determination of surface basicity.

Basicity in MgO (and most metal oxides) arises due to the presence of O^{2-} and OH sites. The polarisability of O^{2-} is directly related to electron density and thus the basicity of the surface. This is influenced by the number of Mg^{2+} nearest neighbours in the ionic lattice and will be affected by the presence of defects (missing cations) or surface termination (Figure 5.36a and b). In this instance, the number of nearest-neighbour Mg^{2+} surrounding each O^{2-} decreases from five in the (100) to four in the (110) to three in the (111) surface, so it would be expected that O^{2-} basicity will increase from (100) < (110) < (111). Indeed, combined TEM and Auger parameter measurements on MgO nanoparticles reveal that the Auger parameter increases linearly with particle size over the range 3–10 nm.[135] Corresponding catalytic data also show that activity in transesterification correlates with base strength, and thus the Auger parameter measurements provide a non-invasive measurement of basicity.

The Auger parameter can also be employed to characterise solid acid materials, as illustrated by the application of Al *KLL* measurements (see Table 5.3) to probe the coordination state of Al in β-zeolites. The chemical state of Al was established by deconvolution of the Al *KLL* Auger transition, which revealed that tetrahedral, octahedral and tricoordinated Al could be differentiated.[136]

XPS chemical shifts of adsorbed probe molecules can also be employed to determine the acid and base properties of mixed oxide catalysts. Typical probe molecules used include pyridine,[137] NH_3,[138] diaminoethane[139] and SO_2. Binding energy shifts are characteristic of the interaction of the probe molecule with the surface acid or base

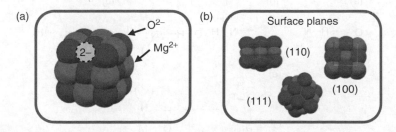

Figure 5.36 (a) Defect and base site formation in MgO. (b) Surface terminations of MgO: reduced $O^{2-}Mg^{2+}$ coordination increases the polarisability and basicity of O^{2-}.

Table 5.3 Al *KLL* Auger parameter values (in units of eV) for different Al coordination environments.

Compound	Al^{3+} (3-coord.)	Al^{3+} (4-coord.)	Al^{3+} (6-coord.)
Zeolite-β[136]	1458.5 ± 0.5	1460.2 ± 0.4	1465.4 ± 0.3
Mordenite[a]	1458.7–1459.3	1459.8–1460.2	1460.9–1465.5

[a]M. Remy, M. Genet, P. Notte, P.F. Lardinois and G. Poncelet, *Microporous Mater.*, **2**, 7 (1993).

site. Interaction of the probe with Lewis or Brønsted acid sites also gives rise to a different surface species, which can be identified by the chemical shift.

When NH_3 binds to metal oxide surfaces, a range of species can form depending on the interaction with Lewis base (O^{2-}), Lewis acid (M^{n+}), hydroxyl (OH) and Brønsted acid (OH^+) sites, as illustrated in Figure 5.37. In the presence of strong acid–base couples, N–H bond cleavage of NH_3 can also occur. N 1s XP binding energies for most stable surface species – $[NH_2]^-$, $[NH_3-M]^{n+}$ and $[NH_4]^+$ – formed on NH_3 adsorption are reported to be in the range 399, 400.5 and 402.5 eV, respectively.[139] Adsorption of pyridine to Brønsted and Lewis sites in a similar manner also gives rise to characteristic N 1s binding energies for pyridinium ion and datively bound species at 405.6 and 398.8 eV, respectively.[138]

Sulfur dioxide can also form a wide range of adsorbed species characteristic of interaction with different surface sites.[139] S^{2-}, SO_2, $[SO_3]^{2-}$ and $[SO_4]^{2-}$ are typically observed at around 162, 166, 167.5 and 169.2 eV. Observation of SO_2 is characteristic of coordination to Lewis

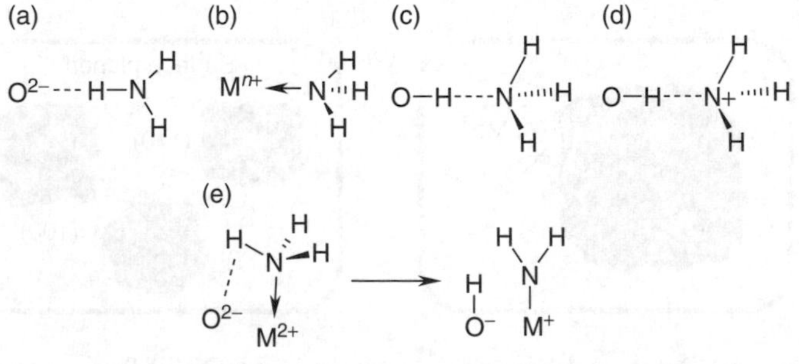

Figure 5.37 Surface species formed upon adsorption of NH_3 on (a) Lewis base, (b) Lewis acid, (c) hydroxyl and (d) Brønsted acid sites of metal oxides. (e) Strong acid–base couples can also lead to NH_3 N–H bond cleavage.

acid sites, whereas formation of $[SO_3]^{2-}$ and $[SO_4]^{2-}$ is characteristic of surface basicity and requires mono- or bidentate interaction with surface O^{2-} or OH sites.

5.6   CONCLUSION

The widespread availability of high-brilliance synchrotron sources, coupled with improved photoelectron detection capabilities, has facilitated *in situ* XPS investigations of diverse catalytic processes over (predominantly) metal single crystals. Despite limited time resolution and pressure restrictions, we hope that this review has shown that such XPS studies can provide important information on the nature, concentration and stability of key surface reaction intermediates, poisons and promoters, which feed into catalyst synthesis and optimisation programmes for a number of selective transformations. The pressure gap remains an important barrier to bridge, with the majority of studies to date conducted under UHV conditions, often at low temperatures in order to stabilise reactive intermediates. However, looking ahead, as the construction of ambient-pressure beamlines continues around the world and the sensitivity of laboratory XPS instruments continues to improve, we anticipate that the coming decade will see HP-XPS on the millisecond timescale become a commonplace, powerful tool in the design of commercial solid catalysts for clean chemical technologies. More widespread use of AP-XPS *via* laboratory systems will also advance the range of applications.[140] Likewise, use of ambient-pressure photoelectron spectroscopy/droplet train apparatus will aid in investigations of the nature and heterogeneous chemistry of liquid–vapour interfaces.[141]

REFERENCES

[1] G. Ertl, H. Knozinger, F. Schuth and J. Weitkamp (Eds), *Handbook of Heterogeneous Catalysis*, 2nd edition, Wiley-VCH, Weinheim, 2008.

[2] K. Seigbahn, C. Nordling, A. Fahlman, R. Nordberg, K. Hamrin, J. Hedman, G. Johansson, T. Bergmark, S.-E. Karlsson, I. Lindgren and B. Lindberg, *ESCA Atomic Molecular and Solid State Structure Studied by Means of Electron Spectroscopy*, Almqvist and Wiksells, Uppsala, 1967.

[3] K. Seigbahn, *Rev. Mod. Phys.*, **54**, 709 (1982).

[4] G. D. Briggs and M. P. Seah, *Practical Surface Analysis*, John Wiley & Sons, Ltd, New York, NY, 1990.

[5] NIST X-Ray Photoelectron Spectroscopy Database, Version 3.5 (National Institute of Standards and Technology, Gaithersburg, 2003), http://srdata.nist.gov/xps/ (last accessed 8 Apr 2013).

[6] M. P. Seah and W. A. Dench, *Surface Interface Anal.*, **1**, 2 (1979).

[7] C. J. Powell, A. Jablonski, I. S. Tilinin, S. Tanuma and D. R. Penn, *J. Electron Spectrosc. Relat. Phenom.*, **98–99**, 1 (1999).

[8] C. S. Fadley, *J. Electron Spectrosc. Relat. Phenom.*, **178–179**, 2 (2010).

[9] S. Tanuma, C. J. Powell and D. R. Penn, *Surf. Interface Anal.*, **43**, 689 (2011).

[10] C. J. Powell and A. Jablonski, *Nucl. Instrum. Methods Phys. Res., Sect. A*, **601**, 53 (2009).

[11] K. Kobayashi, *Nucl. Instrum. Methods Phys. Res., Sect. A*, **601**, 32 (2009).

[12] M. El-Gomati, F. Zaggout, H. Jayacody, S. P. Tear and K. Wilson, *Surf. Interface Anal.*, **37**, 901 (2005).

[13] F. Tao, M. E. Grass, Y. Zhang, D. R. Butcher, J. R. Renzas, Z. Liu, J. Y. Chung, B. S. Mun, M. Salmeron and G. A. Somorjai, *Science*, **322**, 932 (2008).

[14] F. J. Himpsel, F. R. McFeely, A. Taleb-Ibrahimi, J. A. Yarmoff and G. Hollinger, *Phys. Rev. B*, **38**, 6084 (1988).

[15] T. Hattori, *Crit. Rev. Solid State Mater. Sci*, **20**, 339 (1995).

[16] V. I. Nefedov, *XPS of Solid Surfaces*, VSP, Utrecht, Netherlands, 1988.

[17] A. P. Grosvenor, S. D. Wik R. G. Cavell and A. Mar, *Inorg. Chem.*, **44**, 8988 (2005).

[18] K. Wilson and A. F. Lee, in *Handbook of Green Chemistry – Green Catalysis*, P. Anastas and R. Crabtree (Eds), John Wiley & Sons, Ltd., Chichester, 2009.

[19] A. Kotani, *J. Electron Spectrosc. Relat. Phenom.*, **100**, 75 (1999).

[20] A. E. Bocquet, T. Mizokawa, K. Morikawa and A. Fujimori, S. R. Barman, K. Maiti, D. D. Sarma, Y. Tokura and M. Onoda, *Phys. Rev. B*, **53**, 1161 (1996).

[21] G. van der Laan, C. Westra, C. Haas and G. A. Sawatzky, *Phys. Rev. B*, **23**, 4639 (1981).

[22] G. Schön, *Surf. Sci.*, **35**, 96 (1973).

[23] J. Kawai, S. Tsuboyama, K. Ishizu, K. Miyamura and M. Saburi, *Anal. Sci.*, **10**, 853 (1994).

[24] M. C. Biesinger, B. P. Payne, L. W. M. Lau, A. Gerson and R. St. C. Smart, *Surf. Interface Anal.*, **41**, 324 (2009).

[25] K. Takubo, T. Mizokawa, S. Hirata, J.-Y. Son, A. Fujimori, D. Topwal, D. D. Sarma, S. Rayaprol and E.-V. Sampathkumaran, *Phys. Rev. B*, **71**, 073 406 (2005).

[26] T. Fujii, F. M. F. de Groot, G. A. Sawatzky, F. C. Voogt, T. Hibma and K. Okada, *Phys. Rev. B*, **59**, 3195 (1999).

[27] K. Wilson and A. F. Lee, *Applications of XPS to the Study of Inorganic Compounds, Specialist Periodical Reports, Vol. 41 – Spectroscopic Properties of Inorganic and Organometallic Compounds: Techniques, Materials and Applications*, J. Yarwood, R. Douthwaite and S. Duckett (Eds), RSC Publishing, Cambridge, 2010.

[28] G. Moretti, *J. Electron Spectrosc. Relat. Phenom.*, **95**, 95 (1998).

[29] L. Gregoratti, A. Barinov, E. Benfatto, G. Cautero, C. Fava, P. Lacovig, D. Lonza, M. Kiskinova, R. Tommasini, S. Mahl and W. Heichler, *Rev. Sci. Instr.*, **75**, 64 (2004).

[30] A. Barinov, P. Dudin, L. Gregoratti, A. Locatelli, T. O. Mentes, M. A. Nino and M. Kiskinova, *Nucl. Instrum. Methods Phys. Res., Sect. A*, **601**, 195 (2009).

[31] G. A. Somorjai, H. Frei and J. Y. Park, *J. Am. Chem. Soc.*, **131**, 16 589 (2009).

[32] D. F. Ogletree, H. Bluhm, E. D. Hebenstreit and M. Salmeron, *Nucl. Instrum. Methods Phys. Res., Sect. A*, **601**, 151 (2009).

[33] A. F. Lee, K. Wilson, R. L. Middleton, A. Baraldi, A. Goldoni, G. Paolucci and R. M. Lambert, *J. Am. Chem. Soc.*, **121**, 7969 (1999).

[34] A. Baraldi, G. Comelli, S. Lizzit, M. Kiskinova and G. Paolucci, *Surf. Sci. Rep.*, **49**, 169 (2003).

[35] A. Baraldi, *J. Phys.: Condens. Matter*, **20**, 093 001 (2008).

[36] A. Knop-Gericke, E. Kleimenov, M. Havecker, R. Blume, D. Teschner, S. Zafeiratos, R. Schlogl, V. I. Bukhtiyarov, V. V. Kaichev, I. P. Prosvirin, A. I. Nizovskii, H. Bluhm, A. Barinov, P. Dudin and M. Kiskinova, *Adv. Catal.*, **52**, 213 (2009).

[37] E. Antolini, *J. Power Sources*, **170**, 1 (2007).

[38] G. A. Deluga, J. R. Salge, L. D. Schmidt and X. E. Verykios, *Science*, **303**, 993 (2004).

[39] A. Demirbas, *Energy Convers. Manage.*, **50**, 2239 (2009).

[40] C. L. Song, W. M. Zhang, Y. Q. Pei, G. L. Fan and G. P. Xu, *Atmos. Environ.*, **40**, 1957 (2006).

[41] R. R. Davda, J. W. Shabaker, G. W. Huber, R. D. Cortright and J. A. Dumesic, *Appl. Catal. B*, **56**, 171 (2005).

[42] O. Skoplyak, M. A. Barteau and J. G. G. Chen, *Catal. Today*, **147**, 150 (2009).

[43] M. Mavrikakis and M. A. Barteau, *J. Mol. Catal. A: Chem.*, **131**, 135 (1998).

[44] C. J. Weststrate, W. Ludwig, J. W. Bakker, A. C. Gluhoi and B. E. Nieuwenhuys, *J. Phys. Chem. C*, **111**, 7741 (2007).

[45] R. Neubauer, C. M. Whelan, R. Denecke and H. P. Steinruck, *Surf. Sci.*, **507–510**, 832 (2002).

[46] A. F. Lee, D. E. Gawthrope, N. J. Hart and K. Wilson, *Surf. Sci.*, **548**, 200 (2004).

[47] K. I. Gursahani, R. Alcala, R. D. Cortright and J. A. Dumesic, *Appl. Catal., A*, **222**, 369 (2001).

[48] T. Mallat and A. Baiker, *Chem. Rev.*, **104**, 3037 (2004).

[49] A. F. Lee and K. Wilson, *Green Chem.*, **6**, 37 (2004).

[50] A. F. Lee, Z. Chang, P. Ellis, S. F. J. Hackett and K. Wilson, *J. Phys. Chem. C*, **111**, 18 844 (2007).

[51] J. L. Solomon, R. J. Madix and J. Stohr, *J. Chem. Phys.*, **89**, 5316 (1988).

[52] A. F. Lee, S. F. J. Hackett, J. S. J. Hargreaves and K. Wilson, *Green Chem.*, **8**, 549 (2006).

[53] A. F. Lee and K. Wilson, *Green Chem.*, **6**, 37 (2004).

[54] S. F. J. Hackett, R. M. Brydson, M. H. Gass, I. Harvey, A. D. Newman, K. Wilson and A. F. Lee, *Angew. Chem., Int. Ed.*, **46**, 8593 (2007).

[55] M. D. Hughes, Y. J. Xu, P. Jenkins, P. McMorn, P. Landon, D. I. Enache, A. F. Carley, G. A. Attard, G. J. Hutchings, F. King, E. H. Stitt, P. Johnston, K. Griffin and C. J. Kiely, *Nature*, **437**, 1132 (2005).

[56] A. F. Lee, S. F. J. Hackett, G. J. Hutchings, S. Lizzit, J. N. Naughton and K. Wilson, *Catal. Today*, **145**, 251 (2009).

[57] M. A. Keane, *J. Chem. Technol. Biotechnol.*, **80**, 1211 (2005).

[58] S. L. Buchwald, *Acc. Chem. Res.*, **41**, 1439 (2008).

[59] M. X. Yang, M. Xi, H. J Yuan, B. E. Bent, P. Stevens and J. M. White, *Surf. Sci.* **341**, 9 (1995).

[60] X. L. Zhou and J. M. White, *J. Chem. Phys.*, **92**, 5612 (1990).

[61] Y. Song, P. Garner, H. Conrad, A. M. Bradshaw and J. M. White, *Surf. Sci.*, **248**, L279 (1991).

[62] G. J. Szulczewski and J. M. White, *Surf. Sci.*, **399**, 305 (1998).

[63] J. Bushell, A. F. Carley, M. Coughlin, P. R. Davies, D. Edwards, D. J. Morgan and M. Parsons, *J. Phys. Chem. B*, **109**, 9556 (2005).

[64] M. Xi and B. E. Bent, *Surf. Sci.*, **278**, 19 (1992).

[65] M. Xi and B. E. Bent, *J. Am. Chem Soc.*, **115**, 7426 (1993).

[66] J. M. Meyers and A. J. Gellman, *Surf. Sci.*, **337**, 40 (1995).

[67] S. Tjandra and F. Zaera, *J. Catal.*, **164**, 82 (1996).

[68] D. M. Jaramillo, D. E. Hunka and D. P. Land, *Surf. Sci.*, **445**, 23 (2000).

[69] H. von Schenck, J. Weissenrieder, S. Helldén, B. Åkermark and M. Göthelid, *App. Surf. Sci.*, **212**, 508 (2003).

[70] B. E. Bent, *Chem. Rev.*, **96**, 1361 (1996).

[71] G. J. Szulczewski and J. M. White, *Surf. Sci.*, **399**, 305 (1998).

[72] M. Xi and B. E. Bent, *Surf. Sci.*, **278**, 19 (1992).

[73] D. Syomin and B. E. Koel, *Surf. Sci.*, **490**, 265 (2001).

[74] B. M. Haines, D. A. Fischer and J. L. Gland, *Abst. Am. Chem. Soc.*, **230**, U2961 (2005).

[75] N. V. Richardson and N. R. Palmer, *Surf. Sci.*, **114**, L1 (1982).

[76] H. Cabibil, H. Ihm and J. M. White, *Surf. Sci.*, **447**, 91 (2000).

[77] A. F. Lee, Z. Chang, S. F. J. Hackett, A. D. Newman and K. Wilson, *J. Phys. Chem. C*, **111**, 10 455 (2007).

[78] A. F. Lee, P. A. Carr and K. Wilson, *Chem. Commun.*, **23**, 2774 (2004).

[79] A. F. Lee, V. Prabhakaran and K. Wilson, *Chem. Commun.*, **46**, 3827 (2010).

[80] M. Salmeron and R. Schlögl, *Surf. Sci. Rep.*, **63**, 169 (2008).

[81] M. E. Grass, Y. Zhang and D. R. Butcher, J. Y. Park, Y. Li, H. Bluhm, K. M. Bratlie, T. Zhang and G. A. Somorjai, *Angew. Chem. Int. Ed.*, **47**, 8893 (2008).

[82] R. W. Joyner, M. W. Roberts and K. Yates, *Surf. Sci.*, **87**, 501 (1979).

[83] D. F. Ogletree, H. Bluhm, G. Lebedev, C. S. Fadley, Z. Hussain and M. Salmeron, *Rev. Sci. Instrum.*, **73**, 3872 (2002).

[84] H. Bluhm, M. Havecker, A. Knop-Gericke, M. Kiskinova, R. Schlogl and M. Salmeron, *MRS Bull.*, **32**, 1022 (2007).

[85] M. Salmeron and R. Schlogl, *Surf. Sci. Rep.*, **63**, 169 (2008).

[86] J. G. Serafin, A. C. Liu and S. R. Seyedmonir, *J. Mol. Catal. A:Chem.*, **131**, 157 (1998).

[87] R. M. Lambert, F. J. Williams, R. L. Cropley and A. Palermo, *J. Mol. Catal. A: Chem.*, **228**, 27 (2005).

[88] S. Linic and M. A. Barteau, *J. Am. Chem. Soc.*, **124**, 310 (2002).

[89] C. Stegelmann, N. C. Schiodt, C. T. Campbell and P. Stoltze, *J. Catal.*, **221**, 630 (2004).

[90] F. J. Williams, D. P. C. Bird, A. Palermo, A. K. Santra and R. M. Lambert, *J. Am. Chem. Soc.*, **126**, 8509 (2004).

[91] D. Torres, N. Lopez, F. Illas and R. M. Lambert, *Angew. Chem., Int. Ed.*, **46**, 2055 (2007).

[92] V. I. Bukhtiyarov, A. I. Nizovskii, H. Bluhm, M. Havecker, E. Kleimenov, A. Knop-Gericke and R. Schlogl, *J. Catal.*, **238**, 260 (2006).

[93] V. I. Bukhtiyarov, M. Havecker, V. V. Kaichev, A. Knop-Gericke, R. W. Mayer and R. Schlogl, *Phys. Rev. B*, **67**, 235 422 (2003).

[94] G. Ketteler, D. F. Ogletree, H. Bluhm, H. Liu, E. L. D. Hebenstreit and M. Salmeron, *J. Am. Chem. Soc.*, **127**, 18 269 (2005).

[95] H. Gabasch, W. Unterberger, K. Hayek, B Klotzer, E. Kleimenov, D. Teschner, S. Zafeiratos, M. Havecker, A. Knop-Gericke, R. Schlogl, J. Han, F. H. Ribeiro, B. Aszalos-Kiss, T. Curtin and D. Zemlyanov, *Surf. Sci.*, **600**, 2980 (2006).

[96] D. J. Miller, H. Oberg, S. Kaya, H. Sanchez Casalongue, D. Friebel, T. Anniyev, H. Ogasawara, H. Bluhm, L. G. M. Pettersson and A. Nilsson, *Phys. Rev. Lett*, **107**, 195 502 (2011).

[97] H. Gabasch, K. Hayek, B. Klotzer, W. Unterberger, E. Kleimenov, D. Teschner, S. Zafeiratos, M. Havecker, A. Knop-Gericke, R. Schlogl, B. Aszalos-Kiss and D. Zemlyanov, *J. Phys. Chem. C*, **111**, 7957 (2007).

[98] H. Bluhm, M. Havecker, A. Knop-Gericke, E. Kleimenov, R. Schlogl, D. Teschner, V. I. Bukhtiyarov, D. F. Ogletree and M. Salmeron, *J. Phys. Chem. B*, **108**, 14 340 (2004).

[99] R. Blume, M. Hävecker, S. Zafeiratos, D. Teschner, A. Knop-Gericke, R. Schlögl, P. Dudin, A. Barinov and M. Kiskinova, *Catal. Today*, **124**, 71 (2007).

[100] R. Blume, M. Havecker, S. Zafeiratos, D. Teschner, E. Vass, P. Schnorch, A. Knop-Gericke, R. Schlogl, S. Lizzit, P. Dudin, A. Barinovb and M. Kiskinova, *Phys. Chem. Chem. Phys.*, **9**, 3648 (2007).

[101] A. Molnar, A. Sarkany and M. Varga, *J. Mol. Catal. A: Chem.*, **173**, 185 (2001).

[102] D. Teschner, A. Pestryakov, E. Kleimenov, M. Havecker, H. Bluhm, H. Sauer, A. Knop-Gericke and R. Schlogl, *J. Catal.*, **230**, 186 (2005).

[103] D. Teschner, A. Pestryakov, E. Kleimenov, M. Havecker, H. Bluhm, H. Sauer, A. Knop-Gericke and R. Schlogl, *J. Catal.*, **230**, 195 (2005).

[104] D. Teschner, E. Vass, M. Hävecker, S. Zafeiratos, P. Schnörch, H. Sauer, A. Knop-Gericke, R. Schlögl, M. Chamam, A. Wootsch, A. S. Canning, J. J. Gamman, S. D. Jackson, J. McGregor and L. F. Gladden, *J. Catal.*, **242**, 26 (2006).

[105] D. Teschner, J. Borsodi, A. Wootsch, Z. Revay, M. Havecker, A. Knop-Gericke, S. D. Jackson and R. Schlogl, *Science*, **320**, 86 (2008).

[106] D. Teschner, Z. Revay, J. Borsodi, M. Havecker, A. Knop-Gericke, R. Schlogl, D. Milroy, S. D. Jackson, D. Torres and P. Sautet, *Angew. Chem., Int. Ed.*, **47**, 9274 (2008).

[107] A. M. Doyle, S. K. Shaikhutdinov and H. J. Freund, *J. Catal.*, **223**, 444 (2004).

[108] A. M. Doyle, S. K. Shaikhutdinov, H. J. Freund and J. Freund, *Angew. Chem., Int. Ed.*, **44**, 629 (2005).

[109] S. Piccinin, S. Zafeiratos, C. Stampfl, T. W. Hansen, M. Havecker, D. Teschner, V. I. Bukhtiyraov, F. Girgsdies, A. Knop-Gericke, R. Schlogl and M. Scheffler, *Phys. Rev. Lett.*, **104**, 035 503 (2010).

[110] A. F. Lee, C. V. Ellis, K. Wilson and N. S. Hondow, *Catal. Today*, **157**, 243 (2010).

[111] F. Bernardi, G. H. Fecher, M. C. M. Alves and J. Morais, *J. Phys. Chem. Lett.*, **1**, 912 (2010).

[112] I. Tunc, U. K. Demirok, S. S. M. A. Correa-Duatre and L. M. Liz-Marzan, *J. Phys. Chem. B*, **109**, 24 182 (2005).

[113] K. Wilson and J. H. Clark, *Pure Appl. Chem.*, **72**, 1313 (2000).

[114] I. V. Kozhevnikov, *Chem. Rev.*, **98**, 171 (1998).

[115] N. Mizuno and M. Misono, *Chem. Rev.*, **98**, 99 (1998).

[116] I. V. Kozhevnikov, *Catal. Rev. Sci. Eng.*, **37**, 311 (1995).

[117] M. Misono, *Catal. Rev. & Sci. Eng.*, **29**, 269 (1987).

[118] A. D. Newman, A. F. Lee, K. Wilson and N. A. Young, *Catal. Lett.*, **102**, 45 (2005).

[119] A. D. Newman, D. R. Brown, P.F. Siril, A. F. Lee and K. Wilson, *Phys. Chem. Chem. Phys.*, **8**, 2893 (2006).

[120] T. Okuhara, N. Mizuno and M. Misono, *Adv. Catal.*, **41**, 113 (1996).

[121] T. Okuhara, T. Nishimura, H. Watanabe and M. Misono, *J. Mol. Catal.*, **74**, 247 (1992).

[122] T. Okuhara, T. Arai, T. Ichiki, K. Y. Lee and M. Misono, *J. Mol. Catal.*, **55**, 293 (1989).

[123] T. Okuhara and T. Nakato, *Catal. Surv. Jpn.*, **2**, 31 (1998).

[124] K. Narasimharao, D. R. Brown, A. F. Lee, A. D. Newman, P. F. Siril, S. J. Tavener and K. Wilson, *J. Catal.*, **248**, 226 (2007).

[125] A. Corma, *Chem. Rev.*, **95**, 559 (1995).

[126] A. Corma and H. Garcia, *Catal. Today*, **38**, 257 (1997).

[127] M. Stöcker, *Micro & Mesoporous Materials*, **6**, 235 (1996).

[128] M. C. G. Albuquerque, D. C. S. Azevedo, C. L. Cavalcante Jr, J. S. Gonsalez, J. M. M. Robles, R. M. Tost, E. R. Castellon, A. J. Lopez and P. M. Torres, *J. Mol. Catal. A*, **300**, 19 (2009).

[129] D. G. Cantrell, L. J. Gillie, A. F. Lee and K. Wilson, *Appl. Catal. A*, **287**, 183 (2005).

[130] M. C. G. Albuquerque, I. J. Urbistondo, J. S. Gonzalez, J. M. M. Robles, R. M. Tost, E. R. Castellon, A. J. Lopez, D. C. S. Azevedo, C. L. Cavalcante Jr, and P. M. Torres, *Appl. Catal. A*, **334**, 35 (2008).

[131] C. S. Caetano, L. Guerreiro, I. M. Fonseca, A. M. Ramos, J. Vital and J. E. Castanheiro, *Appl. Catal. A*, **359**, 41 (2009).

[132] K. Wilson, C. Hardacre, A. F. Lee, J. M. Montero and L. Shellard, *Green Chem.*, **10**, 654 (2008).

[133] A. Corma and S. Iborra, *Advances in Catalysis*, **49**, 239 (2006).

[134] J. A. D. Matthew and S. Parker, *J. Electron Spectrosc. Relat. Phenom.*, **85**, 175 (1997).

[135] J. M. Montero, P. L. Gai, K. Wilson and A. F. Lee, *Green Chem.*, **11**, 265 (2009).

[136] F. Collignon, P. A. Jacobs, P. Grobet and G. Poncelet, *J. Phys. Chem. B*, **105**, 6812 (2001).

[137] M. Johansson and K. Klier, *Top. Catal.*, **4**, 99 (1997).

[138] C. Guimon, A. Gervasini and A. Auroux, *J. Phys. Chem. B*, **105**, 10 316 (2001).

[139] G. Ballerini, K. Ogle and M.-G. Barthes-Labrousse, *Appl. Surf. Sci.*, **253**, 6860 (2007).

[140] F. Tao, *Chem. Commun.*, **48**, 3812 (2012).

[141] D. E. Starr, E. K. Wong, D. R. Worsnop, K. R. Wilson and H. Bluhm, *Phys. Chem. Chem. Phys.*, **10**, 3093 (2008).

Index

Figures are denoted by *italic page numbers*, Tables by **emboldened numbers**.
Abbreviations: EPR = electron paramagnetic resonance; INS = inelastic neutron scattering; NMR = nuclear magnetic resonance; XAS = X-ray absorption spectroscopy; XES = X-ray emission spectroscopy; XPS = X-ray photoelectron spectroscopy

Absorption coefficient, of photons/X-rays *90–91*, 161

Absorption cross-sections, of neutrons 180

Ag–Cu bimetallic catalysts 335

Ag_2O, ethene oxidation promoted by 329, *330*

AgCl host, Cr^{3+} centres in *259, 260*

Akimotoite ($MgSiO_3$ polymorph), ^{17}O MAS NMR spectrum *37*, 76

L-Alanine, solid, ^{13}C MAS NMR spectrum *9*

Alcohols
catalytic selective oxidation of 319–321
oxidation dehydrogenation of 330–331

Alkynes, selective hydrogenation of 333–5

Allylic alcohols, selective oxidation of 322–4

γ-Al_2O_3, ^{27}Al MAS NMR spectrum *47, 48*

AlPO-5, $^{27}Al/^{19}F$ REDOR 43–4

AlPO-14
^{27}Al MAS NMR spectra *35, 36, 38, 54, 70*
^{31}P MAS NMR spectra *35, 36, 70*

AlPO-53(A), ^{27}Al and ^{31}P NMR spectra *70, 71*

Aluminium (^{27}Al) chemical shifts, effects of coordination number 74

Aluminium (^{27}Al) NMR *35, 36*, 38, 42, 47, 48, 49–50, *54*, **60**, 64, 70, *71*, 74, 76, 79

Aluminophosphates
NMR studies *35, 36*, 38, 49–50, 70–72
substitution by heteroatoms 71–2
see also AlPO-5; AlPO-14; AlPO-53(A); SIZ-4; STA-2

Aluminosilicate glasses, NMR investigations 79–80

Aluminosilicates, NMR investigations of Si/Al ratios 39, 46–7, 69–70, 74–5

Ambient-pressure X-ray photoelectron spectroscopy (AP-XPS) 329–31

Ammonia, species formed on adsorption of NH_3 on metal oxide surfaces 344

Analcime, ^{29}Si MAS NMR spectra *38, 39*, 46

Andalusite (Al_2SiO_5), ^{27}Al MAS NMR spectrum 42

Angular resolved X-ray photoelectron spectroscopy 309–11

Anion disorder, in rock-forming minerals 74

Local Structural Characterisation, First Edition. Edited by Duncan W. Bruce, Dermot O'Hare and Richard I. Walton.
© 2014 John Wiley & Sons, Ltd. Published 2014 by John Wiley & Sons, Ltd.